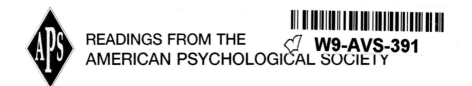

**READINGS FROM THE
AMERICAN PSYCHOLOGICAL SOCIETY**

W9-AVS-391

Current
Directions
in
COGNITIVE
SCIENCE

EDITED BY

Barbara A. Spellman
University of Virginia

Daniel T. Willingham
University of Virginia

PEARSON

Prentice
Hall

Upper Saddle River, New Jersey 07458

© 2005 by PEARSON EDUCATION, INC.
Upper Saddle River, New Jersey 07458

Current Directions © American Psychological Society
1010 Vermont Avenue, NW
Suite 1100
Washington, D.C. 20005-4907

10 9 8 7 6

ISBN 0-13-191991-1

Printed in the United States of America

Contents

Readings from
Current Directions in Psychological Science

Overview of the Book

Picking articles for this reader was, in many ways, a pleasure. As we looked back through the years of *Current Directions* to find articles relevant to cognitive science, we were struck by how many were interesting, well-written, or both. The hard part then was how to narrow down the selection to the 20-25 for the reader.

We assumed that most users of this book would be concurrently taking a course in cognitive psychology or cognitive science. Thus, we decided not to pick articles that were on issues likely to be well-covered in the textbooks for those courses. This decision also led to another decision—not to pick articles that were more than 6 years old. For one thing, such topics were likely to already be in the textbooks; for another, the reader is titled *Current* (not *Historic*) *Directions*.

Beyond that, the reader reflects our shared views as to topics that are particularly timely in cognitive psychology and topics that are applicable beyond cognitive psychology. Indeed, over the past years we have each had reason to answer the question: What are the current trends in cognitive psychology and where do you see the field going? In sum we would say that the directions are micro, macro, and down under. By "micro" we mean the current interest in neuroscience: What is the relation between mind and brain? As the tools for looking at the brain get better, we can learn more about how the mind works. But as Miller and Keller (2000) point out, don't be fooled, one is not reducible to the other. By "macro" we mean two things: first, ties to systems bigger than our individual minds (e.g., to evolution and culture) and second, to human affairs such as law, business, and economics, whose practioners increasingly look to cognitive scientists for information about how people think and behave in the "real world" outside the laboratory. And by "down under" we mean that interest in the unconscious and the role of emotion and affect in influencing cognition is booming. We find all of these directions very exciting and have tried to select articles that illustrate them and, we hope, bring the excitement to you.

Textbooks are good for explaining the theories in a field and for detailing a few central research findings. These articles give you a chance to read recent reviews by experts that will show you something beyond the basics, by taking you to the cutting edge of the field. We hope that these readings will expand your understanding of what cognitive science is and can do. Enjoy.

Visual Perception

Cognitive psychology can be frustrating because it seems so splintered. Beginning students study perception, then attention, then working memory, and so on. They have the sense that each researcher has fenced off a small domain of the mind for study, and scarcely cares or even acknowledges that the mind performs other functions, let alone that these functions might have some influence on the researcher's cherished domain.

There is a grain of truth in this perception. Cognitive psychologists do believe that progress can be made by examining cognitive processes in isolation, because they believe that it is useful to consider the mind as *modular*. That means that the mind is composed of processes, each of which performs a particular cognitive function (e.g., calculating the motion of objects, or maintaining auditory information in working memory). The modules are not entirely independent of one another—they communicate—but they have a certain independence in that their function does not radically change based on what other modules are doing. In addition, most introductory cognitive psychology books don't discuss interaction because it is a more advanced topic—you need to know the basics before discussing how the basics interact. This section includes three articles that consider the interaction of visual perception with other cognitive processes: attention, imagery, and other perceptual systems.

Our intuition tells us that attention increases the processing of visual information, but vision still proceeds to some extent without the benefit of attention. For example, suppose you had walked past a painting in a museum but your friend told you that it was her favorite, so you went back and inspected it more closely. You would likely say that you had *seen* the painting when you first passed it, but had not noticed many of the details until you had directed attention to it. This observation indicates that attention enables deeper or more complex visual perception, but that some perceptual processes operate independently of attention. These processes allow you to know that it's a painting on a wall, and not a mirror or a window. Research reviewed by Mack in "Inattentional Blindness: Looking Without Seeing" suggests that our intuition is not quite right, and that visual perception is more dependent on attention than we would guess. It appears that we do not see objects to which we don't attend. But the interesting twist is that this generalization holds true only for conscious perception. Visible objects can and do affect behavior, although we may not be aware of how they influence us.

Attention relates to perception by making certain types of perception possible. Perception has an altogether different relationship to visual imagery. Again using our intuition as a guide, we might guess that the two have something in common. When asked "does the Statue of Liberty hold her torch in her right hand or her left hand?" most people report that they answer this question by generating a mental picture of the Statue and then inspecting it. This mental picture seems to have something in common with perception—it entails a visual experience—but further reflection reveals important differences between perception and visual imagery. Pause from your reading for a moment and image a tiger. Once you have done so, try to count the number of stripes on the tiger's torso. Most people find this task impossible, although you could of course complete the task if you were perceiving a tiger. Images are less detailed than percepts. Is imagery a watered-down version of perception? No, but it is now known that there is considerable overlap in the two systems. In "The Mind's Eye Mapped onto the Brain's Matter", Behrmann describes in detail the current knowledge of what these systems have in common, and how they differ.

It is easy for us to appreciate how vision and mental imagery overlap; it is more difficult to imagine significant overlap of vision and other sensory systems, such as audition. One might describe the tone of Stan Getz's saxophone as "warm" or even refer to it as "golden" but we mean that as a metaphor, not literally. "Synesthesia: Strong and Weak" by Martino and Marks reports on the small number of people for whom such descriptions are not metaphoric. Synesthetes experience cross talk among the senses such that stimulation of one sensory modality (e.g., a sound) leads to a strong sensation in another modality (e.g., a visual image). Although true synesthesia is rare, it can inform us about basic mechanisms of how perceptions are coded.

The articles in this section illustrate the intricate relationships among cognitive processes. Although we may speak of processes such as "attention" as though they operate in isolation, cognitive psychologists are mindful that this is a convenient simplification, and that even as we study individual cognitive processes we must bear in mind how they interact.

Inattentional Blindness: Looking Without Seeing

Arien Mack[1]

Psychology Department, New School University, New York, New York

Abstract

Surprising as it may seem, research shows that we rarely see what we are looking at unless our attention is directed to it. This phenomenon can have serious life-and-death consequences. Although the inextricable link between perceiving and attending was noted long ago by Aristotle, this phenomenon, now called inattentional blindness (IB), only recently has been named and carefully studied. Among the many questions that have been raised about IB are questions about the fate of the clearly visible, yet unseen stimuli, whether any stimuli reliably capture attention, and, if so, what they have in common. Finally, is IB an instance of rapid forgetting, or is it a failure to perceive?

Keywords

inattention; perception; awareness

Imagine an experienced pilot attempting to land an airplane on a busy runway. He pays close attention to his display console, carefully watching the airspeed indicator on his windshield to make sure he does not stall, yet he never sees that another airplane is blocking his runway!

Intuitively, one might think (and hope) that an attentive pilot would notice the airplane in time. However, in a study by Haines (1991), a few experienced pilots training in flight simulators proceeded with their landing when a clearly visible airplane was blocking the runway, unaware of the second airplane until it was too late to avoid a collision.

As it turns out, such events are not uncommon and even may account for many car accidents resulting from distraction and inattention. This is why talking on cell telephones while driving is a distinctly bad idea. However, the pervasive assumption that the eye functions like a camera and our subjective impression of a coherent and richly detailed world lead most of us to assume that we see what there is to be seen by merely opening our eyes and looking. Perhaps this is why we are so astonished by events like the airplane scenario, although less potentially damaging instances occur every day, such as when we pass by a friend without seeing her.

These scenarios are examples of what psychologists call inattentional blindness (IB; Mack & Rock, 1998). IB denotes the failure to see highly visible objects we may be looking at directly when our attention is elsewhere. Although IB is a visual phenomenon, it has auditory and tactile counterparts as well; for example, we often do not hear something said to us if we are "not listening."

INATTENTIONAL BLINDNESS

The idea that we miss a substantial amount of the visual world at any given time is startling even though evidence for such selective seeing was first reported in the 1970s by Neisser (1979). In one of several experiments, he asked participants to view a video of two superimposed ball-passing games in which one group of players wore white uniforms and another group wore black uniforms. Participants counted the number of passes between members of one of the groups. When the participants were subsequently asked to report what they had seen, only 21% reported the presence of a woman who had unexpectedly strolled though the basketball court carrying an open umbrella, even though she was clearly in view some of the time. Researchers recently replicated this finding with a study in which a man dressed in a gorilla costume stopped to thump his chest while walking through the court and remained visible for between 5 and 9 s (Simons & Chabris, 1999).

Although it is possible that some failures to see the gorilla or the umbrella-carrying woman might have resulted from not looking directly at them, another body of work supports the alternative explanation that the observers were so intent on counting ball passes that they missed the unexpected object that appeared in plain view. Research I have conducted with my colleagues (Mack & Rock, 1998) conclusively demonstrates that, with rare exceptions, observers generally do not see what they are looking directly at when they are attending to something else. In many of these experiments, observers fixated on specified locations while simultaneously attending to a demanding perceptual task, the object of which might be elsewhere. Under these conditions, observers often failed to perceive a clearly visible stimulus that was located exactly where they were looking.

INATTENTIONAL BLINDNESS OR INATTENTIONAL AMNESIA?

Not surprisingly, there is a controversy over whether the types of failures documented in such experiments are really evidence that the observers did not see the stimulus, or whether they in fact saw the stimulus but then quickly forgot it. In other words, is IB more correctly described as *inattentional amnesia* (Wolfe, 1999)? Although this controversy may not lend itself to an empirical resolution, many researchers find it difficult to believe that a thumping gorilla appearing in the midst of a ball game is noticed and then immediately forgotten. What makes the argument for inattentional amnesia even more difficult to sustain is evidence that unseen stimuli are capable of priming, that is, of affecting some subsequent act. (For example, if a subject is shown some object too quickly to identify it and is then shown it again so that it is clearly visible, the subject is likely to identify it more quickly than if it had not been previously flashed. This is evidence of priming: The first exposure speeded the response to the second.) Priming can occur only if there is some memory of the stimulus, even if that memory is inaccessible.

UNCONSCIOUS PERCEPTION

A considerable amount of research has investigated unconscious, or *implicit*, perception and those perceptual processes that occur outside of awareness. This work has led many researchers to conclude that events in the environment, even if not consciously perceived, may direct later behavior. If stimuli not seen because of IB are in fact processed but encoded outside of awareness, then it should be possible to demonstrate that they prime subsequent behavior.

The typical method for documenting implicit perception entails measuring reaction time over multiple trials. Such studies are based on the assumption that an implicitly perceived stimulus will either speed up or retard subsequent responses to relevant stimuli depending on whether the priming produces facilitation or inhibition.[2] However, because subjects in IB experiments cannot be made aware of the critical stimulus, unlike in many kinds of priming studies, only one trial with that stimulus is possible. This requirement rules out reaction time procedures, which demand hundreds of trials because reaction time differences tend to be small and therefore require stable response rates that can be achieved only with many trials. Fortunately, an alternate procedure, stem completion, can be used when the critical stimuli are words. In this method, some observers (experimental group) are exposed to a word in an IB procedure, and other observers (control group) are not. Then, the initial few letters of the unseen word are presented to all the observers, who are asked to complete the string of letters with one or two English words. If the members of the experimental group complete the string with the unseen word more frequently than do the members of the control group, this is taken as evidence that the experimental group implicitly perceived and encoded the word.

IB experiments using this method have demonstrated significant priming (Mack & Rock, 1998), as well as other kinds of evidence that visual information undergoes substantial processing prior to the engagement of attention. For example, evidence that aspects of visual processing take place before attention is allocated has been provided by a series of ingenious IB experiments by Moore and her collaborators (e.g., Moore & Egeth, 1997). This work has shown that under conditions of inattention, basic perceptual processes, such as those responsible for the grouping of elements in the visual field into objects, are carried out and influence task responses even though observers are unable to report seeing the percepts that result from those processes. For example, in one study using a modification of the IB procedure, Moore and Egeth investigated the Müller-Lyer illusion, in which two lines of equal length look unequal because one has outgoing fins, which make it look longer, and the other has ingoing fins, which make it look shorter. In this case, the fins were formed by the grouping of background dots: Dots forming the fins were closer together than the other dots in the background. Moore and Egeth demonstrated that subjects saw the illusion even when, because of inattention, the fins were not consciously perceived. Whatever processes priming entails, the fact that it occurs is evidence of implicit perception and the encoding of a stimulus in memory. Thus, the fact that the critical stimulus in the IB paradigm can prime subsequent responses is evidence that this stimulus is implicitly perceived and encoded.

When Do Stimuli Capture Attention and Why?

That unconsciously perceived stimuli in IB experiments undergo substantial processing in the brain is also supported by evidence that the select few stimuli able to capture attention when attention is elsewhere are complex and meaningful (e.g., the observer's name, an iconic image of a happy face) rather than simple features like color or motion. This fact suggests that attention is captured only after the meaning of a stimulus has been analyzed. There are psychologists who believe that attention operates much earlier in the processing of sensory input, before meaning has been analyzed (e.g., Treisman, 1969). These accounts, however, do not easily explain why modest changes, such as inverting a happy face and changing one internal letter in the observer's name, which alter the apparent meaning of the stimuli but not their overall shape, cause a very large increase in IB (Mack & Rock, 1998).

Meaning and the Capture of Attention

If meaning is what captures attention, then it follows axiomatically that meaning must be analyzed before attention is captured, which is thought to occur at the end stage of the processing of sensory input. This therefore implies that even those stimuli that we are not intending to see and that do not capture our attention must be fully processed by the brain, for otherwise their meanings would be lost before they had a chance of capturing our attention and being perceived. If this is the case, then we are left with some yet-unanswered, very difficult questions. Are all the innumerable stimuli imaged on our retinas really processed for meaning and encoded into memory, and if not, which stimuli are and which are not?

Although we do not yet have answers to these questions, an unpublished doctoral dissertation by Silverman, at New School University, has demonstrated that there can be priming by more than one element in a multielement display, even when these elements cannot be reported by the subject. This finding is relevant to the question whether all elements in the visual field are processed and stored because up to now there has been scarcely any evidence of priming by more than one unreportable element in the field. The fact of multielement priming begins to suggest that unattended or unseen elements are processed and stored, although it says nothing about how many elements are processed and whether the meaning of all the elements is analyzed.

One answer to the question of how much of what is not seen is encoded into memory comes from an account of perceptual processing based on the assumption that perception is a limited-capacity process and that processing is mandatory up to the point that this capacity is exhausted (Lavie, 1995). According to this analysis, the extent to which unattended objects are processed is a function of the difficulty of the perceptual task (i.e., the perceptual load). When the perceptual load is high, only attended stimuli are encoded. When it is low, unattended stimuli are also processed. This account faces some difficulty because it is not clear how perceptual load should be estimated. Beyond this, however, it is difficult to reconcile this account with evidence suggesting that observers are likely to see their own names even when they occur among the stimuli that must be ignored in order to perform a demanding perceptual task (Mack, Pappas, Silverman, & Gay, 2002). It should be noted, however, that these latter results are at odds with a published report (Rees, Russell, Firth, & Driver, 1999) I describe in the next section.

EVIDENCE FROM NEURAL IMAGING

Researchers have used magnetic imaging techniques to try to determine what happens in the brain when observers fail to detect a visual stimulus because their attention is elsewhere. Neural recording techniques may be able to show whether visual stimuli that are unconsciously perceived arouse the same areas of the brain to the same extent as visual stimuli that are seen. This is an important question because it bears directly on the nature of the processing that occurs outside of awareness prior to the engagement of attention and on the difference between the processing of attended and unattended stimuli.

In one study, Scholte, Spekreijse, and Lamme (2001) found similar neural activity related to the segregation of unattended target stimuli from their backgrounds (i.e., the grouping of the unattended stimuli so they stood out from the background on which they appeared), an operation that is thought to occur early in the processing of visual input. This activation was found regardless of whether the stimuli were attended and seen or unattended and not seen, although there was increased activation for targets that were attended and seen. This finding is consistent with the behavioral findings of Moore and Egeth (1997), cited earlier, showing that unattended, unseen stimuli undergo lower-level processing such as grouping, although the additional neural activity associated with awareness suggests that there may be important differences in processing of attended versus unattended stimuli.

In another study, Rees and his colleagues (Rees et al., 1999) used functional magnetic resonance imaging (fMRI) to picture brain activity while observers were engaged in a perceptual task. They found no evidence of any difference between the neural processing of meaningful and meaningless lexical stimuli when they were ignored, although when the same stimuli were attended to and seen, the neural processing of meaningful and meaningless stimuli did differ. These results suggest that unattended stimuli are not processed for meaning. However, in another study that repeated the procedure used by Rees et al. (without fMRI recordings) but included the subject's own name among the ignored stimuli, many subjects saw their names, suggesting that meaning was in fact analyzed (Mack et al., 2002). Thus, one study shows that ignored stimuli are not semantically processed, and the other suggests that they are. This conflict remains unresolved. Are unattended, unseen words deeply processed outside of awareness, despite these fMRI results, which show no evidence of semantic neural activation by ignored words? How can one reconcile behavioral evidence of priming by lexical stimuli under conditions of inattention (Mack & Rock, 1998) with evidence that these stimuli are not semantically processed?

NEUROLOGICAL DISORDER RELATED TO INATTENTIONAL BLINDNESS

People who have experienced brain injuries that cause lesions in the parietal cortex (an area of the brain associated with attention) often exhibit what is called unilateral visual neglect, meaning that they fail to see objects located in the visual field opposite the site of the lesion. That is, for example, if the lesion is on the right, they fail to eat food on the left side of their plates or to shave the left half of their faces. Because these lesions do not cause any sensory deficits, the apparent blindness cannot be attributed to sensory causes and has been explained in terms of the role of the parietal cortex in attentional processing

7

(Rafal, 1998). Visual neglect therefore seems to share important similarities with IB. Both phenomena are attributed to inattention, and there is evidence that in both visual neglect (Rafal, 1998) and IB, unseen stimuli are capable of priming. In IB and visual neglect, the failure to see objects shares a common cause, namely inattention, even though in one case the inattention is produced by brain damage, and in the other the inattention is produced by the task. Thus, evidence of priming by neglected stimuli appears to be additional evidence of the processing and encoding of unattended stimuli.

ATTENTION AND PERCEPTION

IB highlights the intimate link between perception and attention, which is further underscored by recent evidence showing that unattended stimuli that share features with task-relevant stimuli are less likely to suffer IB than those that do not (Most et al., 2001). This new evidence illustrates the power of our intentions in determining what we see and what we do not.

CONCLUDING REMARKS

Although the phenomenon of IB is now well established, it remains surrounded by many unanswered questions. In addition to the almost completely unexplored question concerning whether all unattended, unseen stimuli in a complex scene are fully processed outside of awareness (and if not, which are and which are not), there is the question of whether the observer can locate where in the visual field the information extracted from a single unseen stimulus came from, despite the fact that the observer has failed to perceive it. This possibility is suggested by the proposal that there are two separate visual systems, one dedicated to action, which does not entail consciousness, and the other dedicated to perception, which does entail consciousness (Milner & Goodale, 1995). That is, the action stream may process an unseen stimulus, including its location information, although the perception stream does not. An answer to this question would be informative about the fate of the unseen stimuli.

The pervasiveness of IB raises another unresolved question. Given that people see much less than they think they do, is the visual world a mere illusion? According to one provocative answer to this question, most recently defended by O'Regan and Noe (2001), the outcome of perceptual processing is not the construction of some internal representation; rather, seeing is a way of exploring the environment, and the outside world serves as its own external representation, eliminating the need for internal representations. Whether or not this account turns out to be viable, the phenomenon of IB has raised a host of questions, the answers to which promise to change scientists' understanding of the nature of perception. The phenomenon itself points to the serious dangers of inattention.

Recommended Reading

Mack, A., & Rock, I. (1998). (See References)
Rensink, R. (2002). Change blindness. *Annual Review of Psychology, 53,* 245–277.

Simons, D. (2000). Current approaches to change blindness. *Visual Cognition, 7,* 1–15.
Wilkens, P. (Ed.). (2000). Symposium on Mack and Rock's *Inattentional Blindness. Psyche,*
6 and 7. Retrieved from http://psyche.cs. monash.edu.au/psyche-indexv7.html#ib

Acknowledgments—I am grateful for the comments and suggestions of Bill Prinzmetal
and Michael Silverman.

Notes

1. Address correspondence to Arien Mack, Psychology Department, New School
University, 65 Fifth Ave., New York, NY 10003.
2. An example of a speeded-up response (facilitation, or positive priming) has
already been given. Negative, or inhibition, priming occurs when a stimulus that has
been actively ignored is subsequently presented. For example, if a series of superimposed
red and green shapes is rapidly presented and subjects are asked to report a feature of
the red shapes, later on it is likely to take them longer to identify the green shapes than
a shape that has not previously appeared, suggesting that the mental representation of
the green shapes has been associated with something like an "ignore me" tag.

References

Haines, R.F. (1991). A breakdown in simultaneous information processing. In G. Obrecht & L.W.
Stark (Eds.), *Presbyopia research* (pp. 171–175). New York: Plenum Press.
Lavie, N. (1995). Perceptual load as a necessary condition for selective attention. *Journal of Experimental Psychology: Human Perception and Performance, 21,* 451–468.
Mack, A., Pappas, Z., Silverman, M., & Gay, R. (2002). What we see: Inattention and the capture
of attention by meaning. *Consciousness and Cognition, 11,* 488–506.
Mack, A., & Rock, I. (1998). *Inattentional blindness.* Cambridge, MA: MIT Press.
Milner, D., & Goodale, M.A. (1995). *The visual brain in action.* Oxford, England: Oxford University Press.
Moore, C.M., & Egeth, H. (1997). Perception without attention: Evidence of grouping under conditions of inattention. *Journal of Experimental Psychology: Human Perception and Performance,
23,* 339–352.
Most, S.B., Simons, D.J., Scholl, B.J., Jimenez, R., Clifford, E., & Chabris, C.F. (2001). How not
to be seen: The contribution of similarity and selective ignoring to sustained inattentional
blindness. *Psychological Science, 12,* 9–17.
Neisser, U. (1979). The control of information pickup in selective looking. In A.D. Pick (Ed.), *Perception and its development: A tribute to Eleanor Gibson* (pp. 201–219). Hillsdale, NJ: Erlbaum.
O'Regan, K., & Noe, A. (2001). A sensorimotor account of vision and visual consciousness. *Behavioral and Brain Sciences, 25,* 5.
Rafal, R. (1998). Neglect. In R. Parasuraman (Ed.), *The attentive brain* (pp. 489–526). Cambridge,
MA: MIT Press.
Rees, G., Russell, C., Firth, C., & Driver, J. (1999). Inattentional blindness versus inattentional
amnesia. *Science, 286,* 849–860.
Scholte, H.S., Spekreijse, H., & Lamme, V.A. (2001). Neural correlates of global scene segmentation are present during inattentional blindness [Abstract]. *Journal of Vision, 1*(3), Article 346.
Retrieved from http://journalofvision. org/1/3/346
Simons, D.J., & Chabris, C.F. (1999). Gorillas in our midst: Sustained inattentional blindness for
dynamic events. *Perception, 28,* 1059–1074.
Treisman, A. (1969). Strategies and models of selective attention. *Psychological Review, 76,* 282–299.
Wolfe, J. (1999). Inattentional amnesia. In V. Coltheart (Ed.), *Fleeting memories* (pp. 71–94). Cambridge, MA: MIT Press.

9

Critical Thinking Questions

1. This article notes that some traffic accidents may well be caused by inattentional blindness, but you have likely had the experience of daydreaming as you drove without incident. In fact, your inattention may be so complete that you might drive to a familiar destination (e.g., your home) and feel surprised when you arrive. Does this phenomenon mean that the low-level processing that occurs with inattention is sufficient to support successful driving?

2. This article describes evidence that priming is supported by low-level processes that occur even in the absence of attention. When attention is directed to stimuli, you are aware of them. Other than awareness, what cognitive processes do you think are possible only with attention?

3. Mack cites the neurological phenomenon of unilateral neglect as sharing important similarities with inattentional blindness. Neglect patients fail to process stimuli in *all* modalities, not just vision. Do you think that lack of attention yields "blindness" in other modalities? In other words, if you are not attending, will you not hear an auditory stimulus? Will you not feel a tactile stimulus such as someone touching you?

The Mind's Eye Mapped Onto the Brain's Matter

Marlene Behrmann[1]

Department of Psychology, Carnegie Mellon University, Pittsburgh, Pennsylvania

Abstract

Research on visual mental imagery has been fueled recently by the development of new behavioral and neuroscientific techniques. This review focuses on two major topics in light of these developments. The first concerns the extent to which visual mental imagery and visual perception share common psychological and neural mechanisms; although the research findings largely support convergence between these two processes, there are data that qualify the degree of overlap between them. The second issue involves the neural substrate mediating the process of imagery generation. The data suggest a slight left-hemisphere advantage for this process, although there is considerable variability across and within subjects. There also remain many unanswered questions in this field, including what the relationship is between imagery and working memory and what representational differences, if any, exist between imagery and perception.

Keywords

mental imagery; visual perception; cognitive neuroscience

Consider sitting in your office and answering the question "How many windows do you have in your living room?" To decide how to answer, you might construct an internal visual representation of your living room from the stored information you possess about your home, inspect this image so as to locate the windows, and then count them. This type of internal visual representation (or "seeing with the mind's eye"), derived in the absence of retinal stimulation, is known as visual mental imagery, and is thought to be engaged in a range of cognitive tasks including learning, reasoning, problem solving, and language. Although much of the research on mental imagery has been concerned specifically with visual mental imagery, and hence the scope of this review is restricted to this topic, similar internal mental representations exist in the auditory and tactile modalities and in the motor domain.

The past decade has witnessed considerable progress in our understanding of the psychological and neural mechanisms underlying mental imagery. This is particularly dramatic because, in the not-too-distant past, during the heyday of behaviorism, discussions of mental imagery were almost banished from scientific discourse: Given that there was no obvious way of measuring so private an event as a mental image and there was no homunculus available for viewing the pictures in the head even if they did exist, the study of mental imagery fell into disrepute. Indeed, through the 1940s and 1950s, Psychological Abstracts recorded only five references to imagery. The study of mental imagery was revived in the 1970s, through advances such as the experiments of Shepard and colleagues (Shepard & Cooper, 1982) and of

11

Kosslyn and colleagues (see Kosslyn, 1994), and the dual coding theory of Paivio (1979).

Although a general consensus endorsing the existence of mental imagery began to emerge, it was still not fully accepted as a legitimate cognitive process. Some researchers queried whether subjects were simply carrying out simulations of their internal representations in symbolic, nonvisual ways rather than using a visual, spatially organized code. Other researchers suggested that subjects were simply conforming to the experimenters' expectations and that the data that appeared to support a visual-array format for mental imagery merely reflected the experimenters' belief in this format rather than the true outcome of a mental imagery process (Pylyshyn, 1981). In recent years, powerful behavioral and neuroscientific techniques have largely put these controversies to rest. This review presents some of this recent work.

RELATIONSHIP BETWEEN VISUAL IMAGERY AND VISUAL PERCEPTION

Perhaps the most hotly debated issue is whether mental imagery exploits the same underlying mechanisms as visual perception. If so, generating a visual mental image might be roughly conceived of as running perception backward. In perception, an external stimulus delivered to the eye activates visual areas of the brain, and is mapped onto a long-term representation that captures some of the critical and invariant properties of the stimulus. During mental imagery, the same long-term representations of the visual appearance of an object are used to activate earlier representations in a top-down fashion through the influence of preexisting knowledge. This bidirectional flow of information is mediated by direct connections between higher-level visual areas (more anterior areas dealing with more abstract information) and lower-level visual areas (more posterior areas with representations closer to the input).

Mental Imagery and Perception Involve Spatially Organized Representations

Rather than being based on propositional or symbolic representations, mental images appear to embody spatial layout and topography, as does visual perception. For example, many experiments have shown that the distance that a subject travels in mental imagery is equivalent to that traveled in perceptual performance (e.g., imagining the distance between New York and Los Angeles vs. looking at a real map to judge the distance). Recent neuroimaging studies have also provided support for the involvement of spatially organized representations in visual imagery (see Kosslyn et al., 1999); for example, when subjects form a high-resolution, depictive mental image, primary and secondary visual areas of the occipital lobe (areas 17 and 18, also known as V1 and V2), which are spatially organized, are activated.[2] Additionally, when subjects perform imagery, larger images activate relatively more anterior parts of the visual areas of the brain than smaller images, a finding consistent with the known mapping of how visual information from the world is mediated by different areas of primary visual cortex. Moreover, when repetitive transcranial magnetic stimulation[3] is applied and disrupts the normal function of area 17, response times in both perceptual and imagery tasks increase, further supporting the involvement of primary visual areas in mental imagery (Kosslyn et al., 1999).

Shared Visual and Imagery Areas Revealed Through Functional Imaging

Not only early visual areas but also more anterior cortical areas can be activated by imagined stimuli; for example, when subjects imagine previously seen motion stimuli (such as moving dots or rotating gratings), area MT/MST, which is motion sensitive during perception, is activated (Goebel, Khorram-Sefat, Muckli, Hacker, & Singer, 1998). Color perception and imagery also appear to involve some (but not all) overlapping cortical regions (Howard et al., 1998), and areas of the brain that are selectively activated during the perception of faces or places are also activated during imagery of the same stimuli (O'Craven & Kanwisher, in press). Higher-level areas involved in spatial perception, including a bilateral parieto-occipital network, are activated during spatial mental imagery, and areas involved in navigation are activated during mental simulation of previously learned routes (Ghaem et al., 1997). As is evident, there is considerable overlap in neural mechanisms implicated in imagery and in perception both at lower and at higher levels of the visual processing pathways.

Neuropsychological Data for Common Systems

There is also neuropsychological evidence supporting the shared-systems view. For example, many patients with cortical blindness (i.e., blindness due to damage to primary visual areas of the brain) or with scotomas (blind spots) due to destruction of the occipital lobe have an associated loss of imagery, and many patients with visual agnosia (a deficit in recognizing objects) have parallel imagery deficits. Interestingly, in some of these latter cases, the imagery and perception deficits are both restricted to a particular domain; for example, there are patients who are unable both to perceive and to image only faces and colors, only facial emotions, only spatial relations, only object shapes and colors, or only living things. The equivalence between imagery and perception is also noted in patients who, for example, fail both to report and to image information on the left side of space following damage to the right parietal lobe.

There are, however, also reports of patients who have a selective deficit in either imagery or perception. This segregation of function is consistent with the functional imaging studies showing that roughly two thirds of visual areas of the brain are activated during both imagery and perception (Kosslyn, Thompson, & Alpert, 1997). That is, selective deficits in imagery or perception may be explained as arising from damage to the nonoverlapping regions. Selective deficits are particularly informative and might suggest what constitutes the nonoverlapping regions. Unfortunately, because the lesions in the neuropsychological patients are rather large, one cannot determine precise anatomical areas for these nonoverlapping regions, but insights into the behaviors selectively associated with imagery or perception have been obtained, as discussed next.

Patients with impaired imagery but intact perception are unable to draw or describe objects from memory, to dream, or to verify propositions based on memory ("does the letter W have three strokes?"). It has been suggested that in these cases, the process of imagery generation (which does not overlap with perception) may be selectively affected without any adverse consequences for recognition. I review this generation process in further detail in a later section. It has also been suggested that low- or intermediate-level processes may play a greater (but not exclusive) role in perception than they do in imagery. For example, when asked to image a "kangaroo," one accesses the intact long-term representation of a kangaroo, and this is then instantiated and available for inspection. When one is perceiving a kangaroo, however, featural analysis, as well as perceptual organization such as figure-ground segregation and feature grouping or integration, is required. If a patient has a perceptual deficit because of damage to these low- or intermediate-level processes, the patient will be unable to perceive the display, but imagery might well be spared because it relies less on these very processes (Behrmann, Moscovitch, & Winocur, 1994).

Summary and Challenges

There is substantial evidence that imagery and perception share many (although not all) psychological and neural mechanisms. However, some results suggest that the early visual regions in the occipital lobe are not part of the shared network. For example, it has been shown that reliable occipital activation is not always observed in neuroimaging studies when subjects carry out mental imagery tasks. A possible explanation for these null results concerns the task demands: When the task does not require that the subjects form a highly depictive image, no occipital activation is obtained. A second explanation might have to do with subject sampling: There are considerable individual differences in mental imagery ability, and subjects who score poorly on a mental imagery vividness questionnaire show less blood flow in area 17 than those who score higher. If a study includes only low-imagery individuals, no occipital lobe activation might be obtained. A final explanation concerns the nature of the baseline, or control, condition used in neuroimaging experiments: If subjects are instructed to rest but instead continue to activate internal representations, when the activation obtained in the baseline is subtracted from that obtained in the imagery condition, no primary visual cortex activation will be observed.

Another challenge to the conclusion that primary visual areas are involved in imagery comes from studies of patients who have bilateral occipital lesions and complete cortical blindness but preserved imagery. Indeed, some of these patients are capable of generating such vivid images that they believe these to be veridical perceptions. For example, when a set of keys was held up before one such subject, she correctly identified the stimulus based on the auditory signal but then went on to provide an elaborate visual description of the keys, convinced that she could actually see them (Goldenberg, Müllbacher, & Nowak, 1995). Another subject with bilateral occipital lesions whose perception was so impaired that he could not even differentiate light from dark was still able to draw well from memory and performed exceptionally well on standard imagery tasks (Chatterjee & Southwood, 1995).

In sum, although the data supporting a strong association between imagery and perception are compelling, some findings are not entirely consistent with this conclusion.

GENERATION OF MENTAL IMAGES

A second major debate in the imagery literature concerns the mechanisms involved in generating a mental image. This is often assumed to be a process specific to imagery (or perhaps more involved in imagery than in perception) and involves the active construction of a long-term mental representation. Although there has been debate concerning whether there is such a process at all, several lines of evidence appear to support its existence and role in imagery. There are, for example, reports of neuropsychological patients who have preserved perception but impaired imagery and whose deficit is attributed to image generation. Patient R.M., for example, could copy well and make good shape discriminations of visually presented objects but could not draw even simple shapes nor complete from memory visually presented shapes that were partially complete (Farah, Levine, & Calvanio, 1988).

One controversial and unanswered issue concerns the neural substrate of the generation process. The growing neuropsychological literature has confirmed the preponderance of imagery generation deficits in patients with lesions affecting the left temporo-occipital lobe regions. There is not, however, a perfect relationship between this brain region and imagery generation, as many patients with such lesions do not have an impairment in imagery generation.

In many studies, normal subjects show a left-hemisphere advantage for imagery generation; when asked to image half of an object, subjects are more likely to image the right half, reflecting greater left-hemisphere than right-hemisphere participation in imagery generation. Additionally, right-handed subjects show a greater decrement in tapping with their right than left hand while performing a concurrent imagery task, reflecting the interference encountered by the left hemisphere while tapping and imaging simultaneously. (The left hemisphere controls movement on the right side of the body.) Studies in which information is presented selectively to one visual field (and thereby one hemisphere) have, however, yielded more variable results with normal subjects. Some studies support a left-hemisphere superiority, some support a right-hemisphere superiority, and some find no hemispheric differences at all. Studies with split-brain patients[4] also reveal a trend toward left-hemisphere involvement, but also some variability. Across a set of these rather rare patients, imagery performance is better when the stimulus is presented to the left than to the right hemisphere, although this finding does not hold for every experiment and the results are somewhat variable even within a single subject.

Neuroimaging studies in normal subjects have also provided some support for a left-hemisphere basis for imagery generation. For example, functional magnetic resonance imaging showed more activation of the left inferior occipital-temporal region when subjects generated images of heard words compared with when they were simply listening to these words (D'Esposito et al., 1997). This result had also been observed previously using ERPs[5]; an asymmetry in the waveforms of the two hemispheres implicated the left temporo-occipital regions in imagery generation (see Farah, 1999).

In sum, there is a slight but not overwhelming preponderance of evidence favoring the left hemisphere as mediating the imagery generation process. A conservative conclusion from these studies suggests that there may well be some degree of left-hemisphere specialization, but that many individuals have some capability for imagery generation by the right-hemisphere. Another suggestion is that both hemispheres are capable of imagery generation, albeit in different ways; for example, subjects showed a left-hemisphere advantage in a generation task when they memorized how the parts of a stimulus were arranged but showed a right-hemisphere advantage when they memorized the metric positions of the parts and how they could be "mentally glued" together (see Kosslyn, 1994).

CONCLUSION

Although considerable progress has been made in analyzing the convergence between imagery and perception, there are several outstanding issues. One of these is the relationship between imagery and the activation of internal representations in other cognitive tasks. For example, during visual working memory tasks, a mental representation of an object or spatial location is maintained over a delay period, in the absence of retinal stimulation. In these tasks, areas in the very front of the brain, rather than occipital cortex, are activated despite the similarities between this task and mental imagery. Similarly, in tasks that involve top-down forms of attention, subjects are verbally instructed to search for a target (such as a red triangle) in an upcoming display. Although subjects likely generate an image of a red triangle, this is generally not conceived of as an instance of mental imagery, and in neuro imaging studies the location of activation is not usually sought in occipital cortex.

Another perplexing and unresolved issue concerns the reasons that vivid imagery and hallucinations are not confused with reality, especially given that functional imaging studies show identical activations during hallucinations and perception (ffytche et al., 1998). Several solutions to this dilemma have been proposed, among them the idea that hallucinations derive from a failure to self-monitor an inner voice, with the result that the source of the stimulus is located in the external world. A second explanation suggests that perceptions are deeper and contain more detail than images. As Hume (1739/1963) stated, "The difference betwixt these [imagery-ideas and perception] consists in the degree of force and liveliness, with which they strike upon the mind . . . . Perceptions enter with most force and violence . . . . By ideas I mean the faint images of these in thinking and reasoning" (p. 311). How to verify these claims empirically is not obvious, yet this issue clearly demands resolution.

Recommended Reading

Behrmann, M., Moscovitch, M., & Winocur, G. (1999). Visual mental imagery. In G.W. Humphreys (Ed.), *Case studies in vision* (pp. 81–110). London: Psychology Press.
Farah, M.J. (1999). (See References)
Kosslyn, S.M. (1994). (See References)
Richardson, J.T.E. (1999). *Imagery.* Philadelphia: Psychology Press.

Acknowledgments—This work was supported by grants from the National Institute of Mental Health (MH54246 and MH54766). I thank Martha Farah and Steven Kosslyn for helpful discussions about mental imagery and Nancy Kanwisher for her insightful comments on this manuscript.

Notes

1. Address correspondence to Marlene Behrmann, Department of Psychology, Carnegie Mellon University, Pittsburgh, PA 15213-3890; e-mail: behrmann+@cmu.edu.

2. Visual information is received initially via the retina of the eye and is then transmitted through various visual pathways to the brain. This information is sent initially to the primary visual area of the brain, housed posteriorly in the occipital cortex, and is then sent more anteriorly through secondary visual areas to the temporal lobe of the brain for the purposes of recognition. The primary visual area is also known as area 17 or V1, and the secondary area is known as area 18 or V2. The visual input is also sent from the occipital cortex up to the parietal areas of the brain, which represent and code spatial information.

3. Repetitive transcranial magnetic stimulation is a new method in which magnetic pulses are delivered to the brain from a magnet placed externally on the scalp. The electrical pulses disrupt the function of the underlying brain area temporarily and are thus analogous to a reversible lesion. This method allows investigators to determine the involvement of certain brain areas in particular cognitive processes.

4. Split-brain patients are individuals who have undergone a separation of the two sides of the brain (cerebral hemispheres). This is done in individuals who have intractable and uncontrolled epilepsy in order to prevent the seizure activity from spreading across the entire brain. Unfortunately, it also prevents the transfer of all other forms of information from one hemisphere to the other.

5. ERPs, or evoked response potentials, are recordings of the brain's electrical activity in response to stimuli that are presented to the subject. The potentials are measured over time as waveforms obtained from electrodes placed at specific sites on the scalp, and different waveforms roughly reflect differential participation of some brain sites in the task under examination.

References

Behrmann, M., Moscovitch, M., & Winocur, G. (1994). Intact visual imagery and impaired visual perception in a patient with visual agnosia. *Journal of Experimental Psychology: Human Perception and Performance, 20*, 1068–1087.

Chatterjee, A., & Southwood, M.H. (1995). Cortical blindness and visual imagery. *Neurology, 45*, 2189–2195.

D'Esposito, M., Detre, J.A., Aguirre, G.K., Stallcup, M., Alsop, D.C., Tippett, L.J., & Farah, M.J. (1997). Functional MRI study of mental image generation. *Neuropsychologia, 35*, 725–730.

Farah, M.J. (1999). Mental imagery. In M. Gazzaniga (Ed.), *The cognitive neurosciences* (Vol. 2, pp. 965–974). Cambridge, MA: MIT Press.

Farah, M.J., Levine, D.N., & Calvanio, R. (1988). A case study of a mental imagery deficit. *Brain and Cognition, 8*, 147–164.

ffytche, D.H., Howard, R.J., Brammer, M.J., David, A., Woodruff, P., & Williams, S. (1998). The anatomy of conscious vision: An fMRI study of visual hallucinations. *Nature Neuroscience, 1*, 738–742.

Ghaem, O., Mellet, E., Crivello, F., Tzourio, N., Mazoyer, B., Berthoz, A., & Denis, M. (1997). Mental navigation along memorized routes activates the hippocampus, precuneus and insula. *NeuroReport, 8*, 739–744.

Goebel, R., Khorram-Sefat, D., Muckli, L., Hacker, H., & Singer, W. (1998). The constructive nature of vision: Direct evidence from functional magnetic resonance imaging studies of apparent motion and motion imagery. *European Journal of Neuroscience, 10,* 1563–1573.

Goldenberg, G., Müllbacher, W., & Nowak, A. (1995). Imagery without perception—A case study of anosognosia for cortical blindness. *Neuropsychologia, 33,* 39–48.

Howard, R.J., ffytche, D.H., Barnes, J., McKeefry, D., Ha, Y., Woodruff, P.W., Bullmore, E.T., Simmons, A., Williams, S.C.R., David, A.S., & Brammer, M. (1998). The functional anatomy of imagining and perceiving colour. *NeuroReport, 9,* 1019–1023.

Hume, D. (1963). A treatise of human nature. In V.C. Chappel (Ed.), *The philosophy of David Hume* (pp. 11–311). New York: Modern Library. (Original work published 1739)

Kosslyn, S.M. (1994). *Image and brain.* Cambridge, MA: MIT Press.

Kosslyn, S.M., Pascual-Leone, A., Felician, O., Camposano, S., Keenan, J.P., Thompson, W.L., Ganis, G., Sukel, K.E., & Alpert, N.M. (1999). The role of Area 17 in visual imagery: Convergent evidence from PET and TMS. *Science, 284,* 167–170.

Kosslyn, S.M., Thompson, W.L., & Alpert, N.M. (1997). Neural systems shared by visual imagery and visual perception: A positron emission tomography study. *Neuroimage, 6,* 320–334.

O'Craven, K., & Kanwisher, N. (in press). Mental imagery of faces and places activates corresponding stimulus-specific brain regions. *Journal of Cognitive Neuroscience.*

Paivio, A. (1979). *Imagery and verbal processes.* Hillsdale, NJ: Erlbaum.

Pylyshyn, Z.W. (1981). The imagery debate: Analogue media versus tacit knowledge. *Psychological Review, 88,* 16–45.

Shepard, R.N., & Cooper, L.A. (1982). *Mental images and their transformations.* Cambridge, MA: MIT Press.

Critical Thinking Questions

1. There are individual differences in imagery capabilities. Some people are slow or inaccurate in answering questions that require imagery (e.g., mentally rotating objects) whereas others are fast. If the mechanisms for imagery and perception overlap, would you expect parallel individual differences in perception?

2. If imagery and perception share mechanisms, one might expect that it would be difficult to use both functions simultaneously. For example, one might expect that subjects would close their eyes to reduce perceptual interference when performing an imagery task. Can you think of evidence from your own experience that such interference might take place?

3. How do dreams compare to images and to perception? Dreams seem more like perception because both can be quite vivid, but mental images usually are not. Also, imagery requires constant attention to be maintained, whereas dreams do not. On the other hand, dreams can be likened to imagery because neither entails retinal stimulation. Do you think that dreams are more like imagery or more like perception? Why?

Synesthesia: Strong and Weak

Gail Martino[1] and Lawrence E. Marks

The John B. Pierce Laboratory, New Haven, Connecticut (G.M., L.E.M.), and Department of Diagnostic Radiology (G.M.) and Department of Epidemiology and Public Health (L.E.M.), Yale Medical School, Yale University, New Haven, Connecticut

Abstract

In this review, we distinguish strong and weak forms of synesthesia. Strong synesthesia is characterized by a vivid image in one sensory modality in response to stimulation in another one. Weak synesthesia is characterized by cross-sensory correspondences expressed through language, perceptual similarity, and perceptual interactions during information processing. Despite important phenomenological dissimilarities between strong and weak synesthesia, we maintain that the two forms draw on similar underlying mechanisms. The study of strong and weak synesthetic phenomena provides an opportunity to enrich scientists' understanding of basic mechanisms involved in perceptual coding and cross-modal information processing.

Keywords

synesthesia; cross-modal perception; selective attention

Color is central to Carol's life. As a professional artist, she uses color to create visual impressions in her paintings. Yet unlike most people, Carol also uses color to diagnose her health. She is able to accomplish this by consulting the colored images she sees in connection with pain. For example, a couple of years ago, Carol fell and damaged her leg badly while climbing on rocks at the beach. She diagnosed the severity of her accident not only by the intensity of her pain, but also by the intensity of the orange color that spread across her mind's eye. She said, "When I saw that everything was orange, I knew I should be rushed to the hospital."

Carol's tendency to see colors in response to pain is an example of strong synesthesia. Synesthesia means "to perceive together," and strong synesthesia occurs when a stimulus produces not only the sensory quality typically associated with that modality, but also a quality typically associated with another modality. Strong synesthesia typically arises on its own, although it also can follow the ingestion of drugs such as mescaline and LSD. In this article, we confine our discussion to synesthesia unrelated to drug use.

Over the two centuries since strong synesthesia was first identified in the scientific literature, several heterogeneous phenomena have been labeled as synesthetic. These phenomena range from strong experiences like Carol's, on the one hand, to weaker cross-modal literary expressions, on the other. We believe it is a mistake to label all of these phenomena simply as synesthesia because the underlying mechanisms cannot be identical, although they may overlap. In this review, we distinguish between *strong synesthesia*, which describes the unusual experiences of individuals such as Carol, and *weak synesthesia*, which describes milder forms of cross-sensory connections revealed through language and perception. In both types of synesthesia, cross-modal correspondences are evident, suggesting that the neural processes underlying strong and weak synesthesia, although not wholly identical, nonetheless may have a common core.

STRONG SYNESTHESIA

The Synesthete

Strong synesthesia is an uncommon condition with an unusual demographic profile. Although estimates vary, one recent estimate places the incidence at 1 in 2,000, with females outnumbering males 6 to 1 (Baron-Cohen, Burt, Laittan-Smith, Harrison, & Bolton, 1996). Strong synesthesia clusters in families, leading some researchers to suggest that it has a genetic basis (Baron-Cohen et al., 1996). Empirical evidence for this notion remains sparse.

Except for the finding that there are more female than male synesthetes, few other generalizations characterize strong synesthetes as a group. Attempts have been made to link synesthesia with artistic creativity: Several strong synesthetes described in case studies have worked in the visual arts or music (Cytowic, 1989). Furthermore, several artists who have produced highly creative work—Kandinsky, Rimbaud, and Scriabin—have drawn inspiration from synesthesia. It is unlikely, however, that these artists were themselves strong synesthetes (Dann, 1998). Thus, there are no empirical data to support the idea that strong synesthetes show high artistic creativity.

Strong Synesthetic Correspondence

Case studies offer several characteristics of strong synesthesia (see Cytowic, 1989). In all cases, an association or correspondence exists between an *inducer* in one modality (e.g., pain in Carol's case) and an *induced* percept or image in another (e.g., color). These correspondences have several salient characteristics. For example, they can be idiosyncratic and systematic at the same time. Correspondences are idiosyncratic in that each synesthete has a unique scheme of associations. Middle C on the piano may be blue to one color-music synesthete and green to another. Yet both synesthetes will reveal a systematic relationship between color brightness or lightness and auditory pitch: The higher the pitch of the sound, the lighter or brighter the color of the image (Marks, 1978). Thus C', an octave above middle C, will evoke a correspondingly brighter blue or a lighter green synesthetic color. Besides pitch-brightness and pitch-lightness, auditory-visual synesthesia reveals several other systematic associations. Notable is the association of pitch to shape and size: The higher the pitch of the sound, the sharper, more angular, and smaller the visual image (Marks, 1978).

Synesthetic images are typically simple (e.g., consist of a color or shape), but dynamic (e.g., as the inducer waxes and wanes, so does the image). In some cases, the induced image is so vivid as to be distracting. S, a mnemonist (professional memorizer) and synesthete described by Luria (1968), explained that "crumbly and yellow" images coming from a speaker's mouth were so intense that

S had difficulty attending to the intended message. This observation exemplifies another general principle—induced images tend to be visual, whereas inducing stimuli tend to be auditory, tactile, or gustatory (Cytowic, 1989). The reason for this asymmetry is unknown.

Because strong synesthesias are noticed in early childhood, it is possible they are inborn. Many strong synesthetes claim that their cross-modal experiences "have always been there" (e.g., Cytowic, 1989). Harrison and Baron-Cohen (1997) argued that the higher incidence of synesthesia in females speaks against synesthesia being learned. If synesthesia is learned, why should more synesthetes be female than male?

The connection between the inducer and induced is so entrenched that the image is considered part of the percept's literal identity. For S to refer to a voice as "crumbly and yellow" is to offer a literal rather than a metaphorical description. Given the close connection between the inducer and the induced, one would expect correspondences to be highly memorable and durable. In a study of color-word synesthesia (Baron-Cohen et al., 1996), strong synesthetes and control subjects were asked to report the color induced by many words (the control subjects named the first color that came to mind). One hour later, the strong synesthetes were 97% accurate in recalling pairs, whereas the control subjects were 13% accurate. Baron-Cohen et al. argued that this result means that synesthetic perception is highly memorable and genuine. It is not clear to us, however, whether to attribute the synesthetes' superior performance to their synesthesia or perhaps to better memory for word-color pairings in general.

Processing

Reports of strong synesthetes offer clues about how strong synesthesia is produced and where it arises. With regard to production, the relation between the inducer and the induced is typically unidirectional. That is, although a voice induces a yellow image, a yellow percept need not induce an image of a voice.

Some investigators postulate that strong synesthesia arises from a disorder within low-level sensory mechanisms. The sensory leakage hypothesis claims that information leaks from one sensory channel into another, producing strong synesthesia (Harrison & Baron-Cohen, 1997). Leakage might occur, for example, if nerve cells fail to form or migrate properly during neonatal development.

Another hypothesis places the mechanism within specific regions of the brain. Measurement of cerebral blood flow in a single strong synesthete, M.W., suggests that increased activation of areas involved in memory and emotion (i.e., limbic areas) and simultaneous suppression of areas involved in higher reasoning (i.e., cortical areas) produce synesthetic perceptions (Cytowic, 1989). This "limbic-cortex disconnection" has not yet been replicated (see Frith & Paulesu, 1997). Failures to replicate may be related to the unique nature of strong synesthesia in each individual or due to methodological differences across studies.

Besides knowing where strong synesthesia occurs, it is important to know how and why it takes on its phenomenological form (meaning known through the senses, rather than through thought or intuition). In this regard, case studies of strong synesthetes have provided some insights. Yet case studies are not sufficient to

Table 1 Summary of claims about strong and weak synesthesia

Characteristic	Synesthesia	
	Strong	Weak
Prevalance	Uncommon, gender bias favoring females	Common
Experience of pairings	One stimulus is perceived, the other is experienced as an image	Both stimuli are perceived
Organization of correspondences	Idiosyncratic and systematic	Systematic
Definition of correspondences	Aboslute	Contextual
Role of learning	Some may be unlearned	Some learned, some unlearned
Semantic association	Literal	Metaphorical
Memory	Easily identified and remembered	Easily identified and remembered
Processing	Unidirectional at a low-level sensory locus	Bidirectional at a high-level semantic locus

explain how and why strong synesthetes' perceptions differ from the norm. Toward this end, future research should be guided by a perceptual or cognitive framework. Table 1 offers a summary of characteristics such a framework must address.

WEAK SYNESTHESIA

The phenomenology of strong synesthesia led us to ask whether individuals who lack strong synesthesia nevertheless show analogous cross-modal associations. There is considerable evidence that one can create, identify, and appreciate cross-modal connections or associations even if one is not strongly synesthetic. These abilities constitute weak synesthesia. One form of association is the cross-modal metaphor found in common language (e.g., *warm* color and *sweet* smell) and in literature (e.g., Baudelaire's poem "Correspondences"). Other evidence for weak synesthesia comes from the domain of music. Some people believe that music, like language, contains cross-modal connections. For example, the idea that pitches and colors are associated motivated the invention of the color organ by Castel and inspired the composition *Prometheus* by Scriabin.

Laboratory experiments square with the idea that most people can appreciate cross-modal associations. In such studies, participants are asked to pair a stimulus from one sensory modality to a stimulus from another. These studies show that pairings are systematic. For example, given a set of notes varying in pitch and a set of colors varying in lightness, the higher the pitch, the lighter the color paired with it (see Marks, 1978). This pitch-lightness relation resembles the one observed in strong synesthesia, with one notable difference. In weak synesthesia, the correspondences are defined by context, so that the highest pitch is always associated with the lightest color. Here lies a distinction between strong and weak synesthesia: Although cross-modal correspondences in weak synesthesia are systematic and contextual, those in strong synesthesia are systematic and absolute (display a one-to-one mapping). Despite this difference, it appears that strong and weak synesthetes share an understanding of how visual and auditory dimensions are related (Marks, 1978).

Role of Learning

Are cross-modal correspondences inborn or learned? The answer appears to be, a bit of both. Infants who have not yet learned language show a kind of cross-modal "matching" of loudness-brightness (Lewkowicz & Turkewitz, 1980) and pitch-position (Wagner, Winner, Cicchetti, & Gardner, 1981). Other correspondences develop over time. For example, 4-year-old childrencan match pitch and brightness systematically, but not pitch and visual size. By age 12, children perform these matches as well as do adults (Marks, Hammeal, & Bornstein, 1987).

Processing

What mental processes underlie the ability to form cross-modal associations? The self-reports of strong synesthetes indicate cross-modal interactions are unidirectional and may involve sensory processes. We wondered whether these characteristics of strong synesthesia apply to cross-modal processing more generally. To address this issue, we developed a cross-modal selective attention task. This task measures a person's ability to respond to a stimulus in one modality while receiving concurrent input from a different, "unattended" modality. If an unattended stimulus affects your ability to respond to an attended one, then the two stimuli are said to interact during information processing.

In a typical task, participants may be asked to classify a sound (high or low tone) in the presence of a color (black or white square), so there are four possible combinations of sounds and colors. Participants are faster at classifying high-pitched tones when these are accompanied by white (vs. black) colors, and are faster at identifying low-pitched tones when they are accompanied by black (vs. white) colors. Analogous results are obtained when participants classify the lightness of a square and color is the unattended stimulus (Melara, 1989). This pattern of findings is termed a congruence effect. Congruence effects entail superior performance when attended and unattended stimuli "match" cross-modally (e.g., high pitch + white square; low pitch + black square) rather than "mismatch" (e.g., high pitch + black square; low pitch + white square). Congruence effects suggest that (a) there is cross-modal interaction (unattended signals can affect one's ability to make decisions about attended ones), (b) the cross-modal correspondence between stimuli is important in determining when interactions occur, and (c) interactions are bidirectional (congruence effects can occur when either sounds or colors are attended). (See Martino & Marks, 1999, for converging evidence.)

Why do congruence effects occur? Two accounts predominate. According to *a sensory hypothesis*, congruence effects involve absolute correspondences processed within low-level sensory mechanisms. These correspondences may arise from common properties in underlying neural codes (e.g., temporal properties of neural impulses may link visual brightness to auditory pitch). This account is consistent with the sensory leakage theory of strong synesthesia and with reports that infants show cross-modal correspondences. Alternatively, we propose that congruence effects involve high-level mechanisms, which develop over childhood from experience with percepts and language—an idea we term the *semantic-coding hypothesis* (SCH; e.g., Martino & Marks, 1999).

The SCH makes four claims. First, although cross-modal correspondence may arise from sensory mechanisms in infants, these correspondences reflect postsensory (meaning-based) mechanisms in adults. Second, experience with percepts from various modalities and the language a person uses to describe these percepts produces an abstract semantic network that captures synesthetic correspondence. Third, when synesthetically corresponding stimuli are perceived, they are recoded from sensory representations into abstract ones based on this semantic network. Fourth, the coding of stimuli from different modalities as matching or mismatching depends on the context within which the stimuli are presented. As mentioned previously, cross-modal matches are defined contextually, so the stimuli are perceived as matching or mismatching only when two or more values are presented in each modality.

Critical support for the SCH comes from selective attention studies like the one described earlier. For example, congruence effects occur when tones and colors both vary from trial to trial, but not when either the tone or the color remains constant (Melara, 1989). The sensory account incorrectly predicts that matches should be processed more efficiently than mismatches in both conditions because correspondences are absolute. The SCH explains the result as a context effect: Trial-by-trial variation provides a context in which to define stimulus values relative to one another, thus highlighting a synesthetic association.

Even stronger evidence for the SCH is that linguistic stimuli are sufficient to drive congruence effects. That is, like the colors black and white, the words *black* and *white* produce congruence effects when paired with low- and high-pitched tones (see Martino & Marks, 1999). The sensory hypothesis cannot account for these findings because interaction is claimed to occur at a sensory level. The SCH accounts for them by proposing that all stimuli (sensory and linguistic) are recoded postperceptually into a single abstract representation that captures the synesthetic correspondence between them.

CONCLUSIONS

Synesthesia is not a unitary phenomenon, but instead takes on strong and weak forms. Strong and weak synesthesia differ in phenomenology, prevalence, and perhaps even some mechanisms underlying their expression. Whereas strong synesthesia expresses itself in perceptual experience proper, weak synesthesia is most clearly evident in cross-modal metaphorical language and in cross-modal matching and selective attention.

Several questions about the nature of strong and weak synesthesia await further investigation. Some concern cognitive and neurological underpinnings: Is strong synesthesia mediated by semantic codes, as weak synesthesia appears to be? Are the brain regions involved in the two kinds of synesthesia similar? For example, do both recruit sensory and semantic areas of the brain? Other issues concern development: To what extent may strong synesthesia be learned or embellished over time? The opportunity to tackle such fundamental questions makes synesthesia an exciting topic for future research.

Recommended Reading

Dann, K.T. (1998). (See References) Harrison, J.E., & Baron-Cohen, S. (1997). (See References)

Marks, L. (1978). (See References)

Acknowledgments—Support was provided by National Institutes of Health (NIH) Training Grant T32 DC00025-13 to the first author and by NIH Grant R01 DC02752 to the second author.

Note

1. Address correspondence to Gail Martino, The John B. Pierce Laboratory, 290 Congress Ave., New Haven, CT 06519; e-mail: gmartino@jbpierce.org.

References

Baron-Cohen, S., Burt, L., Laittan-Smith, F., Harrison, J.E., & Bolton, P. (1996). Synaesthesia: Prevalence and familiarity. *Perception, 25,* 1073–1080.

Cytowic, R.E. (1989). *Synaesthesia: A union of the senses.* Berlin: Springer.

Dann, K.T. (1998). *Bright colors falsely seen.* New Haven, CT: Yale University Press.

Frith, C.D., & Paulesu, E. (1997). The physiological basis of synaesthesia. In S. Baron-Cohen & J.E. Harrison (Eds.), *Synaesthesia: Classic and contemporary readings* (pp. 123–147). Cambridge, MA: Blackwell.

Harrison, J.E., & Baron-Cohen, S. (1997). Synaesthesia: A review of psychological theories. In S. Baron-Cohen & J.E. Harrison (Eds.), *Synaesthesia: Classic and contemporary readings* (pp. 109–122). Cambridge, MA: Blackwell.

Lewkowicz, D.J., & Turkewitz, G. (1980). Cross-modal equivalence in early infancy: Auditory-visual intensity matching. *Developmental Psychology, 16,* 597–601.

Luria, A.R. (1968). *The mind of a mnemonist.* New York: Basic Books.

Marks, L. (1978). *The unity of the senses: Interrelations among the modalities.* New York: Academic Press.

Marks, L.E., Hammeal, R.J., & Bornstein, M.H. (1987). Perceiving similarity and comprehending metaphor. *Monographs of the Society for Research in Child Development, 52*(1, Serial No. 215).

Martino, G., & Marks, L.E. (1999). Perceptual and linguistic interactions in speeded classification: Tests of the semantic coding hypothesis. *Perception, 28,* 903–923.

Melara, R.D. (1989). Dimensional interactions between color and pitch. *Journal of Experimental Psychology: Human Perception and Performance, 15,* 69–79.

Wagner, S., Winner, E., Cicchetti, D., & Gardner, H. (1981). "Metaphorical" mapping in human infants. *Child Development, 52,* 728–731.

Critical Thinking Questions

1. How could one use brain imaging data to test the sensory leakage and the semantic coding hypotheses?

2. Is the semantic coding hypothesis for weak synesthesia just another way of describing metaphor? For example, suppose I demonstrate that most people

agree on a mapping between concepts where one of them is *not* sensory (e.g., honesty is warm, but revenge is cold). Does this mean that weak synesthesia is just metaphor applied to sensation?

3. Synesthesia is described as a mapping between different sensations. Could there be a mapping between movement and sensation (e.g., a synesthete experiences a particular color when a particular hand movement is made). If we don't observe that sort of mapping, does its lack tell us anything interesting about synesthesia?

Memory

Memory is usually the most extensively covered topic in Cognitive Psychology textbooks. It is a central cognitive function and is customarily placed in the middle of those books. Without memory, what use would perception be when we would not recognize what we have seen from one instance to the next? Without memory, how good would reasoning and judgment be when we could not base it on information we have already learned? We rely on the accessibility and correctness of our memories every waking moment; we are frustrated when memory fails (i.e., when we forget) and upset when we learn that something we have long remembered is, in fact, not true.

We usually think of our memories as private and personal; we are the only ones with access to them and they represent the truth about our own lives. But what we remember, and whether we remember it accurately, has both causes and consequences in the real world. The four articles in this section each demonstrate how important memory phenomena from the real world can be studied in the laboratory—and how those laboratory studies then give us insight into how to structure situations in the real world to enhance memory accuracy, memory reliability, and creativity. These articles move between the laboratory, the offices of clinical psychologists, the courtroom, and the business conference room.

In "Imagination and Memory", Garry and Polaschek tackle a hot topic in current memory research—how false memories might arise. There are several types of laboratory studies in which subjects are likely to generate false memories; in many of those studies, subjects are asked to try to remember, visualize, or imagine some event that had not actually occurred in their past. The act of imagining an unreal event can increase people's confidence that the event actually happened.

Erroneous memories can have serious real-world consequences both for the rememberer and for others. In "Recovering Memories of Trauma: A View from the Laboratory", McNally describes attempts to discover the mechanisms behind recovered memories—that is, for memories of traumatic childhood events that have been inaccessible for years and then re-appear in adulthood. Studying recovered memories poses many methodological challenges. McNally describes some differences in personality traits and cognitive processing between people with recovered memories and those with continuous or no memories of childhood trauma. Inaccurate memories can also create problems because the legal system relies extensively on people's memories in criminal proceedings. In "The Confidence of Eyewitnesses in Their Identifications from Lineups", Well, Olson, and Charman describe how the accuracy and confidence of an eyewitness to a crime need not be correlated. An unknown, and probably very large, number of people have been mistakenly convicted of crimes based on faulty eyewitness testimony. The results of psychological research on eyewitness identification has finally begun to have some effect on legal policy.

That memory can be conceptualized as an associative network—a mental web of information constructed such that related concepts are likely to activate each other—may help us understand some of its failings and limitations, including the creation of false memories. However, that conceptualization can also give us ideas about how to overcome blocks in memory and creativity. In "Making Group Brainstorming More Effective: Recommendations from and Associative Memory Perspective," Brown and Paulus use the associative network idea to demonstrate and explain why and how different types of interactions can help to overcome blocking in both individuals and group brainstorming.

Imagination and Memory

Maryanne Garry[1] and Devon L.L. Polaschek

School of Psychology, Victoria University of Wellington, Wellington, New Zealand

Abstract

A growing body of literature shows that imagining contrary-to-truth experiences can change memory. Recent experiments are reviewed to show that when people think about or imagine a false event, entire false memories can be implanted. Imagination inflation can occur even when there is no overt social pressure, and when hypothetical events are imagined only briefly. Overall, studies of imagination inflation show that imagining a counterfactual event can make subjects more confident that it actually occurred. We discuss possible mechanisms for imagination inflation and find that, with evidence supporting the involvement of both source confusion and familiarity in creating inflation, the primary mechanism is still to be determined. We briefly review evidence on individual differences in susceptibility to inflation. Finally, the widespread use of imagination-based techniques in self-help and clinical contexts suggests that there may be practical implications when imagination is used as a therapeutic tool.

Keywords

imagination; autobiographical memory

Consistency, said Oscar Wilde, is the last resort of the unimaginative. A poet, playwright, and all-round observer of human behavior, perhaps Wilde knew what psychologists are just beginning to understand: that imagining the past differently from what it was can change the way one remembers it. Psychologists have known, for 15 years or so, that thinking about a different past can change the way people make sense of it. For instance, a person who gets into a car accident may think, "If I hadn't taken this shortcut, I'd never be in this mess now." Perhaps a more common experience for academics is to have a manuscript rejected and think, "If only the editors understood my genius, they would accept this paper."[2] These contrary-to-truth "what if" scenarios, called counterfactual thoughts, can affect the way people judge bad outcomes, personal choices, and accountability (see Roese, 1997, for a review). But an emerging body of literature shows that imagining contrary-to-truth experiences can do more than change the way people make sense of the past. It can also change memories for it.

Loftus (1993) described the first systematic attempt to create in subjects a coherent, detailed memory for things they never did. She simply asked subjects to read detailed descriptions of four events that supposedly happened during their childhood. In actuality, three of these events had been genuinely experienced, and one—about getting lost in a shopping mall—was false. Subjects were asked to write what they remembered about each of these events in separate sessions over the course of a few weeks, and by the last session of the experiment, about 25% of subjects developed a false shopping-mall memory.

It is easy to underestimate the importance of the lost-in-the-mall studies Loftus reported, but even as recently as the early 1990s, there was very little scientific evidence that entire events could systematically be implanted in memory. It was common for nonscientists to charge that laboratory research was not applicable to real events, particularly traumatic ones. It was also common for psychological scientists to argue that it would be difficult to be misled about personally experienced events, especially childhood events.

How did simply reading about being lost in a shopping mall create memories for that event? Loftus (1993) raised a number of possibilities, but it is likely that imagination played a role. Many subjects probably relied on their imagination as a strategy for remembering being lost, because that is what people do when they try to think about an event that they do not remember (Sarbin, 1998).

Hyman and Pentland (1996) used a similar procedure, and compared subjects' performance when they were given a vague instruction to "think about the event" with their performance when they were given an explicit strategic instruction to imagine it. In this research, the unusual false childhood event was about misbehaving at a wedding, knocking over the punch bowl, and spilling punch all over the parents of the bride. One quarter of the "imagine it" subjects and 9% of the "think about it" subjects developed a clear memory of the false experience. Not only do these results demonstrate the effect of imagination in creating memories, but they also suggest that even some "think about it" subjects probably used imagination as a recall strategy.

These studies show that false memories can be created when people think about (and probably imagine) childhood events in an attempt to remember them. However, these studies also relied on social pressure. For example, subjects were told that a family member provided the to-be-remembered events. In some cases, subjects were also interviewed by an experimenter. What if these demands were reduced? Interestingly, research shows that even without them, imagination can still affect memory.

IMAGINATION INFLATION

In a recent study using a less intensive procedure to examine the effect of imagination on memory (Garry, Manning, Loftus, & Sherman, 1996), subjects were pretested on how confident they were that a number of childhood events had happened, asked to imagine some of those events briefly, and then tested again on their confidence that the events had happened. Subjects became more confident they had experienced imagined counterfactual events than nonimagined counterfactual events. This confidence-boosting effect is known as *imagination inflation*. Since this first study, research has consistently shown that briefly imagining the sketchiest details of a counterfactual event is enough to produce imagination inflation (Garry, Frame, & Loftus, 1999; Goff & Roediger, 1998; Heaps & Nash, 1999; Paddock et al., 1998).

What is it about imagining a counterfactual event that causes people to later become more confident that it really happened? The most obvious explanation is that imagining a childhood event does not really do anything interesting at all; it merely reminds people of genuinely experienced events. Indeed, there is no way to tell how much of the imagination inflation is caused by simple reminding in the studies in which social pressures to remember are reduced (Garry et al., 1996). However, Goff and Roediger (1998) addressed the issue by asking subjects first to do some actions but not others. Later, subjects imagined some of those actions anywhere from zero to five times. Goff and Roediger found evidence of imagination inflation, as well as an effect for the number of imaginings.

There are two main explanations for imagination inflation. The first of these is source confusion (Johnson, Hashtroudi, & Lindsay, 1993), a process in which content and source (i.e., the circumstances in which information was learned) become separated. In a source-confusion account of imagination inflation, subjects are said to confuse information from a recently imagined event with information from a genuinely experienced event. The other explanation for imagination inflation is familiarity. In this account of imagination inflation, imagining events makes them more familiar, and subjects incorrectly attribute the increased familiarity of imagined events to their occurrence (see also Jacoby, Kelley, & Dywan, 1989).

Goff and Roediger (1998) argued that imagination inflation is the product of overlapping source confusion and familiarity mechanisms, but a recent experiment (Garry et al., 1999) tried to separate the two. Subjects were asked to imagine either certain childhood events happening to them or those same childhood events happening to another person. Subjects in the two groups showed the same amount of imagination inflation. Moreover, subjects' rating of various imagery qualities and characteristics did not predict the amount of imagination inflation. Both of these findings are evidence that imagination inflation is caused more by familiarity than by source confusion.

Of course, there are alternative explanations for these results. One is that subjects may have pretended to be the other character; for example, if Beth pretended to be an 8-year-old Bill Clinton breaking a window with his hand, the images she has of the event would be from young Bill's perspective, and thus as easily confused with a real event as if Beth imagined breaking the window as a child herself. So the results do not necessarily indicate that source confusion plays no role in imagination inflation. Indeed, evidence for the primary role of source confusion in imagination inflation is found in studies that have varied the time distance of hypothetical events. A source-confusion mechanism predicts greater imagination inflation for long-ago imagined events compared with more recent imagined events, whereas a familiarity mechanism predicts no difference in the amount of imagination inflation. These opposing predictions led to experiments (Garry & Hayes, 1999) in which some subjects imagined hypothetical childhood events and others imagined more recent events from only 5 years ago. The subjects who imagined the long-ago childhood events showed the typical imagination-inflation effect, but those who were asked to imagine the recent events showed no change in confidence.

WHO IS SUSCEPTIBLE TO IMAGINATION INFLATION?

Are some people more likely to inflate than others? This is a question several researchers have asked. Heaps and Nash (1999) found that subjects' tendency toward imagination inflation had nothing to do with their susceptibility to the influence of an authoritative questioner, nor with the vividness of their mental images. Instead, it was their predisposition to both hypnotic suggestion and dissociation (i.e., tendency to lose awareness and confuse fact with fantasy) that predicted imagination inflation, a finding supported by the work of Paddock et al. (1998). Paddock et al. also suggested that age may be inversely related to imagination inflation. It is not certain if age and education somehow modulate the personal characteristics that predispose people to imagination inflation; much research suggests, for instance, that both hypnotic suggestibility and dissociativity decrease with age.

Of course, age is surely the most contentious of the individual differences considered in the false memory literature. Researchers studying the suggestibility of children have long been aware of the dangerous consequences of thinking about a false event. Much research shows that children can adopt misleading suggestions about aspects of genuine events, or come to report entire, elaborate false events (Ceci, Loftus, Leichtman, & Bruck, 1994). Are children susceptible to imagination inflation? When 8- to 10-year-olds imagined hypothetical events from 5 years earlier, the children showed inflated confidence that the events had really happened (Garry & Hayes, 1999).

THEORETICAL IMPLICATIONS

A growing literature shows that imagination can change autobiographies. The most important theoretical implication of this research is the Heisenberg-like suggestion that repeatedly examining one's past can affect the way one remembers it. In the imagination-inflation literature, several important questions remain. It is still not clear, for instance, what the primary mechanism driving imagination inflation is, nor is it known under what circumstances imagining recent autobiographical events might boost confidence that they occurred. Interestingly, Goff and Roediger (1998) actually found no imagination effect when recent events were imagined only once, the condition that most closely parallels the standard imagination-inflation procedure, which examines naturalistic, autobiographical events.

Finally, there are two statistical issues to consider in making sense of imagination inflation. First, although in one sense it is impressive that a seemingly innocuous, brief imagination can affect people's confidence in whether or not an event actually occurred, from a statistical point of view, the imagination-inflation effect is a small one. Future research should look toward increasing the size of the imagination-inflation effect; there are aspects of the current experimental procedures that might be masking or moderating the effect. Second, one might at first glance be tempted to argue that the effect is nothing other than "regression toward the mean," a statistical phenomenon in which people with extremely low scores on some measure (in this case, confidence in whether an event actually occurred) will move toward the population mean on that measure when tested again. In the original imagination-inflation research (Garry et al., 1996), we selected subjects with low confidence scores, and on a retest their scores were higher. However, an interpretation of the results in terms of regression to the mean is easily dismissed, because the same people participated in both the imagination condition and the control, no-imagination condition, and confidence did not rise as much in the control condition as in the imagination condition. Moreover, in both conditions, pretest scores were equally low on the critical events. Thus, regression toward the mean cannot explain the imagination-inflation effect.

CLINICAL IMPLICATIONS

There are practical implications of this research, too. Although the clinical implications of the lost-in-the-mall genre of studies are obvious, what may be less obvious is that the results of the imagination studies (e.g., Garry et al., 1996) also give cause for concern. The imagination-inflation literature may have implications for therapists who are involved in a much broader range of clinical endeavors than working with survivors of sexual abuse. In particular, Goff and Roediger's (1998) findings show that even when people imagine performing certain actions in the present, they can report, mistakenly, that they genuinely did them.

A review of the psychotherapy and self-help literatures suggests that imagined performance features in a variety of therapeutic strategies. For example, cognitive behavioral psychologists use a number of techniques that rely heavily on the use of imagination, particularly in the treatment of anxiety disorders. If Bo Peep is instructed to imagine a series of sheep-related feared situations to reduce her sheep phobia, she will probably become less anxious about sheep, but she may also show inflated confidence that she actually experienced sheep encounters that in fact she only imagined.

The self-help literature is also replete with imagination-based interventions. Books on creative visualization exhort readers to imagine themselves achieving anything from younger-looking skin to becoming president of the United States. In more academic circles, researchers have found that mental simulation can enhance coping with stressful events and improve athletic performance (Orlick & Partington, 1986; Taylor, Pham, Rivkin, & Armor, 1998). These techniques rely on repeatedly imagining situations and actions.

Of course, the unanticipated memory-related consequences suggested by imagination-inflation research may not always be serious. Incorrectly recalling that a particular experience or action actually occurred although it was only imagined may not matter outside of evidential contexts. Nevertheless, good clinical practice requires that clinicians understand both how efficacious treatments work and what their possible side effects are, and further research should be aimed at informing both science and practice. Meanwhile, the growing literature on imagination and memory suggests that there can be some unexpected side effects to using imagination as a therapeutic tool. Inconsistency, at least in the way people remember their own lives, may be the price people pay for being imaginative.

Recommended Reading

Garry, M., Frame, S., & Loftus, E.F. (1999). (See References)
Loftus, E.F. (1997, September). Creating false memories. *Scientific American, 277,* 70–75.

Ofshe, R., & Watters, E. (1994). *Making monsters: False memories, psychotherapy, and sexual hysteria.* Berkeley: University of California.

Notes

1. Address correspondence to Maryanne Garry, School of Psychology, Victoria University of Wellington, Box 600, Wellington, New Zealand.
2. Obviously, we are speaking here about our own experiences.

References

Ceci, S.J., Loftus, E.F., Leichtman, M.D., & Bruck, M. (1994). The possible role of source misattributions in the creation of false beliefs among preschoolers. *International Journal of Clinical and Experimental Hypnosis, 42,* 304–320.

Garry, M., Frame, S., & Loftus, E.F. (1999). Lie down and let me tell you about your childhood. In S. Della Sala (Ed.), *Mind myths: Exploring popular assumptions about the mind and brain* (pp. 113–124). New York: Wiley.

Garry, M., & Hayes, J.H. (1999). *Imagination inflation depends on when the imagined event occurred.* Unpublished manuscript, Victoria University of Wellington, Wellington, New Zealand.

Garry, M., Manning, C.G., Loftus, E.F., & Sherman, S.J. (1996). Imagination inflation: Imagining a childhood event inflates confidence that it occurred. *Psychonomic Bulletin & Review, 3,* 208–214.

Goff, L.M., & Roediger, H.L., III. (1998). Imagination inflation for action events: Repeated imaginings lead to illusory recollections. *Memory & Cognition, 26,* 20–33.

Heaps, C., & Nash, M.R. (1999). Individual differences in imagination inflation. *Psychonomic Bulletin & Review, 6,* 313–318.

Hyman, I.E., & Pentland, J. (1996). The role of mental imagery in the creation of false childhood memories. *Journal of Memory and Language, 35,* 101–117.

Jacoby, L.L., Kelley, C.M., & Dywan, J. (1989). Memory attributions. In H.L. Roediger, III, & F.I.M. Craik (Eds.), *Varieties of memory and consciousness: Essays in honour of Endel Tulving* (pp. 391–422). Hillsdale, NJ: Erlbaum.

Johnson, M.K., Hashtroudi, S., & Lindsay, D.S. (1993). Source monitoring. *Psychological Bulletin, 114,* 3–28.

Loftus, E.F. (1993). The reality of repressed memories. *American Psychologist, 48,* 518–537.

Orlick, T., & Partington, J.T. (1986). *Psyched: Inner views of winning.* Ottawa, Canada: Coaching Association of Canada.

Paddock, J.R., Joseph, A.L., Chan, F.M., Terranova, S., Loftus, E.F., & Manning, C. (1998). When guided visualization procedures may backfire: Imagination inflation and predicting individual differences in suggestibility. *Applied Cognitive Psychology, 12,* S63–S75.

Roese, N.J. (1997). Counterfactual thinking. *Psychological Bulletin, 121,* 133–148.

Sarbin, T.R. (1998). Believed-in imaginings: A narrative approach. In J. de Rivera & T.R. Sarbin (Eds.), *Believed-in imaginings: The narrative construction of reality* (pp. 15–30). Washington, DC: American Psychological Association.

Taylor, S.E., Pham, L.B., Rivkin, I.D., & Armor, D.A. (1998). Harnessing the imagination: Mental simulation, self-regulation, and coping. *American Psychologist, 53,* 429–439.

Critical Thinking Questions

1. Think back to your 4th (or 3rd or 5th) birthday. Close your eyes and try to imagine it. Was there a party? Was there a cake with candles for you to blow out? Presents? (a) Can you envision that thinking about your birthday this way might cause you to believe that you had a birthday party when you really hadn't? Why or why not? (b) Perhaps you do have memories of an early birthday. Are you sure those memories are your own? Might those memories come

from stories that older people (e.g., your parents or siblings) have told you? Might they come from having seen pictures or videos of that birthday? How can you be sure?

2. One of the editors of this book has imagined herself being Scarlett O'Hara (from Gone With the Wind) dozens of times. Yet she does not believe that she has ever owned slaves, picked cotton, or worn a dress made of her mother's curtains. The authors argue that familiarity may be the mechanism behind false memories and imagination inflation, but is familiarity sufficient? In Question 1 above, do you think you might ever believe that: You got a (real) giraffe as a gift? Two Martians came to your party? What do you think the limits are to the kinds of events that for which memory could be affected by imagination? What might that suggest about people's recovered memories for things like satanic rituals?

3. Imagination inflation has been shown to operate more for long-ago events than for recent events. Why might that happen?

Recovering Memories of Trauma:
A View From the Laboratory

Richard J. McNally[1]

Department of Psychology, Harvard University, Cambridge, Massachusetts

Abstract

The controversy over the validity of repressed and recovered memories of childhood sexual abuse (CSA) has been extraordinarily bitter. Yet data on cognitive functioning in people reporting repressed and recovered memories of trauma have been strikingly scarce. Recent laboratory studies have been designed to test hypotheses about cognitive mechanisms that ought to be operative if people can repress and recover memories of trauma or if they can form false memories of trauma. Contrary to clinical lore, these studies have shown that people reporting CSA histories are not characterized by a superior ability to forget trauma-related material. Other studies have shown that individuals reporting recovered memories of either CSA or abduction by space aliens are characterized by heightened proneness to form false memories in certain laboratory tasks. Although cognitive psychology methods cannot distinguish true memories from false ones, these methods can illuminate mechanisms for remembering and forgetting among people reporting histories of trauma.

Keywords

recovered memories; trauma; repression; sexual abuse; dissociation

How victims remember trauma is among the most explosive issues facing psychology today. Most experts agree that combat, rape, and other horrific experiences are unforgettably engraved on the mind (Pope, Oliva, & Hudson, 1999). But some also believe that the mind can defend itself by banishing traumatic memories from awareness, making it difficult for victims to remember them until many years later (Brown, Scheflin, & Hammond, 1998).

This controversy has spilled out of the clinics and cognitive psychology laboratories, fracturing families, triggering legislative change, and determining outcomes in civil suits and criminal trials. Most contentious has been the claim that victims of childhood sexual abuse (CSA) often repress and then recover memories of their trauma in adulthood.[2] Some psychologists believe that at least some of these memories may be false—inadvertently created by risky therapeutic methods (e.g., hypnosis, guided imagery; Ceci & Loftus, 1994).

One striking aspect of this controversy has been the paucity of data on cognitive functioning in people reporting repressed and recovered memories of CSA. Accordingly, my colleagues and I have been conducting studies designed to test hypotheses about mechanisms that might enable people either to repress and recover memories of trauma or to develop false memories of trauma.

For several of our studies, we recruited four groups of women from the community. Subjects in the *repressed-memory group* suspected they had been sexually abused as children, but they had no explicit memories of abuse. Rather, they inferred their hidden abuse history from diverse indicators, such as depressed mood, interpersonal problems with men, dreams, and brief, recurrent visual images (e.g., of a penis), which they interpreted as "flashbacks" of early trauma. Subjects in the *recovered-memory group* reported having remembered their abuse after long periods of not having thought about it.[3] Unable to corroborate their reports, we cannot say whether the memories were true or false. Lack of corroboration, of course, does not mean that a memory is false. Subjects in the *continuous-memory group* said that they had never forgotten their abuse, and subjects in the *control group* reported never having been sexually abused.

PERSONALITY TRAITS AND PSYCHIATRIC SYMPTOMS

To characterize our subjects in terms of personality traits and psychiatric symptoms, we asked them to complete a battery of questionnaires measuring normal personality variation (e.g., differences in absorption, which includes the tendency to fantasize and to become emotionally engaged in movies and literature), depressive symptoms, posttraumatic stress disorder (PTSD) symptoms, and dissociative symptoms (alterations in consciousness, such as memory lapses, feeling disconnected with one's body, or episodes of "spacing out"; McNally, Clancy, Schacter, & Pitman, 2000b).

There were striking similarities and differences among the groups in terms of personality profiles and psychiatric symptoms. Subjects who had always remembered their abuse were indistinguishable from those who said they had never been abused on all personality measures. Moreover, the continuous-memory and control groups did not differ in their symptoms of depression, posttraumatic stress, or dissociation. However, on the measure of negative affectivity—proneness to experience sadness, anxiety, anger, and guilt—the repressed-memory group scored higher than did either the continuous-memory or the control group, whereas the recovered-memory group scored midway between the repressed-memory group on the one hand and the continuous-memory and control groups on the other.

The repressed-memory subjects reported more depressive, dissociative, and PTSD symptoms than did continuous-memory and control subjects. Repressed-memory subjects also reported more depressive and PTSD symptoms than did recovered-memory subjects, who, in turn, reported more dissociative and PTSD symptoms than did control subjects. Finally, the repressed and recovered-memory groups scored higher than the control group on the measure of fantasy proneness, and the repressed-memory group scored higher than the continuous-memory group on this measure.

This psychometric study shows that people who believe they harbor repressed memories of sexual abuse are more psychologically distressed than those who say they have never forgotten their abuse.

FORGETTING TRAUMA RELATED MATERIAL

Some clinical theorists believe that sexually molested children learn to disengage their attention during episodes of abuse and allocate it elsewhere (e.g., Terr,

1991). If CSA survivors possess a heightened ability to disengage attention from threatening cues, impairing their subsequent memory for them, then this ability ought to be evident in the laboratory. In our first experiment, we used directed-forgetting methods to test this hypothesis (McNally, Metzger, Lasko, Clancy, & Pitman, 1998). Our subjects were three groups of adult females: CSA survivors with PTSD, psychiatrically healthy CSA survivors, and nonabused control subjects. Each subject was shown, on a computer screen, a series of words that were either trauma related (e.g., *molested*), positive (e.g., *charming*), or neutral (e.g., *mailbox*). Immediately after each word was presented, the subject received instructions telling her either to remember the word or to forget it. After this encoding phase, she was asked to write down all the words she could remember, irrespective of the original instructions that followed each word.

If CSA survivors, especially those with PTSD, are characterized by heightened ability to disengage attention from threat cues, thereby attenuating memory for them, then the CSA survivors with PTSD in this experiment should have recalled few trauma words, especially those they had been told to forget. Contrary to this hypothesis, this group exhibited memory deficits for positive and neutral words they had been told to remember, while demonstrating excellent memory for trauma words, including those they had been told to forget. Healthy CSA survivors and control subjects recalled remember-words more often than forget-words regardless of the type of word. Rather than possessing a superior ability to forget trauma-related material, the most distressed survivors exhibited difficulty banishing this material from awareness.

In our next experiment, we used this directed-forgetting approach to test whether repressed- and recovered-memory subjects, relative to nonabused control subjects, would exhibit the hypothesized superior ability to forget material related to trauma (McNally, Clancy, & Schacter, 2001). If anyone possesses this ability, it ought to be such individuals. However, the memory performance of the repressed- and recovered-memory groups was entirely normal: They recalled remember-words better than forget-words, regardless of whether the words were positive, neutral, or trauma related.

INTRUSION OF TRAUMATIC MATERIAL

The hallmark of PTSD is involuntary, intrusive recollection of traumatic experiences. Clinicians have typically relied on introspective self-reports as confirming the presence of this symptom. The emotional Stroop color-naming task provides a quantitative, nonintrospective measure of intrusive cognition. In this paradigm, subjects are shown words varying in emotional significance, and are asked to name the colors the words are printed in while ignoring the meanings of the words. When the meaning of a word intrusively captures the subject's attention despite the subject's efforts to attend to its color, Stroop interference—delay in color naming—occurs. Trauma survivors with PTSD take longer to name the colors of words related to trauma than do survivors without the disorder, and also take longer to name the colors of trauma words than to name the colors of positive and neutral words or negative words unrelated to their trauma (for a review, see McNally, 1998).

Using the emotional Stroop task, we tested whether subjects reporting either continuous, repressed, or recovered memories of CSA would exhibit interference for trauma words, relative to nonabused control subjects (McNally, Clancy, Schacter, & Pitman, 2000a). If severity of trauma motivates repression of traumatic memories, then subjects who cannot recall their presumably repressed memories may nevertheless exhibit interference for trauma words. We presented a series of trauma-related, positive, and neutral words on a computer screen, and subjects named the colors of the words as quickly as possible. Unlike patients with PTSD, including children with documented abuse histories (Dubner & Motta, 1999), none of the groups exhibited delayed color naming of trauma words relative to neutral or positive ones.

MEMORY DISTORTION AND FALSE MEMORIES IN THE LABORATORY

Some psychotherapists who believe their patients suffer from repressed memories of abuse will ask them to visualize hypothetical abuse scenarios, hoping that this guided-imagery technique will unblock the presumably repressed memories. Unfortunately, this procedure may foster false memories.

Using Garry, Manning, Loftus, and Sherman's (1996) methods, we tested whether subjects who have recovered memories of abuse are more susceptible than control subjects to this kind of memory distortion (Clancy, McNally, & Schacter, 1999). During an early visit to the laboratory, subjects rated their confidence regarding whether they had experienced a series of unusual, but nontraumatic, childhood events (e.g., getting stuck in a tree). During a later visit, they performed a guided-imagery task requiring them to visualize certain of these events, but not others. They later rerated their confidence that they had experienced each of the childhood events. Nonsignificant trends revealed an inflation in confidence for imagined versus non-imagined events. But the magnitude of this memory distortion was more than twice as large in the control group as in the recovered memory group, contrary to the hypothesis that people who have recovered memories of CSA would be especially vulnerable to the memory-distorting effects of guided imagery.

To use a less-transparent paradigm for assessing proneness to develop false memories, we adapted the procedure of Roediger and McDermott (1995). During the encoding phase in this paradigm, subjects hear word lists, each consisting of semantically related items (e.g., *sour, bitter, candy,* sugar) that converge on a nonpresented word—the *false target*—that captures the gist of the list (e.g., *sweet*). On a subsequent recognition test, subjects are given a list of words and asked to indicate which ones they heard during the previous phase. The false memory effect occurs when subjects "remember" having heard the false target. We found that recovered-memory subjects exhibited greater proneness to this false memory effect than did subjects reporting either repressed memories of CSA, continuous memories of CSA, or no abuse (Clancy, Schacter, McNally, & Pitman, 2000). None of the lists was trauma related, and so we cannot say whether the effect would have been more or less pronounced for words directly related to sexual abuse.

In our next experiment, we tested people whose memories were probably false: individuals reporting having been abducted by space aliens (Clancy, McNally, Schacter, Lenzenweger, & Pitman, 2002). In addition to testing these individuals (and control subjects who denied having been abducted by aliens), we tested individuals who believed they had been abducted, but who had no memories of encountering aliens. Like the repressed-memory subjects in our previous studies, they inferred their histories of trauma from various "indicators" (e.g., a passion for reading science fiction, unexplained marks on their bodies). Like subjects with recovered memories of CSA, those reporting recovered memories of alien abduction exhibited pronounced false memory effects in the laboratory. Subjects who only believed they had been abducted likewise exhibited robust false memory effects.

CONCLUSIONS

The aforementioned experiments illustrate one way of approaching the recovered-memory controversy. Cognitive psychology methods cannot ascertain whether the memories reported by our subjects were true or false, but these methods can enable testing of hypotheses about mechanisms that ought to be operative if people can repress and recover memories of trauma or if they can develop false memories of trauma.

Pressing issues remain unresolved. For example, experimentalists assume that directed forgetting and other laboratory methods engage the same cognitive mechanisms that generate the signs and symptoms of emotional disorder in the real world. Some therapists question the validity of this assumption. Surely, they claim, remembering or forgetting the word *incest* in a laboratory task fails to capture the sensory and narrative complexity of autobiographical memories of abuse. On the one hand, the differences between remembering the word *incest* in a directed-forgetting experiment, for example, and recollecting an episode of molestation do, indeed, seem to outweigh the similarities. On the other hand, laboratory studies may underestimate clinical relevance. For example, if someone cannot expel the word *incest* from awareness during a directed-forgetting experiment, then it seems unlikely that this person would be able to banish autobiographical memories of trauma from consciousness. This intuition notwithstanding, an important empirical issue concerns whether these tasks do, indeed, engage the same mechanisms that figure in the cognitive processing of traumatic memories outside the laboratory.

A second issue concerns attempts to distinguish subjects with genuine memories of abuse from those with false memories of abuse. Our group is currently exploring whether this might be done by classifying trauma narratives in terms of how subjects describe their memory-recovery experience. For example, some of the subjects in our current research describe their recovered memories of abuse by saying, "I had forgotten about that. I hadn't thought about the abuse in years until I was reminded of it recently." The narratives of other recovered-memory subjects differ in their experiential quality. These subjects, as they describe it, suddenly realize that they are abuse survivors, sometimes attributing current life difficulties to these long-repressed memories. That is, they do not say that they have remembered forgotten events they once knew, but rather indicate that they have learned (e.g., through hypnosis) the abuse occurred. It will be important to determine whether these two groups of recovered-memory subjects differ cognitively. For example, are subjects exemplifying the second type of recovered-memory experience more prone to develop false memories in the laboratory than are subjects exemplifying the first type of experience?

Recommended Reading

Lindsay, D.S., & Read, J.D. (1994). Psychotherapy and memories of childhood sexual abuse: A cognitive perspective. *Applied Cognitive Psychology, 8*, 281–338.

McNally, R.J. (2001). The cognitive psychology of repressed and recovered memories of childhood sexual abuse: Clinical implications. *Psychiatric Annals, 31*, 509–514.

McNally, R.J. (2003). Progress and controversy in the study of posttraumatic stress disorder. *Annual Review of Psychology, 54*, 229–252.

McNally, R.J. (2003). *Remembering trauma*. Cambridge, MA: Belknap Press/Harvard University Press.

Piper, A., Jr., Pope, H.G., Jr., & Borowiecki, J.J., III. (2000). Custer's last stand: Brown, Scheflin, and Whitfield's latest attempt to salvage "dissociative amnesia." *Journal of Psychiatry and Law, 28*, 149–213.

Acknowledgments—Preparation of this article was supported in part by National Institute of Mental Health Grant MH61268.

Notes

1. Address correspondence to Richard J. McNally, Department of Psychology, Harvard University, 1230 William James Hall, 33 Kirkland St., Cambridge, MA 02138; e-mail: rjm@wjh.harvard.edu.

2. Some authors prefer the term *dissociation* (or *dissociative amnesia*) to *repression*. Although these terms signify different proposed mechanisms, for practical purposes these variations make little difference in the recovered-memory debate. Each term implies a defensive process that blocks access to disturbing memories.

3. However, not thinking about a disturbing experience for a long period of time must not be equated with an inability to remember it. Amnesia denotes an inability to recall information that has been encoded.

References

Brown, D., Scheflin, A.W., & Hammond, D.C. (1998). *Memory, trauma treatment, and the law*. New York: Norton.

Ceci, S.J., & Loftus, E.F. (1994). 'Memory work': A royal road to false memories? *Applied Cognitive Psychology, 8*, 351–364.

Clancy, S.A., McNally, R.J., & Schacter, D.L. (1999). Effects of guided imagery on memory distortion in women reporting recovered memories of childhood sexual abuse. *Journal of Traumatic Stress, 12*, 559–569.

Clancy, S.A., McNally, R.J., Schacter, D.L., Lenzenweger, M.F., & Pitman, R.K. (2002). Memory distortion in people reporting abduction by aliens. *Journal of Abnormal Psychology, 111*, 455–461.

Clancy, S.A., Schacter, D.L., McNally, R.J., & Pitman, R.K. (2000). False recognition in women reporting recovered memories of sexual abuse. *Psychological Science, 11*, 26–31.

Dubner, A.E., & Motta, R.W. (1999). Sexually and physically abused foster care children and posttraumatic stress disorder. *Journal of Consulting and Clinical Psychology, 67*, 367–373.

Garry, M., Manning, C.G., Loftus, E.F., & Sherman, S.J. (1996). Imagination inflation: Imagining a childhood event inflates confidence that it occurred. *Psychonomic Bulletin & Review, 3,* 208–214.

McNally, R.J. (1998). Experimental approaches to cognitive abnormality in posttraumatic stress disorder. *Clinical Psychology Review, 18,* 971–982.

McNally, R.J., Clancy, S.A., & Schacter, D.L. (2001). Directed forgetting of trauma cues in adults reporting repressed or recovered memories of childhood sexual abuse. *Journal of Abnormal Psychology, 110,* 151–156.

McNally, R.J., Clancy, S.A., Schacter, D.L., & Pitman, R.K. (2000a). Cognitive processing of trauma cues in adults reporting repressed, recovered, or continuous memories of childhood sexual abuse. *Journal of Abnormal Psychology, 109,* 355–359.

McNally, R.J., Clancy, S.A., Schacter, D.L., & Pitman, R.K. (2000b). Personality profiles, dissociation, and absorption in women reporting repressed, recovered, or continuous memories of childhood sexual abuse. *Journal of Consulting and Clinical Psychology, 68,* 1033–1037.

McNally, R.J., Metzger, L.J., Lasko, N.B., Clancy, S.A., & Pitman, R.K. (1998). Directed forgetting of trauma cues in adult survivors of childhood sexual abuse with and without posttraumatic stress disorder. *Journal of Abnormal Psychology, 107,* 596–601.

Pope, H.G., Jr., Oliva, P.S., & Hudson, J.I. (1999). Repressed memories: The scientific status. In D.L. Faigman, D.H. Kaye, M.J. Saks, & J. Sanders (Eds.), *Modern scientific evidence: The law and science of expert testimony* (Vol. 1, pocket part, pp. 115–155). St. Paul, MN: West Publishing.

Roediger, H.L., III, & McDermott, K.B. (1995). Creating false memories: Remembering words not presented in lists. *Journal of Experimental Psychology: Learning, Memory, and Cognition, 21,* 803–814.

Terr, L.C. (1991). Childhood traumas: An outline and overview. *American Journal of Psychiatry, 148,* 10–20.

Critical Thinking Questions

1. Studying recovered memory provides many challenges. A major difficulty, of course, is that the researcher is almost never certain that the remembered event(s) actually occurred. In some of the present research, the groups studied were recovered-memory, repressed-memory, continuous-memory, and control. Why is it important to have each of those groups? Why was it important to do similar studies with the CSA+PTSD, CSA, and control groups? And, silly as it may seem, why was it important to do similar studies on people with memories of alien abduction?

2. Another difficulty is provided by ethical constraints in research. Psychologists can't randomly select a bunch of people to traumatize and then assign them to the recovered-memory condition. What implications does that limitation have for the causal direction of the finding that the recovered-memory and repressed-memory groups differ on some personality dimensions from the continuous-memory and control groups?

3. The article describes four different experimental procedures – directed forgetting, emotional Stroop, imagination inflation, and the false-target list. Only on the last one did the recovered-memory group significantly differ from the others. To what extent do you believe that each of these procedures captures something about the mechanisms behind memory repression? What about the mechanisms behind how false memories might occur for real-life traumatic events that occurred over a long period of time?

4. Suppose you read in the newspaper about three women who claimed to have recently recovered memories for childhood sexual abuse. One remembered the abuse after months of therapy for relationship issues in which the therapist suggested that she try to imagine how she might have been sexually abused as a child. The second remembered the abuse when she realized it made sense as an explanation for a pattern of beliefs and behaviors in her life (e.g., why she was uncomfortable in sexual relationships; why she had body-image problems). The third remembered the abuse out of the blue when she entered a child's bedroom that had the same wallpaper as had been in her childhood bedroom. Of course, all three memories could be true, but which are you most and least likely to believe. Why?

The Confidence of Eyewitnesses in Their Identifications From Lineups

Gary L. Wells,[1] Elizabeth A. Olson,
and Steve D. Charman
Psychology Department, Iowa State University, Ames, Iowa

Abstract

The confidence that eyewitnesses express in their lineup identifications of criminal suspects has a large impact on criminal proceedings. Many convictions of innocent people can be attributed in large part to confident but mistaken eyewitnesses. Although reasonable correlations between confidence and accuracy can be obtained under certain conditions, confidence is governed by some factors that are unrelated to accuracy. An understanding of these confidence factors helps establish the conditions under which confidence and accuracy are related and leads to important practical recommendations for criminal justice proceedings.

Keywords

eyewitness testimony; lineups; eyewitness memory

Mistaken identification by eyewitnesses was the primary evidence used to convict innocent people whose convictions were later overturned by forensic DNA tests (Scheck, Neufeld, & Dwyer, 2000; Wells et al., 1998). The eyewitnesses in these cases were very persuasive because on the witness stand they expressed extremely high confidence that they had identified the actual perpetrator. Long before DNA exoneration cases began unfolding in the 1990s, however, eyewitness researchers in psychology were finding that confidence is not a reliable indicator of accuracy and warning the justice system that heavy reliance on eyewitness's confidence in their identifications might lead to the conviction of innocent people.

Studies have consistently demonstrated that the confidence an eyewitness expresses in an identification is the major factor determining whether people will believe that the eyewitness made an accurate identification. The confidence an eyewitness expresses is also enshrined in the criteria that the U.S. Supreme Court used 30 years ago (and that now guide lower courts) for deciding the accuracy of an eyewitness's identification in a landmark case. Traditionally, much of the experimental work examining the relation between confidence and accuracy in eyewitness identification tended to frame the question as "What is the correlation between confidence and accuracy?" as though there were some single, true correlation value. Today, eyewitness researchers regard the confidence-accuracy relation as something that varies across circumstances. Some of these circumstances are outside the control of the criminal justice system, but some are determined by the procedures that criminal justice personnel control.

A GENERAL FRAMEWORK FOR CONFIDENCE ACCURACY RELATIONS

It has been fruitful to think about eyewitness accuracy and eyewitness confidence as variables that are influenced by numerous factors, some of which are the same and some of which are different. We expect confidence and accuracy to be more closely related when the variables that are influencing accuracy are also influencing confidence than when the variables influencing accuracy are different from those influencing confidence. Consider, for instance, the variable of exposure duration (i.e., how long the eyewitness viewed the culprit while the crime was committed). An eyewitness who viewed the culprit for a long time during the crime should be more accurate than one who had only a brief view. Furthermore, the longer view could be a foundation for the eyewitness to feel more confident in the identification, either because the witness has a more vivid and fluent memory from the longer duration or because the witness infers his or her accuracy from the long exposure duration. Hence, the correlation between confidence and accuracy should be higher the more variation there is in the exposure duration across witnesses (Read, Vokey, & Hammersley, 1990). Suppose, however, that some eyewitnesses were reinforced after their identification decision (e.g., "Good job. You are a good witness."), whereas others were given no such reinforcement. Such postidentification reinforcement does nothing to make witnesses more accurate, but dramatically inflates their confidence (Wells & Bradfield, 1999).

Eyewitness confidence can be construed simply as the eyewitness's belief, which varies in degree, about whether the identification was accurate or not. This belief can have various sources, both internal and external, that need not be related to accuracy. Shaw and his colleagues, for example, have shown that repeated questioning of eyewitnesses about mistaken memories does not make the memories more accurate but does inflate the eyewitnesses' confidence in those memories (Shaw, 1996; Shaw & McLure, 1996). Although the precise mechanisms for the repeated-questioning effect are not clear (e.g., increased commitment to the mistaken memory vs. increased fluidity of the response), these results illustrate a dissociation between variables affecting confidence and variables affecting accuracy.

It is useful to think about broad classes of variables that could be expected to drive confidence and not accuracy, or to drive accuracy and not confidence, or to drive both variables. It is even possible to think about variables that could decrease accuracy while increasing confidence. Consider, for instance, coincidental resemblance. Mistaken identifications from lineups occur primarily when the actual culprit is not in the lineup. Suppose there are two such lineups, one in which the innocent suspect does not highly resemble the real culprit and a second in which the innocent suspect is a near clone (coincidental resemblance) of the real culprit. The second lineup will result not only in an increased rate of mistaken identification compared with the first lineup, but also in higher confidence in that mistake. In this case, a variable that decreases accuracy (resemblance of an innocent suspect to the actual culprit) serves to increase confidence.

THE CORRELATION, CALIBRATION, AND INFLATION OF CONFIDENCE

Although many individual studies have reported little or no relation between eye-witnesses' confidence in their identifications and the accuracy of their identifications, an analysis that statistically combined individual studies indicates that the confidence-accuracy correlation might be as high as +.40 when the analysis is restricted to individuals who make an identification (vs. all witnesses; see Sporer, Penrod, Read, & Cutler, 1995). How useful is this correlation for predicting accuracy from confidence? In some ways, a correlation of .40 could be considered strong. For instance, when overall accuracy is 50%, a .40 correlation would translate into 70% of the witnesses with high confidence being accurate and only 30% of the witnesses with low confidence being accurate. As accuracy deviates from 50%, however, differences in accuracy rates between witnesses with high and low confidence will diminish even though the correlation remains .40.

Another way to think about a .40 correlation is to compare it with something that people experience in daily life, namely the correlation between a person's height and a person's gender. Extrapolating from males' and females' average height and standard deviation (69.1, 63.7, and 5.4 in., respectively; Department of Health and Human Services, n.d.) yields a correlation between height and gender of +.43. Notice that the correlation between height and gender is quite similar to the correlation between eyewitnesses' identification confidence and accuracy. Thus, if eyewitnesses' identifications are accurate 50% of the time, we would expect to encounter a highly confident mistaken eyewitness (or a nonconfident accurate eyewitness) about as often as we would encounter a tall female (or a short male).

Although the eyewitness-identification literature has generally used correlation methods to express the statistical association between confidence and accuracy, it is probably more forensically valid to use calibration and overconfidence/underconfidence measures rather than correlations (Brewer, Keast, & Rishworth, 2002; Juslin, Olson, & Winman, 1996). In effect, the correlation method (specifically, point-biserial correlation) expresses the degree of statistical association by calculating the difference in confidence (expressed in terms of the standard deviation) between accurate and inaccurate witnesses. Calibration, on the other hand, assesses the extent to which an eyewitness's confidence, expressed as a percentage, matches the probability that the eyewitness is correct. Overconfidence reflects the extent to which the percentage confidence exceeds the probability that the eyewitness is correct (e.g., 80% confidence and 60% probability correct), and underconfidence reflects the extent to which the percentage confidence underestimates the probability that the eyewitness is correct (e.g., 40% confidence and 60% probability correct). Juslin et al. pointed out that the confidence-accuracy correlation can be quite low even when calibration is high.

Work by Juslin et al. (1996) indicates that eyewitnesses can be well calibrated at times, but recent experiments (Wells & Bradfield, 1999) illustrate a problem that can arise when trying to use percentage confidence expressed by witnesses to infer the probability that their identifications are accurate. In a series of experiments, eyewitnesses were induced to make mistaken identifications from lineups in which the culprit was absent and were then randomly assigned to receive confirming "feedback" telling them that they identified the actual suspect or to receive no feedback at all. Later, these witnesses were asked how certain they were at the time of their identification (i.e., how certain they were before the feedback). Those who did not receive confirming feedback gave average confidence ratings of less than 50%, but those receiving confirming feedback gave average confidence ratings of over 70%. Because all of these eyewitnesses had made mistaken identifications, even the no-feedback witnesses were overconfident, but the confirming-feedback witnesses were especially overconfident. Confidence inflation is a difficult problem in actual criminal cases because eyewitnesses are commonly given feedback about whether their identification decisions agree with the investigator's theory of the case. In these cases, it is the detective, rather than the eyewitness, who determines the confidence of the eyewitness.

Confirming feedback not only inflates confidence, thereby inducing overconfidence, but also harms the confidence-accuracy correlation. When eyewitnesses are given confirming feedback following their identification decisions, the confidence of inaccurate eyewitnesses is inflated more than is the confidence of accurate eyewitnesses, and the net result is a reduction in the confidence-accuracy correlation (Bradfield, Wells, & Olson, 2002). Hence, although the confidence of an eyewitness can have utility if it is assessed independently of external influences (e.g., comments from the detective, learning about what other eyewitnesses have said), the legal system rarely assesses confidence in this way.

IMPACT ON POLICIES AND PRACTICES

What impact has research on the confidence-accuracy problem had on the legal system? Until relatively recently, the impact has been almost nil. However, when DNA exoneration cases began unfolding in the mid-1990s, U.S. Attorney General Janet Reno initiated a study of the causes of these miscarriages of justice. More than three fourths of these convictions of innocent persons involved mistaken eyewitness identifications, and, in every case, the mistaken eyewitnesses were extremely confident and, therefore, persuasive at trial (Wells et al., 1998). A Department of Justice panel used the psychological literature to issue the first set of national guidelines on collecting eyewitness identification evidence (Technical Working Group for Eyewitness Evidence, 1999). One of the major recommendations was that the confidence of the eyewitness be assessed at the time of the identification, before there is any chance for it to be influenced by external factors.

The state of New Jersey has gone even further in adopting the recommendations of eyewitness researchers. Based on findings from the psychological literature, guidelines from the attorney general of New Jersey now call for double-blind testing with lineups. Double-blind lineup testing means that the person who administers the lineup does not know which person in the lineup is the suspect and which ones are merely fillers. Under the New Jersey procedures, the confidence expressed by the eyewitness will be based primarily on the eyewitness's memory, not on the expectations of or feedback from the lineup administrator.

There is growing evidence that the legal system is now beginning to read and use the psychological literature on eyewitnesses to formulate policies and procedures. The 2002 report of Illinois Governor George Ryan's Commission on Capital Punishment is the latest example of this new reliance on the psychological literature. The commission specifically cited the literature on the problem with confidence inflation and recommended double-blind testing and explicit recording of confidence statements at the time of the identification to prevent or detect confidence inflation (Illinois Commission on Capital Punishment, 2002).

NEW DIRECTIONS

Although the psychological literature on eyewitness identification has done much to clarify the confidence-accuracy issue and specify some conditions under which confidence might be predictive of accuracy, research has started to turn to other indicators that might prove even more predictive of accuracy. One of the most promising examples is the relation between the amount of time an eyewitness takes to make an identification and the accuracy of the identification. Eyewitnesses who make their identification decision quickly (in 10 s or less) are considerably more likely to be accurate than are eyewitnesses who take longer (e.g., Dunning & Perretta, in press). Confidence is a self-report that is subject to distortion (e.g., from postidentification feedback), whereas decision time is a behavior that can be directly observed. Hence, decision time might prove more reliable than confidence as an indicator of eyewitness accuracy. Yet another new direction in eyewitness identification research concerns cases in which there are multiple eyewitnesses. Recent analyses show that the behaviors of eyewitnesses who do not identify the suspect from a lineup can be used to assess the likely accuracy of the eyewitnesses who do identify the suspect from a lineup (Wells & Olson, in press). The future of eyewitness identification research is a bright one, and the legal system now seems to be paying attention.

Recommended Reading

Cutler, B.L., & Penrod, S.D. (1995). *Mistaken identification: The eyewitness, psychology, and the law.* New York: Cambridge University Press.
Scheck, B., Neufeld, P., & Dwyer, J. (2000). (See References)
Wells, G.L., Malpass, R.S., Lindsay, R.C.L., Fisher, R.P., Turtle, J.W., & Fulero, S. (2000). From the lab to the police station: A successful application of eyewitness research. *American Psychologist*, 55, 581–598.

Note

1. Address correspondence to Gary L. Wells, Psychology Department, Iowa State University, Ames, IA 50011; e-mail: glwells@iastate.edu.

References

Bradfield, S.L., Wells, G.L., & Olson, E.A. (2002). The damaging effect of confirming feedback on the relation between eyewitness certainty and identification accuracy. *Journal of Applied Psychology*, 87, 112–120.

Brewer, N., Keast, A., & Rishworth, A. (2002). Improving the confidence-accuracy relation in eye-witness identification: Evidence from correlation and calibration. *Journal of Experimental Psychology: Applied, 8,* 44–56.

Department of Health and Human Services, National Center for Health Statistics. (n.d.). *National Health and Nutrition Examination Survey.* Retrieved May 22, 2002, from http://www.cdc.gov/nchs/about/major/nhanes/datatblelink.htm#unpubtab

Dunning, D., & Perretta, S. (in press). Automaticity and eyewitness accuracy: A 10 - to - 12 second rule for distinguishing accurate from inaccurate positive identifications. *Journal of Applied Psychology.*

Illinois Commission on Capital Punishment. (2002, April). *Report of the Governor's Commission on Capital Punishment.* Retrieved April 23, 2002, from http://www.idoc.state.il.us/ccp/ccp/reports/commission_report/

Juslin, P., Olson, N., & Winman, A. (1996). Calibration and diagnosticity of confidence in eyewitness identification: Comments on what can and cannot be inferred from a low confidence-accuracy correlation. *Journal of Experimental Psychology: Learning, Memory, and Cognition, 5,* 1304–1316.

Read, J.D., Vokey, J.R., & Hammersley, R. (1990). Changing photos of faces: Effects of exposure duration and photo similarity on recognition and the accuracy-confidence relationship. *Journal of Experimental Psychology: Learning, Memory, and Cognition, 16,* 870–882.

Scheck, B., Neufeld, P., & Dwyer, J. (2000). *Actual innocence.* New York: Random House.

Shaw, J.S., III. (1996). Increases in eyewitness confidence resulting from postevent questioning. *Journal of Experimental Psychology: Applied, 2,* 126–146.

Shaw, J.S., III, & McClure, K.A. (1996). Repeated postevent questioning can lead to elevated levels of eyewitness confidence. *Law and Human Behavior, 20,* 629–654.

Sporer, S., Penrod, S., Read, D., & Cutler, B.L. (1995). Choosing, confidence, and accuracy: A meta-analysis of the confidence-accuracy relation in eyewitness identification studies. *Psychological Bulletin, 118,* 315–327.

Technical Working Group for Eyewitness Evidence. (1999). *Eyewitness evidence: A guide for law enforcement.* Washington, DC: U.S. Department of Justice, Office of Justice Programs.

Wells, G.L., & Bradfield, A.L. (1999). Distortions in eyewitnesses' recollections: Can the postidentification-feedback effect be moderated? *Psychological Science, 10,* 138–144.

Wells, G.L., & Olson, E.A. (in press). Eyewitness identification: Information gain from incriminating and exonerating behaviors. *Journal of Experimental Psychology: Applied.*

Wells, G.L., Small, M., Penrod, S., Malpass, R.S., Fulero, S.M., & Brimacombe, C.A.E. (1998). Eyewitness identification procedures: Recommendations for lineups and photospreads. *Law and Human Behavior, 22,* 603–647.

Critical Thinking Questions

1. What factors drive confidence and accuracy in the same direction? What factors might drive one but not the other? Which of these occur at the crime scene (and are not under anyone's direct control) and which are the result of post-crime procedures (and are controllable)?

2. The article mentions that repeated questioning of a witness can lead to increases in confidence (but not accuracy). How does that finding remind you of the issues in the Garry and Polaschek article?

3. Two types of errors can be made by a witness when viewing a line-up: the witness might fail to identify the actual criminal or the witness might improperly identify an innocent suspect. Are the consequences of those two types of error equal? With which type of error is this article more concerned? Why?

4. Current law tells jurors that they may consider witness confidence as relevant to evaluating the witness's accuracy and credibility. Do you believe that jurors should be given instructions regarding the *value* of a witness's confidence on the witness stand? What do you think those instructions should be?

5. In the last sentence, the authors state that "the legal system now seems to be paying attention" to eyewitness identification research. Good research in this area has been around for many years. Why do you think they are paying attention to this research "now"? Is the legal system all of a sudden open to using all social science research or is this a fluke (or maybe a foot in the door)?

6. Suppose you were a witness to a crime. What might you want to do (or not do) to preserve the accuracy of your knowledge/memory? What might you want to do (or not do) to make sure your confidence level reflects your accuracy?

Making Group Brainstorming More Effective: Recommendations From an Associative Memory Perspective

Vincent R. Brown and Paul B. Paulus[1]

Department of Psychology, Hofstra University, Hempstead, New York (V.R.B.), and Department of Psychology, University of Texas at Arlington, Arlington, Texas (P.B.P.)

Abstract

Much literature on group brainstorming has found it to be less effective than individual brainstorming. However, a cognitive perspective suggests that group brainstorming could be an effective technique for generating creative ideas. Computer simulations of an associative memory model of idea generation in groups suggest that groups have the potential to generate ideas that individuals brainstorming alone are less likely to generate. Exchanging ideas by means of writing or computers, alternating solitary and group brainstorming, and using heterogeneous groups appear to be useful approaches for enhancing group brainstorming.

Keywords

brainstorming; cognitive stimulation; groups; group creativity

There is a general belief in the efficacy of collaboration for projects involving innovation or problem solving (Bennis & Biederman, 1997; Sutton & Hargadon, 1996). Although there is some evidence for the effectiveness of collaborative science and teamwork (Paulus, 2000), the enthusiasm for collective work may not always be justified. Controlled studies of idea sharing in groups have shown that groups often overestimate their effectiveness (Paulus, Larey, & Ortega, 1995). Experiments comparing interactive brainstorming groups with sets of individuals who do not interact in performing the same task have found that groups generate fewer ideas and that group members exhibit reduced motivation and do not fully share unique information (e.g., Mullen, Johnson, & Salas, 1991). The strongest inhibitory effect of groups may be production blocking, which is a reduction in productivity due to the fact that group members must take turns in describing their ideas (Diehl & Stroebe, 1991).

One area in which these problems are most evident is the study of group creativity. Most research on creativity has examined individual creativity because it is typically seen as a personal trait or skill. However, today much creative work requires collaboration of people with diverse sets of knowledge and skills. How can such groups overcome the inevitable liabilities of group interaction to reach their creative potential? Is it possible to demonstrate that group interaction can lead to enhanced creativity? Examining these questions has been the aim of our program of research on the cognitive potential of brainstorming groups (Brown, Tumeo, Larey, & Paulus, 1998; Paulus & Brown, in press; Paulus, Dugosh, Dzindolet, Coskun, & Putman, 2002).

COGNITIVE BASES FOR IDEATIONAL CREATIVITY: A MODEL AND SUPPORTING EVIDENCE

Intuitively, the cognitive benefits of brainstorming in a group seem clear: People believe that they come up with ideas in a group that they would not have thought of on their own. The potential for mutual stimulation of ideas is one of the reasons for the popularity of group brainstorming.

Semantic Networks and an Associative Memory Model of Group Brainstorming

Clearly, the retrieval of relevant information from one's long-term conceptual memory is an important part of the brainstorming process because one cannot effectively brainstorm on a topic one knows nothing about. The concepts stored in long-term memory can be thought of as being associated in a semantic network in such a way that related concepts are more strongly connected than unrelated concepts and thus more likely to activate each other (e.g., Collins & Loftus, 1975). Thus, concepts that are more closely connected to those that are currently active should be more accessible than concepts that are less strongly connected to currently active ideas.

To use the semantic network representation as a basis for exploring group brainstorming, many details need to be specified. Rather than explicitly representing four, six, eight, or more semantic networks and the interactions among them—which would be cumbersome—our approach is to represent a brainstormer's knowledge of a given problem as a matrix of category transition probabilities: Each entry in this matrix represents the probability of generating one's next idea from the same category as the previous idea or from a different category (Brown et al., 1998).

A number of individual differences affecting brainstorming performance are captured by this framework. Fluency, or the amount of knowledge one has about the brainstorming problem, is represented by the probabilities in the main body of the matrix relative to the null category, which represents the likelihood of coming up with no idea in a given time interval. Convergent and divergent thinking styles also fit nicely into the framework. On the one hand, a convergent thinker is likely to stick with a category and explore it deeply before moving on to generate ideas from other categories. Thus, a convergent thinker is represented by a matrix with relatively high within-category transition probabilities. On the other hand, a divergent thinker is more likely to jump around between categories, and so is represented by a matrix with lower within-category transition probabilities (and correspondingly higher between-category transition probabilities). Because individual differences can be represented in this framework, the effects of different group compositions can be examined.

Accessibility

The property of the individual's semantic network that is crucial to determining the effectiveness of group brainstorming is category accessibility. People are generally unlikely to explore on their own relevant categories of ideas that have relatively weak connections with other categories in their personal semantic networks. Generation of ideas from these categories requires the spark of input from other brainstormers. For example, a student who lives on campus in a dormitory may be unlikely to generate ideas about parking when brainstorming on ways to improve the university. But if a student who commutes from off campus mentions parking, the dorm dweller may be able to come up with a few thoughts on the matter, perhaps recalling the parking difficulties his or her parents had when they visited. Simulations of the associative memory matrix model show that presenting a brainstormer with ideas from low-accessible categories not only increases the number of ideas generated from those categories, but also increases the total number of ideas generated overall, thus making the individual a more productive brainstormer. This prediction is supported by Leggett (1997), who studied individual brainstormers to evaluate how input of ideas in the absence of a group context (i.e., without inhibitory social influences) might provide cognitive stimulation. Participants listened to audiotapes containing ideas generated by participants who worked on the Thumbs Problem ("What would be the advantages and disadvantages of having an extra thumb on each hand?") in previous studies. Although all brainstormers benefited from being primed with relevant ideas, the amount of benefit depended on whether the ideas came from a category that was frequently or infrequently represented in the responses in the previous studies. Individuals who were primed with ideas from common categories obtained less benefit than those who were primed with ideas from unique categories. In other words, priming categories that were already likely to be utilized did not enhance performance as much as priming categories unlikely to be utilized by someone brainstorming on his or her own.

Attention

Individuals will be influenced by other group members to the extent that they pay attention to those other members' ideas. In the framework of the model, attention is represented as the probability that an individual group member uses the current speaker's idea as the basis for generating his or her next idea (as opposed to simply continuing his or her own internal train of thought).

Simulations predict that, in general, the more attention each individual pays to fellow group members, the better the performance of the group. Conversely, the more each individual's attention is distracted from the ideas of others, perhaps by concern for the social aspects of group brainstorming, the more the performance of the group declines. In particular, the more one attends to fellow brainstormers, the more one is likely to be primed to consider ideas from one's own low-accessible categories. In fact, the model predicts that, in general, if it were not for production blocking, the number of ideas generated by each group member would increase (at least up to a point) as group size increases (Brown et al., 1998).

One way to enhance the effects of attention on brainstorming performance is to instruct brainstormers that at the end of the brainstorming session they will be asked to recall the ideas that were presented during the session. Without memorization instructions, participants may be more likely to focus on the generation of their own ideas and to some extent ignore the ideas being presented by others. Interestingly, the effectiveness of memory instructions appears to be mixed. When participants are listening to audiotapes or exchanging ideas by computer, instructions to memorize facilitate idea generation (Dugosh, Paulus, Roland, & Yang, 2000). However, when participants are asked to exchange written ideas in a round-robin format (Paulus & Yang, 2000), memory instructions inhibit performance. Because the instructions to read the ideas as they are passed from person to person may already ensure that participants attend to the ideas, instructions to memorize may simply add an unnecessary additional processing demand and impede the brainstorming effort.

ENHANCING GROUP BRAINSTORMING

The goal of creating circumstances that optimize group performance requires maximizing the benefits of cognitive facilitation while at the same time minimizing the inhibitory processes that reduce group productivity. We have studied three brainstorming procedures that appear promising for theoretical reasons, and that have garnered some empirical support. These are combining group and solitary brainstorming, having group brainstormers interact by writing instead of speaking ("brainwriting"), and using networked computers on which individuals type their ideas and read the ideas of others (electronic brainstorming).

Individual and Group Brainstorming

At face value, the goal of maximizing the benefits of group exchange while minimizing inhibitory group processes suggests literally combining group and individual brainstorming. Of course, a person cannot brainstorm alone while at the same time brainstorming in a group. But one can alternate group and solitary idea-generation sessions. Preliminary data from our laboratory show that brainstorming in a group before brainstorming alone on the same topic produces more ideas over the course of the two sessions than does brainstorming alone in the first session and then brainstorming in a group the following session (Leggett, Putman, Roland, & Paulus, 1996).

Model simulations make clear the mechanisms that produce this advantage for the group-solitary sequence. The cognitive facilitation that occurs in the group session carries over into the solitary session, during which the brainstormer continues generating ideas without being hindered by production blocking. This effect shows up as a large "productivity spike" for solitary brainstormers in the second session in both model simulations and empirical studies. This order effect should be particularly strong when the initial group consists of heterogeneous members whose knowledge of the task differs. Simulations also indicate that a solitary brainstormer whose idea generation takes place following a group session is likely to sample more categories from the brainstorming topic than a similar brainstormer working in two solitary sessions. This suggests that the group-solitary sequence has an advantage over and above possible increases in overall productivity.

Brainwriting

Another way to take advantage of group priming effects while reducing production blocking would be to have group members interact by writing and reading rather than speaking and listening. This does not seem to be a technique that is often attempted. Perhaps people are so used to communicating orally when they are face to face that researchers do not consider the alternatives. In a study of group brainwriting (Paulus & Yang, 2000), group members wrote their ideas on a piece of paper and passed them on to the next group member, who read the ideas, added his or her own ideas, and passed the paper on. These interactive groups of brainwriters outproduced sets of equal numbers of writers who did not interact in performing the task. This result may be the first laboratory example of face-to-face interactive groups outperforming an equal number of solitary brainstormers.

Although model simulations support the observation that interactive brainwriters can outperform an equal number of solitary brainwriters (Paulus & Brown, in press), the simulation results are complex in some interesting ways. First, simulations predict that interactive brainwriting is not universally superior to individual brainwriting, but is most effective for heterogeneous groups whose members have differing knowledge of the brainstorming problem. Second, up to a point, performance of simulated brainwriting groups improves as the group members pay increasing attention to the written ideas, but performance decreases when attention to the written ideas becomes excessive. Obviously, brainwriters who do not read any of the ideas that are passed along to them will not benefit from the thoughts of their fellow brainwriters. Brainwriters who attend predominantly to the ideas of others will benefit from them to some extent, but not as greatly as those who optimally balance the two goals of attending to the ideas of others and following their own internal train of thought.

Electronic Brainstorming

The effectiveness of brainwriting suggests that using computers for exchanging ideas might be another useful way of tapping the creative potential of groups. Using computers, individuals can be exposed to a broad range of ideas without the verbal "traffic jams" that are problematic for face-to-face groups. In fact, because production blocking is greatly reduced, the larger the group the better—larger groups will increase one's exposure to a broad range of ideas. One of the most consistent findings in the electronic-brainstorming literature is that the benefits are most evident for groups of eight or more (Dennis & Williams, in press).

Unfortunately, it is also easier to ignore the inputs of others in the electronic format than in face-to-face situations. Instructing individuals to attend carefully to ideas shared electronically (e.g., because of an impending memory test) increases the impact of the shared information (Dugosh et al., 2000). Although electronic brainstormers do not have to apply extensive intellectual resources to try to remember the shared ideas because they are available on the computer, for large groups the number of available ideas can become rather overwhelming. It may be important to provide an opportunity for individuals to continue processing the ideas after the interactive session in order to gain full associative advantage of the shared information. The benefits of idea sharing in electronic groups are in fact most evident if individuals are provided such a solitary session after group interaction.

CONCLUSIONS

It is clear that unstructured groups left to their own devices will not be very effective in developing creative ideas. However, a cognitive perspective points to methods that can be used so that group exchange of ideas enhances idea generation. Groups of individuals with diverse sets of knowledge are most likely to benefit from the social exchange of ideas. Although face-to-face interaction is seen as a natural modality for group interaction, using writing or computers can enhance the exchange of ideas. The interaction should be structured to ensure careful attention to the shared ideas. Alternating between individual and group ideation is helpful because it allows for careful reflection on and processing of shared ideas.

There are still a number of significant empirical gaps that need to be addressed. Given that much group exchange consists of verbal interaction in face-to-face groups, studies need to determine the specific extent to which the performance of these groups can be enhanced by using insights from the associative memory perspective. In particular, it will be important to demonstrate that groups that contain members with diverse knowledge bases can effectively use this knowledge interactively for creative purposes. There are also no controlled studies of creativity in groups or teams in organizations outside the laboratory, so that it is not possible to draw definitive conclusions about the effectiveness of groups in the real world. Because group interaction can be a source of social and cognitive interference as well as social and cognitive stimulation, one of the main theoretical challenges will be to integrate the cognitive and social perspectives of group brainstorming. A careful delineation of how these processes interact will be of great benefit to practitioners.

Recommended Reading

Brown, V., Tumeo, M., Larey, T.S., & Paulus, P.B. (1998). (See References)
Osborn, A.F. (1957). *Applied imagination* (1st ed.). New York: Scribner.
Paulus, P.B., & Nijstad, B.A. (Eds.). (in press). *Group creativity.* New York: Oxford University Press.

Note

1. Address correspondence to Paul B. Paulus, Department of Psychology, University of Texas at Arlington, Arlington, TX 76019; e-mail: paulus@uta. edu.

References

Bennis, W., & Biederman, P.W. (1997). *Organizing genius: The secrets of creative collaboration.* Reading, MA: Addison Wesley.

Brown, V., Tumeo, M., Larey, T.S., & Paulus, P.B. (1998). Modeling cognitive interactions during group brainstorming. *Small Group Research, 29,* 495–526.

Collins, A.M., & Loftus, L.F. (1975). A spreading-activation theory of semantic processing. *Psychological Review, 82,* 407–428.

Dennis, A.R., & Williams, M.L. (in press). Electronic brainstorming: Theory, research, and future directions. In P.B. Paulus & B.A. Nijstad (Eds.), *Group creativity.* New York: Oxford University Press.

Diehl, M., & Stroebe, W. (1991). Productivity loss in idea-generating groups: Tracking down the blocking effect. *Journal of Personality and Social Psychology, 61,* 392–403.

Dugosh, K.L., Paulus, P.B., Roland, E.J., & Yang, H.-C. (2000). Cognitive stimulation in brainstorming. *Journal of Personality and Social Psychology, 79,* 722–735.

Leggett, K.L. (1997). *The effectiveness of categorical priming in brainstorming.* Unpublished master's thesis, University of Texas at Arlington.

Leggett, K.L., Putman, V.L., Roland, E.J., & Paulus, P.B. (1996, April). *The effects of training on performance in group brainstorming.* Paper presented at the annual meeting of the Southwestern Psychological Association, Houston, TX.

Mullen, B., Johnson, C., & Salas, E. (1991). Productivity loss in brainstorming groups: A meta-analytic integration. *Basic and Applied Social Psychology, 12,* 3–23.

Paulus, P.B. (2000). Groups, teams and creativity: The creative potential of idea generating groups. *Applied Psychology: An International Review, 49,* 237–262.

Paulus, P.B., & Brown, V. (in press). Ideational creativity in groups: Lessons from research on brainstorming. In P.B. Paulus & B.A. Nijstad (Eds.), *Group creativity.* New York: Oxford University Press.

Paulus, P.B., Dugosh, K.L., Dzindolet, M.T., Coskun, H., & Putman, V.L. (2002). Social and cognitive influences in group brainstorming: Predicting production gains and losses. *European Review of Social Psychology, 12,* 299–325.

Paulus, P.B., Larey, T.S., & Ortega, A.H. (1995). Performance and perceptions of brainstormers in an organizational setting. *Basic and Applied Social Psychology, 18,* 3–14.

Paulus, P.B., & Yang, H.-C. (2000). Idea generation in groups: A basis for creativity in organizations. *Organizational Behavior and Human Decision Processes, 82,* 76–87.

Sutton, R.I., & Hargadon, A. (1996). Brainstorming groups in context. *Administrative Science Quarterly, 41,* 685–718.

Critical Thinking Questions

1. This article relies on the notion of memory as an associative network. What other memory effects—within individual, not group, memory—have you studied that are easily explainable in terms of semantic or associative networks?

2. This article is reminiscent of work on part-set cuing. In part-set cuing studies, subjects may be asked to write down all of the names of U. S. states that they can think of. The "cued" subjects get a list of several states in front of them to "help them start"; the "non-cued" subjects do not. The interesting finding is that non-cued subjects remember more of the unlisted states than the cued subjects. One explanation is that the list keeps bringing the cued subjects' thoughts back to the same items. Those subjects then repeatedly recollect the same states, and the states they have mentally stored as associated to those states, but they can't break out of those mental loops. How can the techniques of group brainstorming solve that problem? Why might it depend on the method of information-sharing used by the group?

3. Why might people working in groups overestimate their own effectiveness in brainstorming more than individuals working alone?

4. Suppose you were in charge of designing an advertising campaign for a new type of product. You want to select several people from your company to get together for a brainstorming session. Who do you pick? A bunch of people who have worked together many times before and who get along well? People from different previous teams who have not worked together before and don't personally like each other? What are the cognitive and social trade-offs to each of these selections?

   (How and why would your personnel selection for this task differ from your personnel selection if you were seeking to evaluate a product or a marketing strategy? See Yaniv in Part 4.)

Associative Learning
and Causal Reasoning

An essential skill for survival is being able to learn about contingencies in the world. Humans have learned which berries are edible, which metals make good weapons, and which medicines cure diseases. How do we pick up such knowledge? Do we all always use it accurately—or similarly? The four articles in this section look at different factors that may affect such learning and reasoning including: evolutionary pre-dispositions, personality traits, imagination, and culture.

The first two articles consider what might be called "associative learning". They describe studies in which subjects (sometimes human, sometimes other species) are exposed over and over to some stimulus; later a response is measured. In "The Malicious Serpent: Snakes as a Prototypical Stimulus for an Evolved Module of Fear," Öhman and Mineka describe both classical conditioning and illusory correlation paradigms for examining fear reactions to snakes. In each paradigm, subjects are shown pictures of various items (e.g., snakes, flowers) that are sometimes paired with an unpleasant shock. In the conditioning paradigm, subjects exhibit stronger fear conditioning to snakes than to flowers. In the illusory correlation paradigm, subjects believe that the snakes were paired more often with the shocks than were the flowers (even when the proportions of such pairings were equal). The authors provide an evolutionary account for why such differential learning might take place.

In "Illusions of Control: How We Overestimate Our Personal Influence," Thompson describes several operant conditioning experiments in which subjects are led to believe that they have control over some outcome (e.g., if they push a button a light will turn on). The outcome may occur totally by chance or it may be a function of both the subjects' actions and chance. When asked to judge how much control they have over the outcome, subjects often overestimate their own causal power. Thompson suggests which kinds of people are most likely to overestimate their causal influence on events and when they are most likely to do so. She ends by speculating on the adaptiveness of the illusory control bias in judgment.

The next two articles reflect on how people determine causality for one-time events—like a car accident or a murder. In "When Possibility Informs Reality: Counterfactual Thinking as a Cue to Causality," Spellman and Mandel consider how counterfactual reasoning might affect a person's causal reasoning about such events. In counterfactual reasoning, we imagine the world different from how it actually is (i.e., counter-to-fact) and play out the consequences. For example, you might imagine that if Timothy McVeigh (the Oklahoma City bomber) had had a different upbringing, he would not have blown up the building. The article describes the controversy over whether such a counterfactual relationship means it is correct to say that the bad upbringing *caused* him to blow up the building.

Whether people will assign causality for a person's behavior to a disposition (i.e., a personal trait) or a situation (i.e., the circumstances) is the topic addressed by Norenzayan and Nisbett in "Culture and Causal Cognition". Whereas Westerners tend to assign causality to an individual's traits (he was always a destructive child), East Asians tend to assign causality to a situation (he was always pressured by others to do so). Thus, different people may make different causal attributions for the same set of events.

All four articles demonstrate, therefore, that our judgments about the relationship between events in the world are not solely based on the contingencies in the world. Rather, our causal judgments are influenced by pre-existing knowledge and beliefs—both conscious and unconscious, both innate and learned.

The Malicious Serpent: Snakes as a Prototypical Stimulus for an Evolved Module of Fear

Arne Öhman[1] and Susan Mineka

Department of Clinical Neuroscience, Karolinska Institute, Stockholm, Sweden (A.Ö.), and Department of Psychology, Northwestern University, Evanston, Illinois (S.M.)

Abstract

As reptiles, snakes may have signified deadly threats in the environment of early mammals. We review findings suggesting that snakes remain special stimuli for humans. Intense snake fear is prevalent in both humans and other primates. Humans and monkeys learn snake fear more easily than fear of most other stimuli through direct or vicarious conditioning. Neither the elicitation nor the conditioning of snake fear in humans requires that snakes be consciously perceived; rather, both processes can occur with masked stimuli. Humans tend to perceive illusory correlations between snakes and aversive stimuli, and their attention is automatically captured by snakes in complex visual displays. Together, these and other findings delineate an evolved fear module in the brain. This module is selectively and automatically activated by once-threatening stimuli, is relatively encapsulated from cognition, and derives from specialized neural circuitry.

Keywords

evolution; snake fear; fear module

Snakes are commonly regarded as shiny, slithering creatures worthy of fear and disgust. If one were to believe the Book of Genesis, humans' dislike for snakes resulted from a divine intervention: To avenge the snake's luring of Eve to taste the fruit of knowledge, God instituted eternal enmity between their descendants. Alternatively, the human dislike of snakes and the common appearances of reptiles as the embodiment of evil in myths and art might reflect an evolutionary heritage. Indeed, Sagan (1977) speculated that human fear of snakes and other reptiles may be a distant effect of the conditions under which early mammals evolved. In the world they inhabited, the animal kingdom was dominated by awesome reptiles, the dinosaurs, and so a prerequisite for early mammals to deliver genes to future generations was to avoid getting caught in the fangs of Tyrannosaurus rex and its relatives. Thus, fear and respect for reptiles is a likely core mammalian heritage. From this perspective, snakes and other reptiles may continue to have a special psychological significance even for humans, and considerable evidence suggests this is indeed true. Furthermore, the pattern of findings appears consistent with the evolutionary premise.

THE PREVALENCE OF SNAKE FEARS IN PRIMATES

Snakes are obviously fearsome creatures to many humans. Agras, Sylvester, and Oliveau (1969) interviewed a sample of New Englanders about fears, and found snakes to be clearly the most prevalent object of intense fear, reported by 38% of females and 12% of males.

Fear of snakes is also common among other primates. According to an exhaustive review of field data (King, 1997), 11 genera of primates showed fear-related responses (alarm calls, avoidance, mobbing) in virtually all instances in which they were observed confronting large snakes. For studies of captive primates, King did not find consistent evidence of snake fear. However, in direct comparisons, rhesus (and squirrel) monkeys reared in the wild were far more likely than lab-reared monkeys to show strong phobiclike fear responses to snakes (e.g., Mineka, Keir, & Price, 1980). That this fear is adaptive in the wild is further supported by independent field reports of large snakes attacking primates (M. Cook & Mineka, 1991).

This high prevalence of snake fear in humans as well as in our primate relatives suggests that it is a result of an ancient evolutionary history. Genetic variability might explain why not all individuals show fear of snakes. Alternatively, the variability could stem from differences in how easily individuals learn to fear reptilian stimuli when they are encountered in aversive contexts. This latter possibility would be consistent with the differences in snake fear between wild- and lab-reared monkeys.

LEARNING TO FEAR SNAKES

Experiments with lab-reared monkeys have shown that they can acquire a fear of snakes vicariously, that is, by observing other monkeys expressing fear of snakes. When nonfearful lab-reared monkeys were given the opportunity to observe a wild-reared "model" monkey displaying fear of live and toy snakes, they were rapidly conditioned to fear snakes, and this conditioning was strong and persistent. The fear response was learned even when the fearful model monkey was shown on videotape (M. Cook & Mineka, 1990).

When videos were spliced so that identical displays of fear were modeled in response to toy snakes and flowers, or to toy crocodiles and rabbits (M. Cook & Mineka, 1991), the lab-reared monkeys showed substantial conditioning to toy snakes and crocodiles, but not to flowers and toy rabbits. Toy snakes and flowers served equally well as signals for food rewards (M. Cook & Mineka, 1990), so the selective effect of snakes appears to be restricted to aversive contexts. Because these monkeys had never seen any of the stimuli used prior to these experiments, the results provide strong support for an evolutionary basis to the selective learning.

A series of studies published in the 1970s (see Öhman & Mineka, 2001) tested the hypothesis that humans are predisposed to easily learn to fear snakes. These studies used a discriminative Pavlovian conditioning procedure in which various pictures served as conditioned stimuli (CSs) that predicted the presence and absence of mildly aversive shock, the unconditioned stimulus (US). Participants for whom snakes (or spiders) consistently signaled shocks showed stronger and more lasting conditioned skin conductance responses (SCRs; palmar sweat responses that index emotional activation) than control participants for whom flowers or mushrooms signaled shocks. When a nonaversive US was used, however, this difference disappeared. E.W. Cook, Hodes, and Lang (1986) demonstrated that qualitatively different responses were conditioned to snakes (heart rate acceleration, indexing fear) than to flowers and mushrooms (heart rate deceleration, indexing attention to the eliciting stimulus). They also reported superior conditioning to snakes than to gun stimuli paired with loud noises. Such results suggest that the selective association between snakes and aversive USs reflects evolutionary history rather than cultural conditioning.

NONCONSCIOUS CONTROL OF RESPONSES TO SNAKES

If the prevalence and ease of learning snake fear represents a core mammalian heritage, its neural machinery must be found in brain structures that evolved in early mammals. Accordingly, the fear circuit of the mammalian brain relies heavily on limbic structures such as the amygdala, a collection of neural nuclei in the anterior temporal lobe. Limbic structures emerged in the evolutionary transition from reptiles to mammals and use preexisting structures in the "reptilian brain" to control emotional output such as flight/fight behavior and cardiovascular changes (see Öhman & Mineka, 2001).

From this neuroevolutionary perspective, one would expect the limbically controlled fear of snakes to be relatively independent of the most recently evolved control level in the brain, the neocortex, which is the site of advanced cognition. This hypothesis is consistent with the often strikingly irrational quality of snake phobia. For example, phobias may be activated by seeing mere pictures of snakes. Backward masking is a promising methodology for examining whether phobic responses can be activated without involvement of the cortex. In this method, a brief visual stimulus is blanked from conscious perception by an immediately following masking stimulus. Because backward masking disrupts visual processing in the primary visual cortex, responses to backward-masked stimuli reflect activation of pathways in the brain that may access the fear circuit without involving cortical areas mediating visual awareness of the stimulus.

In one study (Öhman & Soares, 1994), pictures of snakes, spiders, flowers, and mushrooms were presented very briefly (30 ms), each time immediately followed by a masking stimulus (a randomly cut and reassembled picture). Although the participants could not recognize the intact pictures, participants who were afraid of snakes showed enhanced SCRs only to masked snakes, whereas participants who were afraid of spiders responded only to spiders. Similar results were obtained (Öhman & Soares, 1993) when nonfearful participants, who had been conditioned to unmasked snake pictures by shock USs, were exposed to masked pictures without the US. Thus, responses to conditioned snake pictures survived backward masking; in contrast, masking eliminated conditioning effects in another group of participants conditioned to neutral stimuli such as flowers or mushrooms.

Furthermore, subsequent experiments (Öhman & Soares, 1998) also demonstrated conditioning to masked stimuli when masked snakes or spiders (but not masked flowers or mushrooms) were used as CSs followed by shock USs. Thus, these masking studies show that fear responses (as indexed by SCRs) can be learned and elicited when backward masking prevents visually presented snake stimuli from accessing cortical processing. This is consistent with the notion that responses to snakes are organized by a specifically evolved primitive neural circuit that emerged with the first mammals long before the evolution of neocortex.

ILLUSORY CORRELATIONS BETWEEN SNAKES AND AVERSIVE STIMULI

If expression and learning of snake fear do not require cortical processing, are people's cognitions about snakes and their relationships to other events biased and irrational? One example of such biased processing occurred in experiments on illusory correlations: Participants (especially those who were afraid of snakes) were more likely to perceive that slides of fear-relevant stimuli (such as snakes) were paired with shock than to perceive that slides of control stimuli (flowers and mushrooms) were paired with shock. This occurred even though there were no such relationships in the extensive random sequence of slide stimuli and aversive and nonaversive outcomes (tones or nothing) participants had experienced (Tomarken, Sutton, & Mineka, 1995).

Similar illusory correlations were not observed for pictures of damaged electrical equipment and shock even though they were rated as belonging together better than snakes and shock (Tomarken et al., 1995). In another experiment, participants showed exaggerated expectancies for shock to follow both snakes and damaged electrical equipment before the experiment began (Kennedy, Rapee, & Mazurski, 1997), but reported only the illusory correlation between snakes and shock after experiencing the random stimulus series. Thus, it appears that snakes have a cognitive affinity with aversiveness and danger that is resistant to modification by experience.

AUTOMATIC CAPTURE OF ATTENTION BY SNAKE STIMULI

People who encounter snakes in the wild may report that they first froze in fear, only a split second later realizing that they were about to step on a snake. Thus, snakes may automatically capture attention. A study supporting this hypothesis (Öhman, Flykt, & Esteves, 2001) demonstrated shorter detection latencies for a discrepant snake picture among an array of many neutral distractor stimuli (e.g., flower pictures) than vice versa. Furthermore, "finding the snake in the grass" was not affected by the number of distractor stimuli, whereas it took longer to detect discrepant flowers and mushrooms among many than among few snakes when the latter served as distractor stimuli. This suggests that snakes, but not flowers and mushrooms, were located by an automatic perceptual routine that effortlessly found target stimuli that appeared to "pop out" from the matrix independently of the number of distractor stimuli. Participants who were highly fearful of snakes showed even superior performance in detecting snakes. Thus, when snakes elicited fear in participants, this fear state sensitized the perceptual apparatus to detect snakes even more efficiently.

THE CONCEPT OF A FEAR MODULE

The evidence we have reviewed shows that snake stimuli are strongly and widely associated with fear in humans and other primates and that fear of snakes is relatively independent of conscious cognition. We have proposed the concept of an evolved fear module to explain these and many related findings (Öhman & Mineka, 2001). The fear module is a relatively independent behavioral, mental, and neural system that has evolved to assist mammals in defending against threats such as snakes. The module is selectively sensitive to, and automatically activated by, stimuli related to recurrent survival threats, it is relatively encapsulated from more advanced human cognition, and it relies on specialized neural circuitry.

This specialized behavioral module did not evolve primarily from survival threats provided by snakes during human evolution, but rather from the threat that reptiles have provided through mammalian evolution. Because reptiles have been associated with danger throughout evolution, it is likely that snakes represent a prototypical stimulus for activating the fear module. However, we are not arguing that the human brain has a specialized module for automatically generating fear of snakes. Rather, we propose that the blueprint for the fear module was built around the deadly threat that ancestors of snakes provided to our distant ancestors, the early mammals. During further mammalian evolution, this blueprint was modified, elaborated, and specialized for the ecological niches occupied by different species. Some mammals may even prey on snakes, and new stimuli and stimulus features have been added to reptiles as preferential activators of the module. For example, facial threat is similar to snakes when it comes to activating the fear module in social primates (Öhman & Mineka, 2001). Through Pavlovian conditioning, the fear module may come under the control of a very wide range of stimuli signaling pain and danger. Nevertheless, evolutionarily derived constraints have afforded stimuli once related to recurrent survival threats easier access for gaining control of the module through fear conditioning (Öhman & Mineka, 2001).

ISSUES FOR FURTHER RESEARCH

The claim that the fear module can be conditioned without awareness is a bold one given that there is a relative consensus in the field of human conditioning that awareness of the CS-US contingency is required for acquiring conditioned responses. However, as we have extensively argued elsewhere (Öhman & Mineka, 2001; Wiens & Öhman, 2002), there is good evidence that conditioning to nonconsciously presented CSs is possible if they are evolutionarily fear relevant. Other factors that might promote such nonconscious learning include intense USs, short CS-US intervals, and perhaps temporal overlap between the CS and the US. However, little research on these factors has been reported, and there is a pressing need to elaborate their relative effectiveness in promoting conditioning of the fear module outside of awareness.

One of the appeals of the fear module concept is that it is consistent with the current understanding of the neurobiology of fear conditioning, which gives a central role to the amygdala (e.g., Öhman & Mineka, 2001). However, this understanding is primarily based on animal data. Even though the emerging brain-imaging literature on human fear conditioning is consistent with this database, systematic efforts are needed in order to tie the fear module more convincingly to human brain mechanisms. For example, a conspicuous gap in knowledge concerns whether the amygdala is indeed specially tuned to conditioning contingencies involving evolutionarily fear-relevant CSs such as snakes.

An interesting question that can be addressed both at a psychological and at a neurobiological level concerns the perceptual mechanisms that give snake stimuli privileged access to the fear module. For example, are snakes detected at a lower perceptual threshold relative to non-fear-relevant objects? Are they identified faster than other objects once detected? Are they quicker to activate the fear module and attract attention once identified? Regardless of the locus of perceptual privilege, what visual features of snakes make them such powerful fear elicitors and attention captors? Because the visual processing in pathways preceding the cortical level is crude, the hypothesis that masked presentations of snakes directly access the amygdala implies that the effect is mediated by simple features of snakes rather than by the complex configuration of features defining a snake. Delineating these features would allow the construction of a "super fear stimulus." It could be argued that such a stimulus would depict "the archetypical evil" as represented in the human brain.

Recommended Reading

Mineka, S. (1992). Evolutionary memories, emotional processing, and the emotional disorders. *The Psychology of Learning and Motivation, 28*, 161–206.

Öhman, A., Dimberg, U., & Öst, L.-G. (1985). Animal and social phobias: Biological constraints on learned fear responses. In S. Reiss & R.R. Bootzin (Eds.), *Theoretical issues in behavior therapy* (pp. 123–178). New York: Academic Press.

Öhman, A., & Mineka, S. (2001). (See References)

Note

1. Address correspondence to Arne Öhman, Psychology Section, Department of Clinical Neuroscience, Karolinska Institute and Hospital, Z6:6, S-171 76 Stockholm, Sweden; e-mail: arne.ohman@cns.ki.se.

References

Agras, S., Sylvester, D., & Oliveau, D. (1969). The epidemiology of common fears and phobias. *Comprehensive Psychiatry, 10*, 151–156.

Cook, E.W., Hodes, R.L., & Lang, P.J. (1986). Preparedness and phobia: Effects of stimulus content on human visceral conditioning. *Journal of Abnormal Psychology, 95*, 195–207.

Cook, M., & Mineka, S. (1990). Selective associations in the observational conditioning of fear in rhesus monkeys. *Journal of Experimental Psychology: Animal Behavior Processes, 16*, 372–389.

Cook, M., & Mineka, S. (1991). Selective associations in the origins of phobic fears and their implications for behavior therapy. In P. Martin (Ed.), *Handbook of behavior therapy and psychological science: An integrative approach* (pp. 413–434). Oxford, England: Pergamon Press.

Kennedy, S.J., Rapee, R.M., & Mazurski, E.J. (1997). Covariation bias for phylogenetic versus ontogenetic fear-relevant stimuli. *Behaviour Research and Therapy, 35*, 415–422.

King, G.E. (1997, June). *The attentional basis for primate responses to snakes.* Paper presented at the annual meeting of the American Society of Primatologists, San Diego, CA.

Mineka, S., Keir, R., & Price, V. (1980). Fear of snakes in wild- and laboratory-reared rhesus monkeys (*Macaca mulatta*). *Animal Learning and Behavior, 8*, 653–663.

Öhman, A., Flykt, A., & Esteves, F. (2001). Emotion drives attention: Detecting the snake in the grass. *Journal of Experimental Psychology: General, 131*, 466–478.

Öhman, A., & Mineka, S. (2001). Fear, phobias and preparedness: Toward an evolved module of fear and fear learning. *Psychological Review, 108*, 483–522.

Öhman, A., & Soares, J.J.F. (1993). On the automatic nature of phobic fear: Conditioned electrodermal responses to masked fear-relevant stimuli. *Journal of Abnormal Psychology, 102*, 121–132.

Öhman, A., & Soares, J.J.F. (1994). "Unconscious anxiety": Phobic responses to masked stimuli. *Journal of Abnormal Psychology, 103*, 231–240.

Öhman, A., & Soares, J.J.F. (1998). Emotional conditioning to masked stimuli: Expectancies for aversive outcomes following nonrecognized fear-irrelevant stimuli. *Journal of Experimental Psychology: General, 127*, 69–82.

Sagan, C. (1977). *The dragons of Eden: Speculations on the evolution of human intelligence.* London: Hodder and Stoughton.

Tomarken, A.J., Sutton, S.K., & Mineka, S. (1995). Fear-relevant illusory correlations: What types of associations promote judgmental bias? *Journal of Abnormal Psychology, 104*, 312–326.

Wiens, S., & Öhman, A. (2002). Unawareness is more than a chance event: Comment on Lovibond and Shanks (2002). *Journal of Experimental Psychology: Animal Behavior Processes, 28*, 27–31.

Critical Thinking Questions

1. In many of the studies described here, subjects were shown the same pairings of snakes (or spiders) with shocks (or food) and flowers (or mushrooms) with shocks (or food). Subjects showed more fear, and indicated a higher correlation with shock, to the snakes. Do you think that comparing snakes/spiders to flowers/mushrooms represents a strong test of the authors' hypothesis? What about when they compare snakes to guns or to electrical equipment? What other kinds of things would you want to use in these experiments and why?

2. If the strong association between snakes and fear is the result of evolution, why don't all humans (or all monkeys) show it? If it goes back as far as the authors suggest, what else might you hypothesize that people should be predisposed to fear?

3. What role does consciousness play in evoking fear reactions to snakes?

Illusions of Control:
How We Overestimate Our Personal Influence

Suzanne C. Thompson[1]

Department of Psychology, Pomona College, Claremont, California

Abstract

Illusions of control are common even in purely chance situations. They are particularly likely to occur in settings that are characterized by personal involvement, familiarity, foreknowledge of the desired outcome, and a focus on success. Person-based factors that affect illusions of control include depressive mood and need for control. One explanation of illusory control is that it is due to a control heuristic that is used to estimate control by assessing the factors of intentionality and connection to the outcome. Motivational influences on illusory control and consequences of overestimating one's control are also covered.

Keywords

illusions of control; perceived control

In an intriguing set of studies, Langer (1975) showed that people often overestimate their control even in situations governed purely by chance. In one of Langer's studies, some people were allowed to pick their own lottery ticket, and others got a ticket picked for them. Later, participants were given the option of exchanging their ticket for one in a lottery with more favorable odds. Despite the fact that exchanging the ticket increased the odds of winning, those who had selected their own lottery ticket number did not choose to exchange this personally chosen ticket. People seemed to think that choosing their own ticket increased their odds of winning as if their action of choosing their own ticket gave them some control over the outcome of the lottery. Similar *illusions of control* have been demonstrated in a variety of circumstances.

Alloy and Abramson (1979) extended the study of illusory control by manipulating factors that affect illusory control and measuring participants judgments of control. In these studies, people tried to get a green light to come on by pushing a button. They were told that the button might control the light, but, in actuality, there was no relationship between participants actions and whether the light came on. The light came on 25% of the time for participants in the low-reinforcement condition and 75% of the time for those in the high-reinforcement condition, regardless of how often or when the button was pushed. Estimates of personal control over the light were quite high in the high-reinforcement condition equivalent to a moderate to intermediate degree of control.

WHEN DO PEOPLE OVERESTIMATE CONTROL?

Do people always overestimate their control? Evidently not. You may have noted in the studies just described that the illusions of control occurred for people who selected their own lottery tickets or who were in the high-reinforcement condition. Participants in the other conditions were not so susceptible to thinking they had control. Both situational and person-based factors influence whether or not people will overestimate their control.

Situational factors include personal involvement, familiarity, foreknowledge of the desired outcome, and success at the task. *Personal involvement* refers to someone being the active agent as opposed to having others act for him or her. When people act for themselves, illusory control is likely, but when no personal action is involved for example, when someone else selects their tickets or throws the dice for them the sense of being able to control the outcome is greatly diminished. *Familiarity* is another important influence on illusions of control. When circumstances or the materials being worked with are familiar, it is easier to have inflated judgments of personal control than when the situation or task is new. *Foreknowledge of the desired outcome*, the extent to which people know the outcome they want when they act, is a third condition that affects whether or not personal control will be overestimated. In a prediction situation, actors know which outcome they want to achieve when they act. In contrast, in a postdiction situation, the action is taken before the actors know the outcome they would like. Illusions of control are stronger in prediction than in postdiction situations. *Success at the task* has a major influence on whether or not personal control is overestimated. Success versus failure has sometimes been manipulated experimentally by the number of times people get the outcome they are trying to achieve (e.g., the light comes on 75% vs. 25% of the time). Although the outcome is random and not in response to participants behavior, participants who get the outcome they want at a high rate have significantly and substantially higher estimates of their control than those who get the desired outcome at a low rate. People are also more likely to overestimate their control if the task involves acting to get a desired outcome than if the point is to avoid losing what they have. Evidently, the former type of task leads to a focus on successes and the latter highlights the losses. Even the patterning of successes and failures can affect the focus on success or failure. A string of early successes leads to higher estimates of personal control than does a string of early failures, even if the total number of successes over the entire session is the same.

The person factors that have an effect on illusory control include mood and need for control. In many circumstances, nondepressed individuals tend to overestimate their control, whereas those who are in a *depressed mood* have a more realistic assessment of their ability to control an outcome (Alloy & Abramson, 1979). These effects are found even when mood is manipulated experimentally. Nondepressed participants who are put in a depressed mood through a mood-induction technique make control judgments similar to those of participants who are chronically depressed (Alloy, Abramson, & Viscusi, 1981). Although the topic has not been studied extensively, there is some indication that a strong *need for control* can also lead people to overestimate their potential to affect outcomes. In an experimental demonstration of the effects of need for control, participants who were randomly assigned to a hungry condition were more likely than participants in a sated condition to think they could control a chance task for which a hamburger was a prize (Biner, Angle, Park, Mellinger, & Barber, 1995).

To sum up, people do not always overestimate their control, but when the situation is one that is familiar and focused on success, when the desired outcome is known and people are taking action for themselves, the setup is ripe for illusory control. In addition, nondepressed individuals and those with a need for the outcome are particularly prone to overestimate their control.

WHY DO PEOPLE OVERESTIMATE THEIR CONTROL?

The question of why illusions of control occur was first addressed by Langer (1975), who proposed that skill and chance situations are easily confused. This confusion is especially likely to happen when chance situations have the trappings that are characteristic of skill-based situations (e.g., familiarity, choice, involvement). However, there are several reasons why a confusion of skill and chance situations is not by itself a good explanation for the variety of conditions under which illusions of control appear. For example, the factors that increase illusory control are broader than ones that reflect a confusion between chance and controllable situations. Illusory control is also more likely in situations that involve a focus on success rather than failure, as well as among individuals who are in a positive rather than depressed mood and who have a need for control. Also, feedback that highlights failures and negative moods can eliminate or reverse illusory judgments of control. None of these effects is easily explained by a confusion of skill and chance elements.

My colleagues and I have offered a more complete explanation of why illusions of control occur (Thompson, Armstrong, & Thomas, 1998). We propose that people use a control heuristic to judge their degree of influence over an outcome. A heuristic is a shortcut or simple rule that can be used to reach a judgment, in this case, an estimate of one's control over achieving an outcome. The control heuristic involves two elements: one's intention to achieve the outcome and the perceived connection between one's action and the desired outcome. Both intention and connection are cues that are used to judge how much control one has. If one intended an outcome and can see a connection between one's own action and the outcome, then perceptions of personal control are high.

Like most heuristics, this simple rule will often lead to accurate judgments because intentionality and a perceived connection (e.g., seeing that the outcome immediately follows one's action) often occur in situations in which one does, in fact, have control. However, the heuristic can also lead to overestimations of control because intentionality and connection can occur in situations in which a person has little or no control. For example, gamblers playing the slot machines in Las Vegas pull the handles with the intention of getting a winning combination to slide into place. When their handle pulling is followed by the desired outcome, a connection is established between their action and the outcome. Because intentionality and connection are strong, the stage is set for gamblers to think that they have some control in this situation.

The control heuristic provides a unifying explanation for all of the various factors that have been found to affect illusions of control. Each is influential because it affects perceived intentionality or the connection between one's behavior and the outcome. For example, personal involvement is essential for the illusion of control because connection between an actor's actions and an event cannot be observed or imagined unless the actor acts. Foreknowledge is important for illusory control because without it the actor cannot intend to effect a particular outcome. Success-focused tasks increase illusions of control because these circumstances lead actors to overestimate the connection between their action and the successful outcome. They do this because compared with tasks focusing on failure, success-oriented tasks produce more instances in which actor's actions are followed by the desired outcome and because these types of tasks direct attention to success; both of these factors increase perceptions of the number of positive confirming cases. Failure experiences and a focus on losing have the opposite effect by highlighting the times when actors' behavior is not followed by the desired outcome, thereby weakening the perception of connection. Depressed mood may be associated with lower assessments of personal control because people who are depressed focus on failure, not on success. Because of this focus on failure, they are less likely than nondepressed people to notice and overestimate the connection between their own behavior and successes.

DOES WANTING TO HAVE CONTROL LEAD TO ILLUSORY CONTROL?

Because the benefits of believing oneself to have control (e.g., positive mood and increased motivation) may be realized even if one's control is illusory, it seems reasonable to suggest that people are often motivated to overestimate their control. Some recent studies in my lab have examined the effects of a motive for control on illusory control judgments. Participants worked on a computer task that was similar to the light task described earlier. Motive for control was manipulated by paying some participants for each time a green computer screen appeared. Despite the fact that participants had no control, the motive for control had a strong effect on control judgments: Participants who were paid for each appearance of the green screen had significantly higher perceptions of control than participants lacking this motive.

According to the control-heuristic model, motives for control affect judgments of control by biasing either judgments of intentionality or judgments of connection. For example, believing that one could have foreseen the outcome heightens a sense of acting intentionally. Therefore, for individuals who are motivated to have control, judgments of intentionality can be heightened through the hindsight bias, in which individuals overestimate the degree to which they could have anticipated an outcome. Counterfactuals, or imagined alternative outcomes to an event, are another route for a control motive to increase illusions of control. Perceptions of connection can be strengthened through the use of controllable counterfactuals because imagining undoing antecedents that are controllable by the actor heightens the connection between action and outcome. For example, a lottery winner who thinks, "If I hadn't thought to buy that ticket, I wouldn't have won $1,000" highlights the connection between her action and the desired outcome, whereas imagining a counterfactual with an uncontrollable element would lessen the connection.

THE CONSEQUENCES OF ILLUSORY CONTROL

Not all people overestimate their personal control. Moderately depressed individuals tend to have a realistic sense of how much they are contributing to an outcome. Does that mean that people are better off if they overestimate their personal control?

Only a few studies have addressed the issue of the adaptiveness of illusory control per se. Some of these studies have found that illusory control enhances adaptive functioning. A study by Alloy and Clements (1992) examined college students who varied in the degree to which they exhibited the illusion of control. Compared with students who tended not to have illusions of control, students who displayed greater illusions of control had less negative mood after a failure on a lab task, were less likely to become discouraged when they subsequently experienced negative life events, and were less likely to get depressed a month later, given the occurrence of a high number of negative life events. Thus, individuals who are susceptible to an illusion of control may be at a decreased risk for depression and discouragement in comparison to those individuals who are not.

In contrast to this positive finding, there is also evidence that overestimating one's control is not adaptive. For example, Donovan, Leavitt, and Walsh (1990) investigated the influence of illusory control on child-care outcomes. The degree of illusory control was measured with a simulated child-care task in which mothers tried to terminate the crying of an audio-taped baby. Mothers judged their control over stopping the tape by pushing a button. In actuality, their responses were not connected to the operation of the tape. A subsequent simulation assessed the mothers ability to learn effective responses for getting an infant to stop crying. Mothers with a high illusion of control on the first task were more susceptible to helplessness on the second task, for example, by not responding even when control was possible.

Which is the correct view: that illusory thinking is generally useful because it leads to positive emotions and motivates people to try challenging tasks, or that people are better off if they have an accurate assessment of themselves and their situation? Perhaps a third possibility is correct: Sometimes illusory control is adaptive, and at other times it is not. For example, illusions of control may be reassuring in stressful situations, but lead people to take unnecessary risks when they occur in a gambling context. The challenge for researchers is to examine the consequences of illusory control in a variety of situations to answer this important question.

Recommended Reading

Alloy, L.B., & Abramson, L.Y. (1979). (See References)
Langer, E.J. (1975). (See References)
Thompson, S.C., Armstrong, W., & Thomas, C. (1998). (See References)

Note

1. Address correspondence to Suzanne Thompson, Department of Psychology, Pomona College, Claremont, CA 91711.

References

Alloy, L.B., & Abramson, L.Y. (1979). Judgment of contingency in depressed and nondepressed students: Sadder but wiser? *Journal of Experimental Psychology: General, 108,* 441–485.

Alloy, L.B., Abramson, L.Y., & Viscusi, D. (1981). Induced mood and the illusion of control. *Journal of Personality and Social Psychology, 41,* 1129–1140.

Alloy, L.B., & Clements, C.M. (1992). Illusion of control: Invulnerability to negative affect and depressive symptoms after laboratory and natural stressors. *Journal of Abnormal Psychology, 307,* 234–245.

Biner, P.M., Angle, S.T., Park, J.H., Mellinger, A.E., & Barber, B.C. (1995). Need and the illusion of control. *Personality and Social Psychology Bulletin, 23,* 899–907.

Donovan, W.L., Leavitt, L.A., & Walsh, R.O. (1990). Maternal self-efficacy: Illusory control and its effect on susceptibility to learned helplessness. *Child Development, 61,* 1638–1647.

Langer, E.J. (1975). The illusion of control. *Journal of Personality and Social Psychology, 32,* 311–328.

Thompson, S.C., Armstrong, W., & Thomas, C. (1998). Illusions of control, underestimations, and accuracy: A control heuristic explanation. *Psychological Bulletin, 123,* 143–161.

Critical Thinking Questions

1. What kinds of people are more or less likely to show the illusion of control? If depressed people believe that they have less control, how do we know that the depression is causing the belief and not that the belief is (at least in part) causing the depression?

2. Besides deciding to become depressed, what could you do to decrease the illusion of control and increase your accuracy when you have to make these kinds of judgments?

3. How would you currently answer the question: Is illusory thinking adaptive? What else would you want to know before you would be happy with your answer? Do you think that the question is ultimately necessarily answerable?

When Possibility Informs Reality: Counterfactual Thinking as a Cue to Causality

Barbara A. Spellman[1] and David R. Mandel

Department of Psychology, University of Virginia, Charlottesville, Virginia (B.A.S.), and Department of Psychology, University of Hertfordshire, Hatfield, Hertfordshire, England (D.R.M.)

Abstract

People often engage in counterfactual thinking, that is, imagining alternatives to the real world and mentally playing out the consequences. Yet the counterfactuals people tend to imagine are a small subset of those that could possibly be imagined. There is some debate as to the relation between counterfactual thinking and causal beliefs. Some researchers argue that counterfactual thinking is the key to causal judgments; current research suggests, however, that the relation is rather complex. When people think about counterfactuals, they focus on ways to prevent bad or uncommon outcomes; when people think about causes, they focus on things that covary with outcomes. Counterfactual thinking may affect causality judgments by changing beliefs about the probabilities of possible alternatives to what actually happened, thereby changing beliefs as to whether a cause and effect actually covary. The way in which counterfactual thinking affects causal attributions may have practical consequences for mental health and the legal system.

Keywords

counterfactuals; causality; reasoning

If I *were* in New York right now, and if I happened to be standing on a sidewalk on my way to lunch, waiting for the light to change, and if a car happened to jump the curb, I might be struck dead. By not being there, I may have freed that space on the sidewalk for someone else who might be standing there and get run over. And in that event you might say that I'm partially responsible for that death. In a weak sense I'd be responsible at the end of a long causal chain. We're all linked by these causal chains to everyone around us. . . .

McInerney (1992, p. 120)

Most of us have outlandish suppositional "what if" thoughts like this from time to time. We may wonder, "What if my parents had never met?" (*Back to the Future*) or "What if I had never been born?" (*It's a Wonderful Life*). Seemingly less outlandish and far more ubiquitous are our thoughts about the ways in which things might have been different "if only" an event leading up to a particular outcome had been different. For instance, suppose that you get into a car accident while taking a scenic route home that you rarely use but chose to take today because it was particularly sunny. If you are like many other people, you might "undo" the accident by thinking something like "If only I had taken my usual route, I wouldn't have been in the accident" (Kahneman & Tversky, 1982; Mandel & Lehman, 1996). These "if only" thoughts are termed *counterfactual conditionals* because they focus on a changed (or *mutated*) version of reality (i.e., one that is counter to fact) and because they represent a conditional if-then relation between an antecedent event (e.g., the route taken) and a consequent event (e.g., getting into an accident).

CAUSAL CONSEQUENCES OF COUNTERFACTUAL THINKING

Counterfactual thinking has consequences beyond mere daydreams or entertainment. Engaging in counterfactual thinking may influence people's causal attributions.

Practical consequences of this relation can be seen for mental health and the legal system.

For example, counterfactual thinking can produce what Kahneman and Miller (1986) termed *emotional amplification*—a heightening of affect brought about by realizing that an outcome was not inevitable because it easily could have been undone. Counterfactual thinking can amplify feelings of regret, distress and self-blame, and shame and guilt, as well as satisfaction and happiness (see Roese & Olson, 1995, for reviews). Perhaps recurrent counterfactual thinking, in which people obsessively mutate their own actions, leads people to exaggerate their own causal role for both their misfortunes and their successes, resulting in those heightened emotions.

Counterfactual thinking can also affect liability and guilt judgments in legal settings. For example, in one study, subjects read a story about a date rape and then listened to a mock lawyer's closing argument suggesting possible mutations to the story (Branscombe, Owen, Garstka, & Coleman, 1996). If the argument mutated the defendant's actions so that the rape would be undone, the rapist was assigned more fault (cause, blame, and responsibility) than if his actions were mutated but the rape still would have occurred. Similarly, when the victim's actions were mutated, she was assigned more fault if the mutation would undo the rape than if the rape still would have occurred.

THE METHOD BEHIND OUR MUSINGS

Although in our counterfactual musings there are an infinite number of ways we can mutate antecedents so that an outcome would be different, we tend to introduce relatively minor mutations that are systematically constrained (see Roese, 1997). People commonly mutate abnormal events by restoring them to their normal default (Kahneman & Miller, 1986). For example, in the car-accident story, people undo the accident by restoring the abnormal antecedent of taking the unusual route to its normal default—taking the usual route (Kahneman & Tversky, 1982). People also tend to mutate antecedents that are personally controllable (see Roese, 1997). For instance, one is more likely to think, "If only I had decided to take the usual route . . ." than to think, "If only it wasn't sunny" or even "If only the other driver was not reckless" (Mandel & Lehman, 1996). More generally, research on counterfactual thinking demonstrates that people entertain a very small and nonrandom sample from the universe of possible mutations.

Just as various antecedent events leading up to an outcome are differentially likely to be mutated, various outcomes are differentially likely to trigger counterfactual thinking. Negative events and abnormal events are the most likely triggers (see Roese, 1997; Roese & Olson, 1995). Consequently, research on counterfactual thinking has focused on subjects' counterfactual thoughts about negative events like car accidents, food poisonings, and stock-market losses, and unexpected positive events like winning a lottery.

COUNTERFACTUAL THINKING AS TESTS OF CAUSALITY

Kahneman and Tversky's (1982) highly influential chapter on the *simulation heuristic* described the basic logic by which if-only thinking could be used to explore the causes of past events. They proposed that people often run an if-then simulation in their minds to make various sorts of judgments, including predictions of future events and causal explanations of past events. Accordingly, to understand whether X caused Y, a person may imagine that X had not occurred and then imagine what events would follow. The easier it is to imagine that Y would not now follow, the more likely the person is to view X as a cause of Y.

Kahneman and Tversky's (1982) proposal is certainly plausible: They were not suggesting that all counterfactual-conditional thinking is necessarily directed at understanding the causes of past events, but simply making the point that such thinking could be useful toward that end. Nor did they argue that all forms of causal reasoning are equally likely to rely on counterfactual thinking; their examples suggested that counterfactual thinking would be most useful for understanding the causes of a single case (as in law) rather than for understanding causal relations governing a set of events (as in science). Indeed, their proposal echoed earlier work on causal reasoning by the philosopher Mackie (1974) and the legal philosophers Hart and Honoré (1959/1985), and amounted to what in political science and law is assumed generally to be an important criterion for attributing causality: In order to ascribe causal status to an actor or event, one should believe that *but for* the causal candidate in question, the outcome would not have occurred. In other words, one must demonstrate that the causal candidate was *necessary*, under the circumstances, for the outcome to occur. Despite the modest scope of Kahneman and Tversky's proposal, however, many attribution researchers subsequently posited that counterfactual thinking plays a primary role in causal attributions.

EMPIRICAL EVIDENCE

The first empirical study to measure both counterfactual thinking and judgments of causality in the same story (Wells & Gavanski, 1989) supported the idea that mutability was the key to causality. Subjects read about a woman who was taken to dinner by her boss. The boss ordered for both of them, but the dish he ordered contained wine—to which the woman was allergic. She ate the food and died. The boss had considered ordering something else: In one version, the alternative dish also contained wine; in the other version, the alternative dish did not contain wine. Subjects were asked to list four things that could have been different in the story to prevent the woman's death (mutation task) and to rate how much of a causal role the boss's ordering decision played in her death. Relative to subjects who read that both dishes contained wine, subjects who read that the other dish did not contain wine (a) were more likely to mutate the boss's decision and (b) rated the boss's decision as more causal. These results were interpreted as showing two things: that for causality to be assigned to an event (the boss's decision), the event must have a counterfactual that would have undone the outcome (i.e., mutability is necessary for causality) and that the more available a counterfactual alternative is, the more causal that event seems (i.e., ease of mutability affects perceived causal impact).

Later research, however, has found different patterns for mutability and causality in other, richer, scenarios. For example, in another study (Mandel & Lehman, 1996), subjects read a story about a car accident like the one alluded to earlier (based on Kahneman & Tversky, 1982): On a sunny day, Mr. Jones decided to take an unusual route home; being indecisive, he braked hard to stop at a yellow light; he began driving when the light turned green; he was hit by a drunk teenager who charged through the red light. One group of subjects was asked how Jones would finish the thought "If only . . . ," and another group was asked what Jones would think caused the accident. The subjects tended to mutate Jones's decision to take the unusual route and his indecisive driving but to assign causality to the teenager's drunk driving. Hence, the most mutated event need not be perceived as the most causal event.

Therefore, although sometimes mutability and causality converge (as in the food decision), sometimes they differ (as in the car accident). Yet the latter experiment does not disconfirm that mutability is necessary for causality: The drunk driver's actions—the most causal event—can easily be mutated to undo the outcome (even though other aspects of the story are mutated more often).

However, there is a situation that demonstrates that mutability is *not necessary* for causality. Suppose that Able and Baker shoot Smith at the exact same instant. Smith dies and the coroner reports that either shot alone would have killed him (i.e., there is *causal overdetermination*). Able can argue that he did not kill Smith because even if he had not shot Smith, Smith would have died anyway. Mutating Able's actions alone does not work; in fact, subjects mutate Able's and Baker's actions together to undo Smith's death. But when attributing causality, subjects judge Able and Baker individually to be causes of Smith's death and sentence Able and Baker each to the maximum jail time (Spellman, Maris, & Wynn, 1999). Both philosophers and legal theorists consider such cases of causal overdetermination to be an exception to the but-for causality rule. Because causality is assigned even when mutation does not undo the outcome, mutability is not necessary for causality.

Moreover, mutability is *not sufficient* for causality. For example, one might believe that Jones's car accident would not have occurred if it had not been a sunny day (as Jones would have taken his normal route); however, no one would argue that the sunny day caused the accident. Obviously, therefore, not everything that can be mutated to undo an outcome is judged as a cause of that outcome.

RECENT THEORETICAL PERSPECTIVES

The relation between mutability and causality is not simply one of identity, necessity, or sufficiency; rather, a more complex picture has emerged.

According to a recent proposal (Mandel & Lehman, 1996), everyday counterfactuals are not tests of causality. Rather, for negative outcomes, counterfactuals are after-the-fact realizations of ways that would have been sufficient to prevent the outcomes—and especially ways that actors themselves could have prevented their misfortunes. Data from the car-accident experiment reported earlier support this account. In addition to the counterfactual- and causal-question groups, a third group of subjects was included in this experiment. These subjects answered a question about how Jones would think the accident could have been prevented. Not surprisingly, subjects' answers for the preventability question focused on the same aspects of the story as the answers for the counterfactual question, but were different from the answers for the causality question. Thus, although, logically, sufficient-but-foregone preventors (e.g., taking the usual route) seem like necessary causes, psychologically, attributions of preventability and causality are different. In attributing preventability, people focus on controllable antecedents (e.g., choice of route, stopping at a yellow light); in attributing causality, people focus on antecedents that general knowledge suggests would covary with, and therefore predict, the outcome (e.g., drunk drivers).

The *crediting causality* model (Spellman, 1997) tells us how a covariation analysis of causal candidates might be performed in reasoning about single cases. According to this model, causality attributions in single cases are analogous to causality attributions in science. In science, to find whether something is causal (e.g., whether drunk driving causes accidents), people presumably compare the probability of the effect when the cause is present (e.g., the probability of accidents given drunk drivers) with the probability of the effect when the cause is absent (e.g., the probability of accidents given no drunk drivers); if the former probability is larger, people infer causality. For the single case, people may consider and compare that probability with the probability of an accident before that event occurred (i.e., the baseline probability). If the event (e.g., a drunk driver) raises the probability of the outcome, then it is causal; if it does not (e.g., a sunny day), it is not.

Counterfactual thinking might affect causal attributions by acting as input to that probability comparison. If one can counterfactually imagine many alternatives to an outcome like an accident, then the baseline probability of the accident seems low, and, therefore, the causality of the event in question (e.g., drunk driver) seems high. If there are no alternatives to an outcome (e.g., all foods contain wine and will cause illness), then the baseline probability of the outcome is high, and the causality of the event in question (e.g., ordering a particular dish) seems low. Similarly, in the experiment involving counterfactual thinking about the date rape (Branscombe et al., 1996), thinking about mutations that would undo the rape may have changed estimates of the baseline probability of a rape, and thus affected the subjects' legal judgments. Note that when it seems that nothing can be mutated to undo the outcome, people feel the outcome was inevitable; that is, it was due to fate (Mandel & Lehman, 1996).

CONCLUDING REMARKS

The philosopher David Hume described causation—or, more accurately, causal thinking—as the "cement of the universe" because it binds together people's perceptions of events that would otherwise appear unrelated. It seems, however, that our beliefs about the universe of the actual, to which Hume referred, are affected by our considerations of the merely possible—created by the "what ifs" and "if onlys" of counterfactual thinking.

Recommended Reading

Kahneman, D., & Miller, D.T. (1986). (See References)
Mackie, J.L. (1974). (See References)
Mandel, D.R., & Lehman, D.R. (1996). (See References)
Roese, N.J. (1999). *Counterfactual Research News* [On-line]. Available: http://www.psych.nwu.edu/psych/people/faculty/roese/research/cf/cfnews.htm
Roese, N.J., & Olson, J.M. (Eds.). (1995). (See References)

Acknowledgments—This work was supported in part by a NIMH FIRST Award to the first author and by Research Grant No. 95-3054 from the Natural Sciences and Engineering Research Council of Canada to the second author. We thank Darrin Lehman and Angeline Lillard for helpful comments on an earlier draft.

Note

1. Address correspondence to Barbara A. Spellman, Department of Psychology, 102 Gilmer Hall, University of Virginia, Charlottesville, VA 22903; email: spellman@virginia.edu.

References

Branscombe, N.R., Owen, S., Garstka, T.A., & Coleman, J. (1996). Rape and accident counterfactuals: Who might have done otherwise and would it have changed the outcome? *Journal of Applied Social Psychology, 26*, 1042–1067.

Hart, H.L., & Honoré, A.M. (1985). *Causation in the law* (2nd ed.). Oxford, England: Oxford University Press. (Original work published 1959)

Kahneman, D., & Miller, D.T. (1986). Norm theory: Comparing reality to its alternatives. *Psychological Review, 93*, 136–153.

Kahneman, D., & Tversky, A. (1982). The simulation heuristic. In D. Kahneman, P. Slovic, & A. Tversky (Eds.), *Judgment under uncertainty: Heuristics and biases* (pp. 201–208). New York: Cambridge University Press.

Mackie, J.L. (1974). *The cement of the universe: A study of causation.* Oxford, England: Oxford University Press.

Mandel, D.R., & Lehman, D.R. (1996). Counterfactual thinking and ascriptions of cause and preventability. *Journal of Personality and Social Psychology, 71*, 450–463.

McInerney, J. (1992). *Brightness falls.* New York: Knopf.

Roese, N.J. (1997). Counterfactual thinking. *Psychological Bulletin, 121*, 133–148.

Roese, N.J., & Olson, J.M. (Eds.). (1995). *What might have been: The social psychology of counterfactual thinking.* Mahwah, NJ: Erlbaum.

Spellman, B.A. (1997). Crediting causality. *Journal of Experimental Psychology: General, 126*, 323–348.

Spellman, B.A., Maris, J.R., & Wynn, J.E. (1999). *Causality without mutability.* Unpublished manuscript, University of Virginia, Charlottesville.

Wells, G.L., & Gavanski, I. (1989). Mental simulation of causality. *Journal of Personality and Social Psychology, 56*, 161–169.

Critical Thinking Questions

1. Imagine that you are an attorney defending your client, the X cigarette company, against a law suit by a plaintiff who has smoked X cigarettes for years and has recently developed lung cancer. The plaintiff is also suing companies Y and Z because he smoked their cigarettes, too. Proving causality is essential in this kind of law suit. Think of several ways in which you might defend your client against the claim that "Cigarette X caused the plaintiff's lung cancer". Do you always seem to get to the right or just result?

2. This article claims that "there are an infinite number of ways we can mutate antecedents so that an outcome would be different." Consider the car accident story in the article. Can you think of an infinite number of things to mutate? (Hint: get out of the box.) Why do you think people tend to converge on the same set of minor mutations (e.g., taking the usual rather than unusual route) and do not often consider the lack of Martian intervention?

3. Suppose you convince your friend to skip class and go skiing with you. Your friend hits a patch of ice, the binding does not release, and she breaks her leg. What counterfactuals come to mind? Which make you feel regret? What causes come to mind? Do the lists of counterfactuals and causes overlap? Which do, which don't, and why?

Culture and Causal Cognition

Ara Norenzayan and Richard E. Nisbett[1]
Centre de Récherche en Epistemologie Appliquée, Ecole Polytechnique, Paris, France (A.N.), and Department of Psychology, University of Michigan, Ann Arbor, Michigan (R.E.N.)

Abstract

East Asian and American causal reasoning differs significantly. East Asians understand behavior in terms of complex interactions between dispositions of the person or other object and contextual factors, whereas Americans often view social behavior primarily as the direct unfolding of dispositions. These culturally differing causal theories seem to be rooted in more pervasive, culture-specific mentalities in East Asia and the West. The Western mentality is analytic, focusing attention on the object, categorizing it by reference to its attributes, and ascribing causality based on rules about it. The East Asian mentality is holistic, focusing attention on the field in which the object is located and ascribing causality by reference to the relationship between the object and the field.

Keywords

causal attribution; culture; attention; reasoning

Psychologists within the cognitive science tradition have long believed that fundamental reasoning processes such as causal attribution are the same in all cultures (Gardner, 1985). Although recognizing that the content of causal beliefs can differ widely across cultures, psychologists have assumed that the ways in which people come to make their causal judgments are essentially the same, and therefore that they tend to make the same sorts of inferential errors. A case in point is the fundamental attribution error, or FAE (Ross, 1977), a phenomenon that is of central importance to social psychology and until recently was held to be invariable across cultures.

The FAE refers to people's inclination to see behavior as the result of dispositions corresponding to the apparent nature of the behavior. This tendency often results in error when there are obvious situational constraints that leave little or no role for dispositions in producing the behavior. The classic example of the FAE was demonstrated in a study by Jones and Harris (1967) in which participants read a speech or essay that a target person had allegedly been required to produce by a debate coach or psychology experimenter. The speech or essay favored a particular position on an issue, for example, the legalization of marijuana. Participants' estimates of the target's actual views on the issue reflected to a substantial extent the views expressed in the speech or essay, even when they knew that the target had been explicitly instructed to defend a particular position. Thus, participants inferred an attitude that corresponded to the target person's apparent behavior, without taking into account the situational constraints operating on the behavior. Since that classic study, the FAE has been found in myriad studies in innumerable experimental and naturalistic contexts, and it has been a major focus of theorizing and a continuing source of instructive pedagogy for psychology students.

CULTURE AND THE FAE

It turns out, however, that the FAE is much harder to demonstrate with Asian populations than with European-American populations (Choi, Nisbett, & Norenzayan, 1999). Miller (1984) showed that Hindu Indians preferred to explain ordinary life events in terms of the situational context in which they occurred, whereas Americans were much more inclined to explain similar events in terms of presumed dispositions. Morris and Peng (1994) found that Chinese newspapers and Chinese students living in the United States tended to explain murders (by both Chinese and American perpetrators) in terms of the situation and even the societal context confronting the murderers, whereas American newspapers and American students were more likely to explain the murders in terms of presumed dispositions of the perpetrators.

Recently Jones and Harris's (1967) experiment was repeated with Korean and American participants (Choi et al., 1999). Like Americans, the Koreans tended to assume that the target person held the position he was advocating. But the two groups responded quite differently if they were placed in the same situation themselves before they made judgments about the target. When observers were required to write an essay, using four arguments specified by the experimenter, the Americans were unaffected, but the Koreans were greatly affected. That is, the Americans' judgments about the target's attitudes were just as much influenced by the target's essay as if they themselves had never experienced the constraints inherent in the situation, whereas the Koreans almost never inferred that the target person had the attitude expressed in the essay.

This is not to say that Asians do not use dispositions in causal analysis or are not occasionally susceptible to the FAE. Growing evidence indicates that when situational cues are not salient, Asians rely on dispositions or manifest the FAE to the same extent as Westerners (Choi et al., 1999; Norenzayan, Choi, & Nisbett, 1999). The cultural difference seems to originate primarily from a stronger East Asian tendency to recognize the causal power of situations.

The cultural differences in the FAE seem to be supported by different folk theories about the causes of human behavior. In one study (Norenzayan et al., 1999), we asked participants how much they agreed with paragraph descriptions of three different philosophies about why people behave as they do: (a) a strongly dispositionist philosophy holding that "how people behave is mostly determined by their personality," (b) a strongly situationist view holding that behavior "is mostly determined by the situation" in which people find themselves, and (c) an interactionist view holding that behavior "is always jointly determined by personality and the situation." Korean and American participants endorsed the first position to the same degree, but Koreans endorsed the situationist and interactionist views more strongly than did Americans.

These causal theories are consistent with cultural conceptions of personality as well. In the same study (Norenzayan et al., 1999), we administered a scale designed to measure agreement with two different theories of personality: entity theory, or the belief that behavior is due to relatively fixed dispositions such as traits, intelligence, and moral character, and incremental theory, or the belief that behavior is conditioned on the situation and that any relevant dispositions are subject to change (Dweck, Hong, & Chiu, 1993). Koreans for the most part rejected entity theory, whereas Americans were equally likely to endorse entity theory and incremental theory.

ANALYTIC VERSUS HOLISTIC COGNITION

The cultural differences in causal cognition go beyond interpretations of human behavior. Morris and Peng (1994) showed cartoons of an individual fish moving in a variety of configurations in relation to a group of fish and asked participants why they thought the actions had occurred. Chinese participants were inclined to attribute the behavior of the individual fish to factors external to the fish (i.e., the group), whereas American participants were more inclined to attribute the behavior of the fish to internal factors. In studies by Peng and Nisbett (reported in Nisbett, Peng, Choi, & Norenzayan, in press), Chinese participants were shown to interpret even the behavior of schematically drawn, ambiguous physical events—such as a round object dropping through a surface and returning to the surface—as being due to the relation between the object and the presumed medium (e.g., water), whereas Americans tended to interpret the behavior as being due to the properties of the object alone.

The Intellectual Histories of East Asia and Europe

Why should Asians and Americans perceive causality so differently? Scholars in many fields, including ethnography, history, and philosophy of science, hold that, at least since the 6th century B.C., there has been a very different intellectual tradition in the West than in the East (especially China and those cultures, like the Korean and Japanese, that were heavily influenced by China; Nisbett et al., in press). The ancient Greeks had an *analytic* stance: The focus was on categorizing the object with reference to its attributes and explaining its behavior using rules about its category memberships. The ancient Chinese had a holistic stance, meaning that there was an orientation toward the field in which the object was found and a tendency to explain the behavior of the object in terms of its relations with the field.

In support of these propositions, there is substantial evidence that early Greek and Chinese science and mathematics were quite different in their strengths and weaknesses. Greek science looked for universal rules to explain events and was concerned with categorizing objects with respect to their essences. Chinese science (some people would say it was a technology only, though a technology vastly superior to that of the Greeks) was more pragmatic and concrete and was not concerned with foundations or universal laws. The difference between the Greek and Chinese orientations is well captured by Aristotle's physics, which explained the behavior of an object without reference to the field in which it occurs. Thus, a stone sinks into water because it has the property of gravity, and a piece of wood floats because it has the property of levity. In contrast, the principle that events always occur in some context or field of forces was understood early on in China.

Some writers have suggested that the mentality of East Asians remains more holistic than that of Westerners (e.g., Nakamura, 1960/1988). Thus, modern East Asian laypeople, like the ancient Chinese intelligentsia, are attuned to the field and the overall context in determining events. Western civilization was profoundly shaped by ancient Greece, so one would expect the Greek intellectual stance of object focus to be widespread in the West.

Attention to the Field Versus the Object

If East Asians tend to believe that causality lies in the field, they would be expected to attend to the field. If Westerners are more inclined to believe that causality inheres in the object, they might be expected to pay relatively more attention to the object than to the field. There is substantial evidence that this is the case.

Attention to the field as a whole on the part of East Asians suggests that they might find it relatively difficult to separate the object from the field. This notion rests on the concept of *field dependence* (Witkin, Dyk, Faterson, Goodenough, & Karp, 1974). Field dependence refers to a relative difficulty in separating objects from the context in which they are located. One way of measuring field dependence is by means of the rod-and-frame test. In this test, participants look into a long rectangular box at the end of which is a rod. The rod and the box frame can be rotated independently of one another, and participants are asked to state when the rod is vertical. Field dependence is indicated by the extent to which the orientation of the frame influences judgments of the verticality of the rod. The judgments of East Asian (mostly Chinese) participants have been shown to be more field dependent than those of American participants (Ji, Peng, & Nisbett, in press).

In a direct test of whether East Asians pay more attention to the field than Westerners do (Masuda & Nisbett, 1999), Japanese and American participants saw underwater scenes that included one or more *focal* fish (i.e., fish that were larger and faster moving than other objects in the scene) among many other objects, including smaller fish, small animals, plants, rocks, and coral. When asked to recall what they had just viewed, the Japanese and American participants reported equivalent amounts of detail about the focal fish, but the Japanese reported far more detail about almost everything else in the background and made many more references to interactions between focal fish and background objects. After watching the scenes, the participants were shown a focal fish either on the original background or on a new one. The ability of the Japanese to recognize a particular focal fish was impaired if the fish was shown on the "wrong" background. Americans' recognition was uninfluenced by this manipulation.

ORIGINS OF THE CULTURAL DIFFERENCE IN CAUSAL COGNITION

Most of the cross-cultural comparisons we have reviewed compared participants who were highly similar with respect to key demographic variables, namely, age, gender, socioeconomic status, and educational level. Differences in cognitive abilities were controlled for or ruled out as potential explanations for the data in studies involving a task (e.g., the rod-and-frame test) that might be affected by such abilities. Moreover, the predicted differences emerged regardless of whether the East Asians were tested in their native languages in East Asian countries or tested in English in the United States. Thus, the lack of obvious alternative explanations, combined with positive evidence from intellectual history and the convergence of the data across a diverse set of studies (conducted in laboratory as well as naturalistic contexts), points to culturally shared causal theories as the most likely explanation for the group differences.

But why might ancient societies have differed in the causal theories they produced and passed down to their contemporary successor cultures? Attempts to answer such questions must, of course, be highly speculative because they involve complex historical and sociological issues. Elsewhere, we have summarized the views of scholars who have suggested that fundamental differences between societies may result from ecological and economic factors (Nisbett et al., in press). In China, people engaged in intensive farming many centuries before Europeans did. Farmers need to be cooperative with one another, and their societies tend to be collectivist in nature. A focus on the social field may generalize to a holistic understanding of the world. Greece is a land where the mountains descend to the sea and large-scale agriculture is not possible. People earned a living by keeping animals, fishing, and trading. These occupations do not require so much intensive cooperation, and the Greeks were in fact highly individualistic. Individualism in turn encourages attending only to the object and one's goals with regard to it. The social field can be ignored with relative impunity, and causal perception can focus, often mistakenly, solely on the object. We speculate that contemporary societies continue to display these mentalities because the social psychological factors that gave rise to them persist to this day.

Several findings by Witkin and his colleagues (e.g., Witkin et al., 1974), at different levels of analysis, support this historical argument that holistic and analytic cognition originated in collectivist and individualist orientations, respectively. Contemporary farmers are more field dependent than hunters and industrialized peoples; American ethnic groups that operate under tighter social constraints are more field dependent than other groups; and individuals who are attuned to social relationships are more field dependent than those who are less focused on social relationships.

FUTURE DIRECTIONS

A number of questions seem particularly interesting for further inquiry. Should educational practices take into account the differing attentional foci and causal theories of members of different cultural groups? Can the cognitive skills characteristic of one cultural group be transferred to another group? To what extent can economic changes transform the sort of cultural-cognitive system we have described? These and other questions about causal cognition will provide fertile ground for research in the years to come.

Recommended Reading

Choi, I., Nisbett, R.E., & Norenzayan, A. (1999). (See References)
Fiske, A., Kitayama, S., Markus, H.R., & Nisbett, R.E. (1998). The cultural matrix of social psychology. In D.T. Gilbert, S.T. Fiske, & G. Lindzey (Eds.), The handbook of social psychology (4th ed., Vol. 2, pp. 915–981). Boston: McGraw-Hill.
Lloyd, G.E.R. (1996). Science in antiquity: The Greek and Chinese cases and their relevance to problems of culture and cognition. In D.R. Olson & N. Torrance (Eds.), *Modes of thought: Explorations in culture and cognition* (pp. 15–33). Cambridge, England: Cambridge University Press.
Nisbett, R.E., Peng, K., Choi, I., & Norenzayan, A. (in press). (See References)
Sperber, D., Premack, D., & Premack, A.J. (Eds.). (1995). *Causal cognition: A multidisciplinary debate*. Oxford, England: Oxford University Press.

Note

1. Address correspondence to Richard E. Nisbett, Department of Psychology, University of Michigan, Ann Arbor, MI 48109; e-mail: nisbett@umich.edu.

References

Choi, I., Nisbett, R.E., & Norenzayan, A. (1999). Causal attribution across cultures: Variation and universality. *Psychological Bulletin, 125,* 47–63.
Dweck, C.S., Hong, Y.-Y., & Chiu, C.-Y. (1993). Implicit theories: Individual differences in the likelihood and meaning of dispositional inference. *Personality and Social Psychology Bulletin, 19,* 644–656.
Gardner, H. (1985). *The mind's new science.* New York: Basic Books.
Ji, L., Peng, K., & Nisbett, R.E. (in press). Culture, control, and perception of relationships in the environment. *Journal of Personality and Social Psychology.*
Jones, E.E., & Harris, V.A. (1967). The attribution of attitudes. *Journal of Experimental Social Psychology, 3,* 1–24.
Masuda, T., & Nisbett, R.E. (1999). *Culture and attention to object vs. field.* Unpublished manuscript, University of Michigan, Ann Arbor.
Miller, J.G. (1984). Culture and the development of everyday social explanation. *Journal of Personality and Social Psychology, 46,* 961–978.
Morris, M.W., & Peng, K. (1994). Culture and cause: American and Chinese attributions for social and physical events. *Journal of Personality and Social Psychology, 67,* 949–971.
Nakamura, H. (1988). *The ways of thinking of eastern peoples.* New York: Greenwood Press. (Original work published 1960)
Nisbett, R.E., Peng, K., Choi, I., & Norenzayan, A. (in press). Culture and systems of thought: Holistic vs. analytic cognition. *Psychological Review.*
Norenzayan, A., Choi, I., & Nisbett, R.E. (1999). *Eastern and Western folk psychology and the prediction of behavior.* Unpublished manuscript, University of Michigan, Ann Arbor.
Ross, L. (1977). The intuitive psychologist and his shortcomings. In L. Berkowitz (Ed.), *Advances in experimental social psychology* (Vol. 10, pp. 173–220). New York: Academic Press.
Witkin, H.A., Dyk, R.B., Faterson, H.F., Goodenough, D.R., & Karp, S.A. (1974). *Psychological differentiation.* Potomac, MD: Erlbaum.

Critical Thinking Questions

1. When Westerners and East Asians learn about the same events, they sometimes make different causal attributions – Westerners to the person and East Asians to the situation. However, different attributions may also be made by people within the same culture. For example, if a young man from a poor

neighborhood gets into trouble with drugs, liberals might blame the poverty and conservatives might blame the parents. Do these differences mean that people are, in fact, using different reasoning processes or have different cognitive skills? Or might it mean that they are relying on different information when they reason?

2. East Asians and Westerners also differ in field dependence. Does that necessarily imply that they should make different causal attributions?

3. Would you expect East Asians and Westerners to differ on the illusory control experiments described by Thompson? (If you haven't read Thompson, the general experimental paradigm is described in the overview to this section.)

Solving Problems
and Making Decisions

The topics in the previous section and in this section—reasoning, problem solving, and decision making—are usually lumped together under the term "higher-order cognition". Discussions of these abilities often presuppose that we have gathered, through our senses or out of memory, some kind of information, and that we are going to consciously, logically, and rationally use it to try to understand events, solve a problem, make a decision, or achieve a goal.

Research from the 1970's–90's is replete with examples of how people use "heuristics"—shortcuts in reasoning that don't always lead to the correct answer—and how people demonstrate "biases"—systematic errors in reasoning. One conclusion that might be drawn from that body of research is that humans are "irrational" in our thinking. However, in the last decade, the debate over what counts as "rational" has heated up. Does rationality imply that we must be consistent with the dictates of formal mathematical or logical reasoning? Or that we make the best decisions we can make given our limited cognitive capacities? Or that we make decisions that are, if not perfect, obviously good enough for us to have survived given the structure of the environment in which our cognition evolved?

In the previous section we looked at two possible influences on causal reasoning—evolution and culture. In this section we look at other possible influences on reasoning including our unconscious, our emotions, and the opinions of people around us. Each influence calls into question exactly what "rationality" entails.

In "What Have Psychologists (and Others) Discovered about the Process of Scientific Discovery?," Klahr and Simon describe how researchers study the scientific discovery process. Although scientific *reasoning* is often portrayed as a logical process (especially in undergraduate Research Methods textbooks), scientific *discovery* is far messier, and may depend on a combination of knowledge, perseverance, fortuity, and insight. That last issue is the theme taken up by Siegler in "Unconscious Insights". Using an anecdote and experimental evidence, he argues that we may sometimes solve problems unconsciously before we become aware of our solution strategy.

The last two articles address influences on decision making. In "Anticipated Emotions as Guides to Choice," Mellers and McGraw describe how people's knowledge of alternative outcomes to an event can affect their emotional reactions to that event. They also describe how and when people are not good at predicting their own future emotions. One implication of those studies is that if we make choices based on our anticipated emotions, we could be making bad decisions quite often.

One way to make better decisions might be to ask other people for input. In "The Benefit of Additional Opinions", Yaniv describes how and when asking others for their opinions will enhance our own decision making.

Taken together these articles address the question: How do we solve problems and make decisions—and what *do* we, and *should* we, rely on as we do so? Afterwards, we are still left with the question: are humans "rational" (and by what definition) in these—our "highest-order"— endeavors?

What Have Psychologists (And Others) Discovered About the Process of Scientific Discovery?

David Klahr[1] and Herbert A. Simon

Department of Psychology, Carnegie Mellon University, Pittsburgh, Pennsylvania

Abstract

We describe four major approaches to the study of science—historical accounts of scientific discoveries, psychological experiments with nonscientists working on tasks related to scientific discoveries, direct observation of ongoing scientific laboratories, and computational modeling of scientific discovery processes—by viewing them through the lens of the theory of problem solving. We compare and contrast the different approaches, indicate their complementarities, and provide examples from each approach that converge on a set of principles of scientific discovery.

Keywords

scientific discovery; problem solving

Early in the 20th century, Einstein, in reflecting on his own mental processes leading to the theory of relativity, said, "I am not sure whether there can be a way of really understanding the miracle of thinking" (Wertheimer, 1945, p. 227). However, in the past 25 years, several disciplines, including psychology, history, and artificial intelligence,[2] have produced a substantial body of knowledge about the process of scientific discovery that allows us to say a great deal about it.[3] Although the strengths of one approach are often the weaknesses of another, the work has collectively yielded consistent insights into the scientific discovery process.

ASSESSING THE FOUR MAJOR APPROACHES

Historical accounts of the great scientific discoveries—typically based on diaries, scientific publications, autobiographies, lab notebooks, correspondence, interviews, grant proposals, and memos of famous scientists—have high face validity. That is, it is clear that they are based on what they purport to study: real science. However, such studies have some weaknesses. For one thing, their sources are often subjective and unverifiable. Moreover, the temporal resolution of historical analysis is often coarse, but it can become much finer when laboratory notebooks and correspondence are available. Historical investigations often generate novel results about the discovery process, by focusing on a particular scientist and state of scientific knowledge, as well as by highlighting social and motivational factors not addressed by other approaches.

Although historical studies of discovery focus much more on successes than on failures, laboratory studies are designed to manipulate the discovery context in order to examine differences in processes associated with success and failure. Face validity of lab studies varies widely: from studies only distantly related to real scientific tasks to those that model essential aspects of specific scientific discoveries (e.g., Dunbar's, 1993, simulated molecular genetics laboratory; Schunn & Anderson's, 1999, comparison of experts' and novices' ability to design and interpret memory experiments; Qin & Simon's, 1990, study in which college sophomores rediscovered Kepler's third law of planetary motion). Laboratory studies tend to generate fine-grained data over relatively brief periods and typically ignore or minimize social and motivational factors.

The most direct way to study science is to study scientists in their day-to-day work, but this is extraordinarily difficult and time-consuming. A recent example is Dunbar's (1994) analysis of discovery processes in several world-class molecular genetics research labs. Such studies have high face validity and potential for detecting new phenomena. Moreover, they may achieve much finer-grained temporal resolution of ongoing processes than historical research, and they provide rigor, precision, and objectivity that is lacking in retrospective accounts.

A theory of discovery processes can sometimes be incorporated in a computational model that simulates and reenacts discoveries. Modeling draws upon the same kinds of information as do historical accounts, but goes beyond history to hypothesize cognitive mechanisms that can make the same discoveries, following the same path. Modeling generates theories and tests them against data obtained by the other methods. It tests the sufficiency of the proposed mechanisms to produce a given discovery and allows comparison between case studies, interpreting data in a common language to reveal both similarity and differences of processes. Modeling enables us to express a theory rigorously and to simulate phenomena at whatever temporal resolution and for whatever durations are relevant.

SCIENTIFIC DISCOVERY AS PROBLEM SOLVING

Crick argued that discoveries are major when they produce important knowledge, whether or not they employ unusual thought processes: "The path [to the double helix]. . . was fairly commonplace. What was important was *not the way it was discovered*, but the object discovered—the structure of DNA itself" (Crick, 1988, p. 67; italics added). Psychologists have been making the case for the "nothing special" view of scientific thinking for many years. This does not mean that anyone can walk into a scientist's lab and make discoveries. Practitioners must acquire an extensive portfolio of methods and techniques, and must apply their skills aided by an immense base of shared knowledge about the domain and the profession. These components of expertise constitute the *strong methods*. The equally important *weak methods* scientists use underlie all human problem-solving processes.

A problem consists of an initial state, a goal state, and a set of operators for transforming the initial state into the goal state by a series of intermediate steps. Operators have constraints that must be satisfied before they can be applied. The set of states, operators, and constraints is called a *problem space*, and the problem-solving process can be characterized as a search for a path that links initial state to goal state.

Initial state, goal state, operators, and constraints can each be more or less well-defined. For example, one could have a well-defined initial state and an ill-defined goal state and set of operators (e.g., make "something pretty" with these materials and tools), or an ill-defined initial state and a well-defined final state (e.g., find a vaccine against HIV). But well-definedness depends on the familiarity of the problem-space elements, and this, in turn, depends on an interaction between the problem and the problem solver.

Although scientific problems are much less well-defined than the puzzles commonly studied in the psychology laboratory, they can be characterized in these terms. In both cases, well-definedness and familiarity depend not only on the problem, but also on the knowledge that is available to the scientist. For that reason, much of the training of scientists is aimed at increasing the degree of well-definedness of problems in their domain. The size of a problem space grows exponentially with the number of alternatives at each new step in the problem (e.g., the number of possible paths one must consider at each possible move when planning ahead in chess). Effective problem solving must constrain search to only a few such paths. Strong methods, when available, find solutions with little or no search. For example, in chess, there are many standard openings that allow experts to make their initial moves with little search. Similarly, someone who knows algebra can use simple linear equations to choose between two sets of fixed and variable costs when deciding which car to rent instead of painstakingly considering the implications of driving each car a different distance. But the problem solver must first recognize the fit between the given problem (renting a car) and the strong method (high school algebra).

Weak methods, requiring little knowledge of a problem's structure, do not constrain search as much. One particularly important weak method is analogy, which attempts to map a new problem onto one previously encountered, so that the new problem can be solved by a known procedure. However, the mapping may be quite complex, and it may fail to produce a solution.

Analogy enables the problem solver to shift the search from the given problem space to one in which the search may be more efficient, sometimes making available strong methods that greatly abridge search. Prior knowledge can then be used to plan the next steps of problem solving, replace whole segments of step-by-step search, or even suggest an immediate solution. The recognition mechanism uses this store of knowledge to interpret new situations as instances of previously encountered situations. This is a key weapon in the arsenal of experts and a principal factor in distinguishing expert from novice performance.

In the past 25 years, analogy has assumed prominence in theories of problem solving and scientific discovery. Nersessian (1984) documented its role in several major 19th-century scientific discoveries. Recent studies of contemporary scientists working in their labs have revealed the central role played by analogy in scientific discovery (Dunbar, 1994; Thagard, 1998).

Although many strong methods are applied in scientific practice, weak methods are of special interest for scientific discovery because they are applicable in a wide variety of contexts, and strong methods become less available as the scientist approaches the boundaries of knowledge.

COMPLEMENTARITY OF APPROACHES

Viewing scientific discovery as problem solving provides a common language for describing it and facilitates studying the same discovery using more than one approach.

In the late 1950s, Monod and Jacob discovered how control genes regulate the synthesis of lactose (a sugar found in milk) in bacteria (Jacob & Monod, 1961). The literature explaining this discovery (e.g., Judson, 1996) tends to use terms such as "a gleam of perception," but to characterize a discovery as a gleam of perception is to not describe it at all. One must identify specific and well-understood cognitive processes and then determine their role in the discovery. Among the most important steps for Jacob and Monod in discovering the mechanisms of genetic control were representational changes that enabled them to replace their entrenched idea—that genetic control must involve some kind of activation—with the idea that it employed inhibition instead.

Dunbar (1993) created a laboratory task that captured important elements of Monod and Jacob's problem, while simplifying to eliminate many others. His participants—asked to design and run (simulated) experiments to discover the lactose control mechanism—faced a real scientific task with high face validity. Although the task was simplified, the problem, the "givens," the permissible research methods, and the structure of the solution were all preserved. Dunbar's study cast light on the problem spaces that Monod and Jacob searched, and on some of the conditions of search that were necessary or sufficient for success (e.g., knowing that there was such a thing as a control gene, but not exactly how it worked).

In this example, a historically important scientific discovery provided face validity for the laboratory study, and the laboratory provided information about the discovery processes with fine-grained temporal resolution.

CONVERGENT EVIDENCE OF DISCOVERY PRINCIPLES

In this section, we give a few additional examples of convergent evidence obtained by using two or more approaches to study the same discovery.

Surprise

Recently, reigning theories of the scientific method have generally taken hypotheses as unexplained causes that motivate experiments designed to test them. In this view, the hypotheses derive from scientists' "intuitions," which are beyond explanation. Historians of science have taken a less rigid position with respect to hypotheses, and include their origins within the scope of historical inquiry.

For example, the discovery of radium by the Curies started with their attempt to obtain pure radioactive uranium from pitchblende. As they proceeded, they were surprised to find in pitchblende levels of radioactivity higher than in pure uranium. As a surprise calls for an explanation, they conjectured that the pitchblende contained a second substance (which they named radium) more radioactive than uranium. They succeeded in extracting the radium and determined its key properties. In this case, a phenomenon led to a hypothesis, rather than a hypothesis leading to an experimental phenomenon. This occurs frequently in science. A surprise violates prior expectations. In the face of surprise, scientists frequently divert their path to ascertain the scope and import of the surprising phenomenon and its mechanism.

Response to surprise was investigated in a laboratory study (Klahr, Fay, & Dunbar, 1993) in which participants had to discover the function of an unknown key on a simulated rocket ship. They were given an initial hypothesis about how the key worked. Some participants were given a plausible hypothesis, and others were given an implausible hypothesis. In all cases, the suggested hypothesis was wrong, and the rocket ship produced some unexpected, and sometimes surprising, behavior. Adults reacted to an implausible hypothesis by proposing a competing hypothesis and then generating experiments that could distinguish between them. In contrast, children (third graders) tended to dismiss an implausible hypothesis and ignore evidence that supported it. Instead, they attempted to demonstrate that their favored hypothesis was correct. It seems that an important step in acquiring scientific habits of thinking is coming to accept, rather than deny, surprising results, and to explore further the phenomenon that gave rise to them.

Krebs's biochemical research leading to the discovery of the chain of reactions (the reaction path) by which urea (the end product of protein metabolism) is synthesized in the body has been the topic of convergent studies focusing on response to unexpected results. The discovery has been studied historically by Holmes (1991) and through the formulation of two computational models (Grasshoff & May, 1995; Kulkarni & Simon, 1988), both of which have modeled the discovery. After the models proposed an experiment and were given its outcome, they then proposed another experiment, using the previous outcomes to guide their decision about what sort of experiment would be useful. Using no more knowledge of biochemistry than Krebs possessed at the outset, both programs discovered the reaction path by following the same general lines of experimentation as Krebs followed. One of these models (Kulkarni & Simon, 1988) addressed the surprise issue directly (in this case, surprise in finding a special catalytic role for the amino acid ornithine). The simulated scientist formed expectations (as did Krebs) about experimental outcomes. When the expectations were violated, steps were taken to explain the surprise. Thus, historical studies, simulation models, and laboratory experiments all provide evidence that the scientist's reaction to phenomena—either observational or experimental—that are surprising can lead to generating and testing new theories.

Multiple Search Spaces

This reciprocal relation between hypotheses and phenomena arises in laboratory studies, historical studies, and computational models of discovery, enabling us to characterize scientists' thinking processes as problem-solving search in multiple spaces.

Dual Search

The discovery process can be characterized as a search in two spaces: a hypothesis space and an experiment space. When attempting to discover how a particular control button worked on a programmable device, participants in the "rocket ship" study described earlier (Klahr & Dunbar, 1988) had to negotiate this dual search by (a) designing experiments to disclose the button's functions (searching the experiment space) and (b) proposing rules that explained the device's behavior (searching the hypothesis space). Thus, participants were required to coordinate two kinds of problems, and they approached this dual search with different emphases. Some ("experimenters") focused on the space of possible manipulations, whereas others ("theorists") focused on the space of possible explanations of the responses.

Historical studies usually reveal both hypothesis-space search and experiment-space search. For example, most histories of Faraday's discovery of induction of electricity by magnets place much emphasis on the influence of Ampère's theory of magnetism on Faraday's thought. However, a strong case can be made that Faraday's primary search strategy was in the space of experiments, his discovery path being shaped by phenomena observed through experimentation more than by theory.

The number of search spaces depends on the nature of the scientific problem. For example, in describing the discovery of the bacterial origins of stomach ulcers, Thagard (1998) demonstrated search in three major spaces: hypothesis space, experiment space, and a space of instrumentation.

Analogy in Search for Representations

Bohr used the solar system analogy to arrive at his quantum model of the hydrogen atom. He viewed the electrons in the hydrogen atom as planets orbiting the nucleus, although, according to classical understanding of the solar system, this would mean that the charged electrons would dissipate energy until they fell into the nucleus. Instead of abandoning the analogy, Bohr borrowed Planck's theory that energy could be dissipated only in quantum leaps, then showed that these leaps would produce precisely the spectrum of light frequencies that scientists 30 years previously had demonstrated hydrogen produces when its electrons move from a higher-energy stationary state to a lower-energy one.

Search in the Strategy Space

Finally, changes in strategy, even while the representation of a problem is fixed, may enable discovery. Often the change in strategy results from, or leads to, the invention of new scientific instruments or procedures. Breeding experiments go back to Mendel (and experiments for stock breeding go much further back), but the productivity of such experiments depended on mutation rates. Müller, with the "simple" idea that x-rays could induce higher rates of mutation, substantially raised that productivity.

CREATIVITY AND PROBLEM SOLVING IN SCIENCE AND BEYOND

Scientific discovery is a type of problem solving using both weak methods that are applicable in all disciplines and strong methods that are mainly domain-specific. Scientific discovery is based on heuristic search in problem spaces: spaces of instances, of hypotheses, of representations, of strategies, of instruments, and perhaps others. This heuristic search is controlled by general mechanisms such as trial-and-error, hill-climbing, means-ends analysis, analogy, and response to surprise. Recognition processes, evoked by familiar patterns in phenomena, access knowledge and strong methods in memory, linking the weak methods to the domain-specific mechanisms.

All of these constructs and processes are encountered in problem solving wherever it has been studied. A painter is not a scientist; nor is a scientist a lawyer or a cook. But they all use the same weak methods to help solve their respective problems. When their activity is described as search in a problem space, each can understand the rationale of the other's activity, however abstruse and arcane the content of any special expertise may appear.

At the outer boundaries of creativity, problems become less well structured, recognition becomes less able to evoke prelearned solutions or domain-specific search heuristics, and more reliance has to be placed on weak methods. The more creative the problem solving, the more primitive the tools. Perhaps this is why "childlike" characteristics, such as the propensity to wonder, are so often attributed to creative scientists and artists.

Recommended Reading

Klahr, D. (2000). *Exploring science: The cognition and development of discovery processes.* Cambridge, MA: MIT Press.

Klahr, D., & Simon, H.A. (1999). Studies of scientific discovery: Complementary approaches and convergent findings. *Psychological Bulletin, 125,* 524–543.

Zimmerman, C. (2000). The development of scientific reasoning skills. *Developmental Review, 20,* 99–149.

Acknowledgments—Preparation of this article and some of the work described herein were supported in part by a grant from the National Institute of Child Health and Human Development (HD 25211) to the first author. We thank Jennifer Schnakenberg for a careful reading of the penultimate draft.

Notes

1. Address correspondence to David Klahr, Department of Psychology, Carnegie Mellon University, Pittsburgh, PA 15213.

2. In addition, the sociology of science explains scientific discovery in terms of political, anthropological, or social forces. The mechanisms linking such forces to scientific practice are usually motivational, social-psychological, or psychodynamic, rather than cognitive. Although this literature has provided important insights on how social and professional constraints influence scientific practices, we do not have much to say about it in this brief article.

3. This article summarizes an extensive review listed as the second recommended reading. Full references to historical sources alluded to in the present article can be found there, as well as in the first recommended reading. The third recommended reading focuses on developmental aspects of the discovery process.

References

Crick, F. (1988). *What mad pursuit: A personal view of scientific discovery.* New York: Basic Books.

Dunbar, K. (1993). Concept discovery in a scientific domain. *Cognitive Science, 17,* 397–434.

Dunbar, K. (1994). How scientists really reason: Scientific reasoning in real-world laboratories. In R.J. Sternberg & J. Davidson (Eds.), *The nature of insight* (pp. 365–395). Cambridge, MA: MIT Press.

Grasshoff, G., & May, M. (1995). From historical case studies to systematic methods of discovery: Working notes. In *American Association for Artificial Intelligence Spring Symposium on Systematic Methods of Scientific Discovery* (pp. 45–56). Stanford, CA: AAAI.

Holmes, F.L. (1991). *Hans Krebs: The formation of a scientific life, 1900–1933, Volume 1*. New York: Oxford University Press.

Jacob, F., & Monod, J. (1961). Genetic regulatory mechanisms in the synthesis of proteins. *Journal of Molecular Biology*, 3, 318–356.

Judson, H.F. (1996). *The eighth day of creation: Makers of the revolution in biology* (expanded ed.). Plainview, NY: Cold Spring Harbour Laboratory Press.

Klahr, D., & Dunbar, K. (1988). Dual space search during scientific reasoning. *Cognitive Science*, 12, 1–55.

Klahr, D., Fay, A.L., & Dunbar, K. (1993). Heuristics for scientific experimentation: A developmental study. *Cognitive Psychology*, 24, 111–146.

Kulkarni, D., & Simon, H.A. (1988). The process of scientific discovery: The strategy of experimentation. *Cognitive Science*, 12, 139–176.

Nersessian, N.J. (1984). *Faraday to Einstein: Constructing meaning in scientific theories*. Dordrecht, The Netherlands: Martinus Nijhoff.

Qin, Y., & Simon, H.A. (1990). Laboratory replication of scientific discovery processes. *Cognitive Science*, 14, 281–312.

Schunn, C.D., & Anderson, J.R. (1999). The generality/specificity of expertise in scientific reasoning. *Cognitive Science*, 23, 337–370.

Thagard, P. (1998). Ulcers and bacteria: I. Discovery and acceptance. *Studies in the History and Philosophy of Biology and Biomedical Science*, 9, 107–136.

Wertheimer, M. (1945). *Productive thinking*. New York: Harper & Row.

Critical Thinking Questions

1. This article describes four major approaches to the study of scientific discovery. What are the strengths and weaknesses of each of these techniques? Why is it important to have multiple techniques?

2. What distinguishes a true scientific breakthrough from the kinds of problem solving that we do in everyday life (e.g., discovering a shortcut to school, fixing your computer, etc.)—is it the thought processes involved or the product created?

3. How does the "difficulty" of a problem depend on both the structure of the problem and the knowledge of the problem solver?

4. Many popular games, like Clue, Guess Who, Mastermind, and 20 Questions, involve reasoning that is similar to scientific reasoning. For the games you know—what is the hypothesis space and what is the experiment space? Could these games be used to study scientific reasoning in the laboratory? Why might there be people who say they "hate science" but who still enjoy playing these kinds of games?

Unconscious Insights

Robert S. Siegler[1]

Psychology Department, Carnegie Mellon University, Pittsburgh, Pennsylvania

Abstract

From early in the history of psychology, theorists have argued about whether insights are initially unconscious or whether they are conscious from the start. Empirically identifying unconscious insights has proven difficult, however: How can we tell if people have had an insight if they do not tell us they have had one? Fortunately, although obtaining evidence of unconscious insights is difficult, it is not impossible. The present article describes an experiment in which evidence of unconscious insights was obtained. Almost 90% of second graders generated an arithmetic insight at an unconscious level before they were able to report it. Within five trials of the unconscious discovery, 80% of the children made the discovery consciously, as indicated by their verbal reports. Thus, the initial failure to report the insight could not be attributed to the children lacking the verbal facility to describe it. The results indicate that at least in some cases, insights arise first at an unconscious level, and only later become conscious. Rising activation of the new strategy may be the mechanism that leads children to become conscious of using it.

Keywords

insight; discovery; conscious; unconscious; arithmetic

Do unconscious ideas drive our conscious perceptions, thoughts, and behavior? Over the past decade, advances in cognitive neuropsychology have helped spark a rebirth of interest in this enduring question. To cite one example, there have been documented cases of blindsight, in which patients who have suffered brain damage are unaware of seeing objects but can accurately "guess" the objects' locations (Weiskranz, 1997). The influence of unconscious knowledge on behavior is not limited to brain-damaged patients. College students who are asked to learn artificial grammars, four-way statistical interactions, and other complex rule systems often are unaware of having learned the rules, yet they can use the rule systems to classify novel instances (Lewicki, Hill, & Czyzewska, 1992). Recent findings indicate that even insightful discoveries sometimes arise unconsciously before they reach a conscious (i.e., reportable) level. This article describes some of the research leading to this conclusion.

PERSPECTIVES ON INSIGHT AND CONSCIOUSNESS

Long before there was a scientific field of psychology, philosophers, mathematicians, and scientists described the role of consciousness in their own insights. Archimedes' experience of stepping into the bath, seeing the water rise, and exclaiming "Eureka" is probably the prototypical insight: a sudden change from not knowing a problem's solution to knowing it consciously.

In the Archimedes anecdote, an external event stimulated the insight. Other thinkers have emphasized the contribution of unconscious processes and dreams to their discoveries. Perhaps the prototypical account of this type is Kekule's dream of intertwined snakes, which led him to "see" the structure of the benzene ring.

Although these two cases differ in what led up to the insight, the insight itself emerged suddenly in both cases. Other accounts differ, though. Wittgenstein (1969), for example, compared generation of new ideas to a sunrise: Although our experience is that the new day suddenly "dawns," the amount of light actually grows continuously over a fairly protracted period of time.

These examples suggest two major questions regarding the relation between consciousness and insight: Do insights arise at an unconscious (i.e., nonreportable) level before they arise consciously, and do insights arise suddenly or gradually? These are basic questions about human nature, and they have motivated considerable theorizing over the past century (see Sternberg & Davidson, 1995, and the special section of *American Psychologist* edited by Loftus, 1992, for incisive discussions of classical and current perspectives on these issues). However, the questions have proven resistant to empirical resolution. The main reason is the difficulty of obtaining evidence regarding unconscious insights. Simply put, how can we know that people have an insight if they do not tell us that they had it? Recently, however, Elsbeth Stern and I found a way to obtain independent measures of conscious and unconscious insights and thus to examine the relation between them (Siegler & Stern, 1998).

THE INVERSION TASK

On problems of the form A + B - B (e.g., 18 + 24 - 24), the answer always is A. Such inversion problems are useful for studying insight because they can be solved in either insightful or noninsightful ways. The noninsightful way is to use the standard procedure of adding the first two numbers and subtracting the third. The insightful way involves simply saying the first number.

In addition to allowing both insightful and noninsightful solutions, the inversion task has the unusual property of allowing independent measurement of conscious and unconscious versions of the insight. Immediately retrospective verbal reports provided the measure of conscious use of the insight in the research Stern and I conducted. Young school-age children typically report their arithmetic strategies quite accurately, as indicated by converging evidence from reaction time and error patterns (cf. Siegler, 1987). What made the inversion problems special, however, was that an implicit measure of strategy use, one that did not require any verbal report, also could be obtained: the child's solution time. Ordinarily, solution times are insufficient to infer the strategy that was used on an individual trial. However, they are considerably more useful for inferring strategy use on inversion problems. The reason is that solving the problems via computation generates much slower solution times than solving them by using the arithmetic insight. Consistent with this view, solution times on inversion problems in our study were bimodally distributed: 92% of times were either fast (4 s or less) or slow (8 s or more). Converging evidence from overt behavior supported the view that the fast times reflected use of the insight and that the slow times reflected use of computation. Overt computational activity was observed on 80% of trials classified as computation, versus 0% of trials on which children were classified as using the shortcut. (Methods used to classify strategy use are discussed in the next section.)

CONSCIOUS AND UNCONSCIOUS DISCOVERIES

Having both a verbal report and a solution time on each trial made it possible to define three main strategies: *computation*, *shortcut*, and *unconscious shortcut*. Children were classified as using the computation strategy on each trial on which they took more than 4 s to come up with a solution and said they computed the answer; they were classified as using the shortcut on each trial on which the solution time was 4 s or less and they said they used the shortcut; and they were classified as using the unconscious shortcut on each trial on which their solution time was 4 s or less but they claimed to have computed the answer.

We expected that on the large majority of trials, the verbal-report and solution-time measures of strategy use would converge: Children would either solve the problem quickly and say they used the shortcut or take longer to solve it and say they computed the answer. However, we also expected that sometimes the measures would diverge: The child would solve the problem in 4 s or less but claim to have solved the problem through addition and subtraction. Such trials, if they occurred most often at predicted places in the learning sequence, would indicate unconscious use of the shortcut.

THE UNCONSCIOUS-ACTIVATION HYPOTHESIS

Based on previous research showing unconscious influences on other types of thinking, we formulated the *unconscious-activation hypothesis*: Increasing activation of a strategy leads to people first using it unconsciously; then, as the activation increases further, people become conscious of using it. The straightforward implication of this hypothesis was that the unconscious shortcut would emerge before the conscious version of the strategy.

To further test the unconscious-activation hypothesis, we created two experimental conditions. One was the *blocked-problems condition*. Children in it were presented A + B - B problems, that is, problems that could be solved by the inversion principle, on 100% of the trials. The other experimental condition was the *mixed-problems condition*. In it, children were presented A + B - B problems on 50% of trials, and on the other 50% were presented A + B - C problems (i.e., problems, such as 18 + 24 - 15, in which the three numbers differed and therefore that could not be solved via the shortcut strategy). The unconscious-activation hypothesis predicted that presenting children inversion problems on 100% of the trials would lead to a more rapid increase in activation of the shortcut, which in turn would lead to (a) more rapid discovery of the unconscious-shortcut and shortcut strategies (discovery after fewer inversion problems), (b) a shorter gap between discovery of the unconscious shortcut and discovery of the shortcut, (c) more consistent use of the shortcut on inversion problems once it was discovered, and (d) greater generalization of the strategy to novel problems of similar appearance, such as A - B + B and A + B + B.

AN EXPERIMENT ON CONSCIOUS AND UNCONSCIOUS INSIGHTS

To test these predictions, we presented 31 German second graders with either the blocked problems or the mixed problems. The experiment was conducted over an 8-week period, one session per week.

Each of the predictions of the unconscious-activation hypothesis was borne out. Almost 90% of the children discovered the unconscious version of the short-cut before the conscious version. Relative to children in the mixed-problems condition, children in the blocked-problems condition discovered both the unconscious-shortcut and the shortcut strategies after seeing fewer inversion problems, exhibited a shorter gap between their discovery of the two strategies, used the strategies more often once they discovered them, and generalized the strategies more widely to novel types of problems.

Examination of strategy use just before and after discovery of the unconscious-shortcut and shortcut strategies provided particularly direct support for the unconscious-activation hypothesis. Figure 1 illustrates the circumstances surrounding the first use of the unconscious shortcut among children in the blocked-problems condition. Trial 0 for a given child is the trial on which the

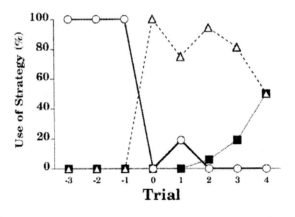

Fig. 1. Percentage use of computation, unconscious-shortcut, and shortcut strategies in the blocked-problems condition on trials immediately preceding and following children's first use of the unconscious shortcut. Each child's first use of the unconscious shortcut is designated Trial 0; the trial just before it is designated Trial -1, the trial just after it is designated Trial 1, and so on.

child first used the unconscious shortcut; thus, by definition, 100% of children used the unconscious shortcut on Trial 0. Trial -1 for a given child is whichever trial immediately preceded that child's Trial 0; Trial 1 is whichever trial immediately followed the child's Trial 0; and so on.

Data from the blocked-problems condition were particularly striking. Figure 1 reveals that just before their first use of the unconscious shortcut, all of these children used the computation strategy. After their initial use of the unconscious shortcut, most of them continued to use the unconscious shortcut over the next three trials. By the fourth trial after the initial use of the unconscious shortcut, half of the children reported using the shortcut. By the fifth trial, 80% of the children reported using it.

Figure 2 shows a parallel analysis centered on first use of the shortcut by the same children. On the three trials immediately preceding its first use, roughly 80% of these children used the unconscious shortcut (as opposed to less than 10% use of this strategy for the study as a whole). Once the children began to report using the shortcut, they continued to use it quite consistently within that session. However, when they returned a week later for the next session, fewer than 35% used the shortcut on any trial before Trial 5. Thereafter, more children rediscovered the shortcut, and by the end of the session, more than 90% of them were again using it.

Changes in solution times from the trials just before the first use of the unconscious shortcut to the trial of discovery suggested that the unconscious shortcut represented a sudden, qualitative shift in thinking. On the three trials immediately before its first use, solution times averaged 12 s; on its first use, the mean solution time was 2.7 s. Solution times on subsequent unconscious-shortcut trials (and on

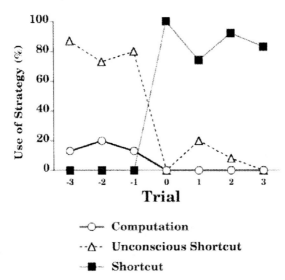

Fig. 2. Percentage use of computation, unconscious-shortcut, and shortcut strategies in the blocked-problems condition on trials immediately preceding and following children's first use of the shortcut. Each child's first use of the shortcut is designated Trial 0; the trial just before it is designated Trial -1, the trial just after it is designated Trial 1, and so on.

shortcut trials as well) continued to average between 2 s and 3 s in all sessions. Thus, although children who used the unconscious shortcut did not report doing anything different, they had already had the insight at a behavioral level.

The lack of reporting of the insight on unconscious-shortcut trials could not be attributed to the children being generally inarticulate, to the insight being difficult to put into words, or to children's perceptions of social desirability preventing them from reporting an approach that they knew they were using. If those were the reasons for children initially not reporting the shortcut, why would the same children have almost invariably reported using it a few trials later in the same session? Further supporting the view that use of the shortcut was at first unconscious, when children rediscovered the shortcut in the session following the one in which they initially used it, most again used the unconscious version just before beginning to report its use.

CONCLUSIONS

These results shed light on both of the questions raised at the outset regarding insights and consciousness. With regard to the first question, the findings demonstrate that insights are not always conscious from the start. At least sometimes, they arise first in unconscious form.

The results also provide an answer to the second question: Insights are abrupt in some senses, but gradual in others. On the one hand, the dramatic reduction in solution times that accompanied the first use of the unconscious shortcut indicates a sense in which insight was abrupt. The fact that solution times on shortcut trials did not decline further indicates that in terms of efficiency of execution, the shortcut emerged full-blown. On the other hand, the insight was gradual in two other senses. First, children initially discovered the shortcut in a nonreportable form and only later became able to report using it. Second, use of the shortcut increased slowly, never extending to more than 60% of trials in a given session.

The results also raise several intriguing questions. Do adults also begin to use new strategies unconsciously before they become conscious of using them, or is unconscious use of strategies unique to children? Are unconscious insights limited to single-step strategies, as in the present case, or do people also discover multistep strategies at an unconscious level before they discover them consciously? Perhaps most important, through what cognitive processes are unconscious insights generated?

Underlying these and other relatively specific questions is the main point of our study, a point consistent with a wide range of previous research: Having a thought, or even an insight, is not the same as being aware of having that thought or insight. Learning how consciousness is related to insight remains one of the basic challenges in understanding human psychology, just as it was in the days of Archimedes.

Recommended Reading

Davidson, J. (1995). The suddenness of insight. In R.J. Sternberg & J.E. Davidson (Eds.), *The nature of insight* (pp. 125–155). Cambridge, MA: MIT Press.

Goldin-Meadow, S., Alibali, M.W., & Church, R.B. (1993). Transitions in concept acquisition: Using the hand to read the mind. *Psychological Review, 100,* 279–297.

Schooler, J.W., Fallshore, M., & Fiore, S.M. (1995). Epilogue: Putting insight into perspective. In R.J. Sternberg & J.E. Davidson (Eds.), *The nature of insight* (pp. 367–402). Cambridge, MA: MIT Press.

Siegler, R.S., & Stern, E. (1998). (See References)

Acknowledgments—I would like to thank the National Institutes of Health (Grant 19011) and the Spencer Foundation for funding that helped make possible much of the research reported in this article.

Note

1. Address correspondence to Robert S. Siegler, Psychology Department, Carnegie Mellon University, Pittsburgh, PA 15213.

References

Lewicki, P., Hill, T., & Czyzewska, M. (1992). Nonconscious acquisition of information. *American Psychologist, 47,* 796–801.

Loftus, E.F. (Ed.). (1992). Science watch [Special section]. *American Psychologist, 47,* 761–809.

Siegler, R.S. (1987). The perils of averaging data over strategies: An example from children's addition. *Journal of Experimental Psychology: General, 116,* 250–264.

Siegler, R.S., & Stern, E. (1998). A microgenetic analysis of conscious and unconscious strategy discoveries. *Journal of Experimental Psychology: General, 127,* 377–397.

Sternberg, R.J., & Davidson, J.E. (Eds.). (1995). *The nature of insight.* Cambridge, MA: MIT Press.

Weiskranz, L. (1997). *Consciousness lost and found.* Oxford, England: Oxford University Press.

Wittgenstein, L. (1969). *On certainty.* New York: Harper & Row.

Critical Thinking Questions

1. This article mentions the famous parable about Kekule—that he discovered the structure of benzene after dreaming about a snake. If someone tells you how he solved a problem or why she had a particular insight, should you necessarily believe them (assuming that they are not lying)? In particular, should you believe an account written long after the event occurred? (Recall what you know about the reconstructive nature of memory.)

2. In the Inversion task, the conscious strategy was measured by asking the children which strategy they used and the unconscious strategy was measured by solution time. Given the two conscious categories and two unconscious categories, there should be four combinations. What is each called? Which one doesn't exist? In what order did the children go through them?

3. Do you believe that you have ever used an unconscious strategy before you became aware of it? Think of solving math, statistics, chemistry, or physics problems. Also think of knowing what to do in a card game or where to move in a sport without being taught when and why to take those specific actions.

Anticipated Emotions as Guides to Choice

Barbara A. Mellers[1] and A. Peter McGraw

Department of Psychology, The Ohio State University, Columbus, Ohio

Abstract

When making decisions, people often anticipate the emotions they might experience as a result of the outcomes of their choices. In the process, they simulate what life would be like with one outcome or another. We examine the anticipated and actual pleasure of outcomes and their relation to choices people make in laboratory studies and real-world studies. We offer a theory of anticipated pleasure that explains why the same outcome can lead to a wide range of emotional experiences. Finally, we show how anticipated pleasure relates to risky choice within the framework of subjective expected pleasure theory.

Keywords

anticipated emotions; choice; pleasure

When making decisions, people often anticipate how they will feel about future outcomes and use those feelings as guides to choice. To understand this process, we have investigated the anticipated and actual pleasure of outcomes that follow decisions in laboratory and real-world studies. In this article, we present a theory of anticipated pleasure called decision affect theory and show how it relates to decision making. We claim that when making decisions, people anticipate the pleasure or pain of future outcomes, weigh those feelings by the chances they will occur, and select the option with greater average pleasure.[2]

Imagine a decision maker who is considering two locations for a summer vacation. The first is perfect in all regards—as long as the weather is nice. Unfortunately, the weather is hard to predict. The second location is quite acceptable, and the weather is almost always good. To make a choice, the decision maker anticipates the pleasure of the first vacation assuming good weather and the displeasure of the vacation assuming bad weather. These feelings are weighted by the perceived likelihood of good or bad weather, respectively, and the resulting feelings are combined to obtain an average feeling of anticipated pleasure. The second location is evaluated in the same manner, and the location with greater average pleasure is selected.

We begin by summarizing several studies and then answer three related questions. First, what variables influence anticipated pleasure? Second, how is anticipated pleasure related to choice? And third, how accurately do people anticipate the pleasure of future outcomes?

EXPERIMENTS

In our laboratory studies, we presented participants with pairs of monetary gambles on a computer screen (Mellers, Schwartz, Ho, & Ritov, 1997; Mellers, Schwartz, & Ritov, 1999). Each gamble was displayed as a pie chart with two regions, representing wins or losses. On each trial, respondents chose the gamble they preferred to play. In some conditions, a spinner appeared in the center of the chosen gamble and began to rotate. Eventually it stopped, and participants learned how much they won or lost. In other conditions, spinners appeared in the center of both gambles. The spinners rotated independently and eventually stopped, at which point participants learned their outcome and that of the other gamble. Outcomes ranged from a $32 win to a $32 loss. In some studies, the outcomes were hypothetical, and in others, the outcomes were real. If the outcome was hypothetical, participants anticipated the pleasure they would have felt had the outcome been real. If the outcome was real, participants rated their actual pleasure. Both types of judgments were made on a category rating scale from "very happy" to "very unhappy."

Within this paradigm, participants are likely to find two counterfactual comparisons particularly salient (Bell, 1982, 1985; Loomes & Sugden, 1982, 1986). Comparisons of the imagined outcome with other outcomes of the chosen gamble are called disappointment or elation. Comparisons of the imagined outcome with an outcome of the unchosen gamble are called regret or rejoicing.

In the real-world studies, we used participants who had already made a choice, but did not yet know the outcome of their choice. We asked them to anticipate their feelings about all possible outcomes of the choice. Later, when they learned what the actual outcome was, they rated their feelings regarding what occurred. The studies involved grades, diets, and pregnancy tests. In the grading study, undergraduates predicted their final grade in introductory psychology and anticipated their emotional reactions to all possible grades. The following quarter, they told us their actual grades and feelings about those grades. In the dieting study, clients participating in a commercial weight-loss program told us their weekly weight-loss goals and anticipated their feelings about various outcomes. They returned the following week, learned about their weight changes, and reported their feelings. Finally, in the pregnancy study, women waiting for a pregnancy test at Planned Parenthood anticipated their emotions about possible test results. Ten minutes later, they learned the results and judged their reactions. We now present selected results from these studies.

WHAT VARIABLES INFLUENCE ANTICIPATED PLEASURE?

Our most important findings about pleasure are shown in Figure 1, which presents results from the gambling studies. Our findings can be summarized in terms of outcome effects, comparison effects, and surprise effects. Outcome effects are illustrated in Figure 1a. As the amount of the imagined outcome increases, so does the anticipated pleasure.

Figures 1b and 1c show comparison effects. Figure 1b plots the anticipated pleasure of an obtained win of $8 or loss of $8, separately for trials on which the unobtained outcome was a loss of $32 or gain of $32. When the unobtained outcome was more desirable, the anticipated pleasure about the obtained outcome declined. This is because people anticipate disappointment when they imagine getting the worse outcome of two outcomes. Figure 1c plots the anticipated pleasure of an obtained win of $8 or a loss of $8, separately for trials on

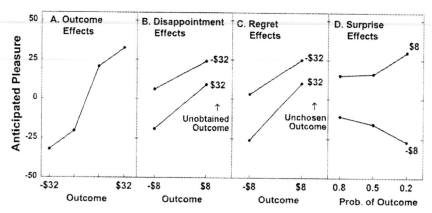

Fig. 1. Selected results from the gambling studies showing the effects of outcomes (a), comparison (b and c), and surprise (d) on anticipated pleasure. Comparison effects are illustrated by anticipated pleasure when the unobtained (b) or unchosen (c) outcome was a loss of $32 versus a gain of $32. Surprise effects (d) are shown for both a gain of $8 and a loss of $8. Prob. = probability.

which the outcome of the other gamble was a loss or gain of $32. A similar pattern appears: When the outcome of the unchosen gamble was more appealing, anticipated pleasure decreased. This is because people anticipate regret when they imagine having made the wrong choice.

Comparison effects of both disappointment and regret on anticipated pleasure are asymmetric. The displeasure of getting the worse of two outcomes is typically greater in magnitude than the pleasure of receiving the better outcome. Comparison effects are powerful enough to make an imagined loss that is the better of two losses more pleasurable than an imagined gain that is the worse of two gains, as we found in other studies (Mellers et al., 1999).

The results shown in Figure 1d illustrate the effects of surprise. Participants anticipated more pleasure with a win of $8 the less likely it was, and they anticipated less pleasure with a loss of $8 the less likely it was. In other words, both positive and negative feelings are stronger when outcomes are surprising. Surprising outcomes have greater intensity than expected outcomes. Surprise amplifies the emotional experience.

Figure 2 presents selected results from our real-world studies. Figure 2a shows the effects of outcome in the dieting study. As imagined weight loss increased, dieters anticipated greater pleasure. Figure 2b presents the comparison effects in the grading study. Students with lower expectations anticipated greater pleasure from all possible grades. Finally, Figure 2c shows the effects of surprise for women in the pregnancy study who preferred not to be pregnant. Surprising outcomes were associated with more intense anticipated feelings than expected outcomes.

The effects of outcomes, comparisons, and surprise shown in Figures 1 and 2 are predicted by an account of anticipated pleasure called decision affect theory. In this theory, anticipated pleasure depends on the utility (or psycho-

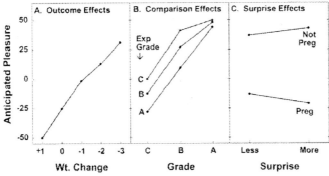

Fig. 2. Selected results from the dieting, grading, and pregnancy studies showing the effects of outcome (a), comparison (b), and surprise (c), respectively, on anticipated pleasure. Comparison effects are illustrated by anticipated pleasure when the expected grade was an A versus a B versus a C. Surprise effects are shown for both women who found out they were pregnant and those who found out they were not pregnant ("Preg"). Wt. = weight.

logical satisfaction) of the outcome and salient comparisons. Comparisons are weighted by how surprising the outcome is. We have provided formal treatments of this theory elsewhere (Mellers et al., 1997, 1999; Mellers & McGraw, 2001).

HOW IS ANTICIPATED PLEASURE RELATED TO CHOICE?

In several studies, we have found that anticipated pleasure is closely connected to choice (Mellers et al., 1999). We assume that decision affect theory predicts the pleasure people anticipate for future outcomes of a given option. Then they weigh those anticipated feelings by the perceived chances of their occurrence, and combine them to form an average anticipated pleasure for each option. The option with greater average pleasure is selected. More detailed descriptions of this theory, called subjective expected pleasure theory, are presented elsewhere (Mellers et al., 1999).

Individuals whose choices are consistent with subjective expected pleasure theory can differ in several respects. For example, if the vacationer we described in the introduction anticipates tremendous pleasure with the first location or is optimistic about good weather, he is more likely to select that location than the alternative location. Greater anticipated pleasure or greater optimism tend to produce greater risk seeking, whereas less anticipated pleasure or more pessimism lead to greater risk aversion.

Subjective expected pleasure theory is similar in some respects to subjective expected utility theory (Savage, 1954). In subjective expected utility theory, decision makers are assumed to consider the utility associated with each outcome, weigh that utility by the perceived chances it will occur, and sum the values for all the outcomes. Utilities are often described in terms of psychological satisfaction, so it seems logical to assume they would not differ from anticipated pleasure. However, utilities do differ from anticipated pleasure. In most theories of choice, utilities depend only on the status quo, but no other reference points. Anticipated pleasure depends on multiple reference points. Furthermore, in most theories of choice, utilities are assumed to be independent of beliefs. In contrast, the anticipated pleasure of outcomes varies systematically with beliefs about their occurrence; anticipated feelings associated with surprising outcomes are amplified relative to anticipated feelings associated with expected outcomes. Because the utility of an outcome differs from the anticipated pleasure of that outcome, the predictions of subjective expected utility theory and subjective expected pleasure theory can differ.

We tested subjective expected pleasure theory by examining whether it could predict participants' actual choices in our gambling studies. To do this, we fit decision affect theory to the anticipated pleasure of outcomes. That is, we estimated parameter values that produced the smallest squared deviations between participants' judgments of anticipated pleasure and the theory's predictions. Then, using these predictions of decision affect theory, we calculated the average anticipated pleasure of each gamble. Finally, we predicted choices based on the assumption that participants select the option with the greater average anticipated pleasure. Predicted choices were correlated with actual choices in five different experiments (Mellers et al., 1999). The correlations ranged from .66 to .86, with an average of .74. These values were remarkably high given the fact that subjective expected pleasure theory was never fit directly to choices. That is, choice predictions were obtained by fitting decision affect theory to judgments of anticipated pleasure.

We further tested decision affect theory by investigating whether anticipated pleasure (which contains utilities, comparisons, and surprise effects) added to the predictability of risky choice over and beyond utilities. To answer this question, we computed the correlations between predicted choices and actual choices after removing the predictions of subjective expected utility theory. These correlations were positive and ranged from .64 to .03, with an average correlation of .33. These analyses show that anticipated pleasure, which is sensitive to comparisons and surprise effects, improves the predictability of choice over and beyond utilities.

HOW WELL CAN DECISION MAKERS
ANTICIPATE PLEASURE?

If people make choices by comparing the average anticipated pleasure of options, the accuracy of their predictions becomes a critical concern. Inaccurate predictions could easily lead to peculiar choices. People who overestimate the pleasure of favorable outcomes, for example, would tend to be overly risk seeking. People who overestimate the displeasure of unfavorable outcomes would tend to be overly risk averse.

We examined the accuracy of affective predictions in both laboratory and real-world studies by comparing anticipated pleasure with actual pleasure (Mellers, 2000). In the laboratory studies, predictions were quite accurate. In the pregnancy and dieting studies, however, participants made systematic prediction errors, and those errors were in the same direction. Specifically, participants overestimated the displeasure of unfavorable outcomes. Women who received bad news from their pregnancy tests actually felt better than they expected. These results are surprising because judgments were made only 10 min apart. Likewise, dieters who gained weight or failed to lose it also felt better than they expected. These results are also surprising given the fact that most dieters are quite familiar with attempts to lose weight, and therefore should have experience with their actual reactions to unsuccessful attempts.

Other errors in affective forecasts have also been found (cf. Loewenstein & Schkade, 1999). Errors can occur from the emotions experienced during the choice process. These experienced emotions influence perceptions, memories, and even decision strategies. Other errors occur because people focus on whatever is salient at the moment, what Schkade and Kahneman (1998) call the focusing illusion. In a fascinating demonstration, Schkade and Kahneman asked students at universities in the Midwest and in California to judge their own happiness and the happiness of students at the other location. The comparison highlighted the advantages of California—better climate, more cultural opportunities, and greater natural beauty. Both students in the Midwest and those in California predicted that Californians were happier, but in fact, students at the two locations were equally happy.

The focusing illusion can also lead people to base affective predictions on transitions rather than final states (Kahneman, 2000). Gilbert, Pinel, Wilson, Blumberg, and Wheatley (1998) asked untenured college professors to anticipate how they would feel about receiving or not receiving tenure. Not surprisingly, the professors expected to be happy if given tenure and extremely unhappy otherwise. Actually, however, the professors who were denied tenure were much happier than they expected to be. Errors in affective forecasting that Gilbert et al. found were in the same direction as those we found in the pregnancy and dieting studies. People anticipated feeling worse about negative outcomes than they actually felt.

FUTURE DIRECTIONS

Research in decision making has demonstrated that anticipated pleasure improves the predictability of choice over and beyond utilities. The effects of comparisons and surprise add valuable information to descriptive theories of choice. Disappointment and regret are by no means the only comparisons that influence anticipated pleasure, however. Many other reference points may be salient. When people make a series of gambling choices, for example, the pleasure of a win or loss is affected by previous wins and losses. In competitive situations, people anticipate the pleasure of their success by comparing their performance with that of others, not to mention their own personal expectations.

Many questions remain. Emotions are far more complex than simple unidimensional ratings of pleasure or pain. People can experience pain from sadness, anger, fear, and disappointment. No one would argue that these emotions should be treated as equivalent. Furthermore, some decision outcomes simultaneously give rise to pleasure and pain. In those cases, people feel ambivalence. Finally, what about the duration of emotional experiences? When is regret a fleeting incident, and when does it last a lifetime? Answers to these questions will deepen social scientists' understanding of emotions, and lead to better tools for guiding choice.

Recommended Reading

Gilbert, D.T., & Wilson, T.D. (2000). Miswanting: Some problems in the forecasting of future affective states. In J. Forgas (Ed.), *Thinking and feeling: The role of affect in social cognition* (pp. 178–197). Cambridge, England: Cambridge University Press.

Kahneman, D., & Varey, C. (1991). Notes on the psychology of utility. In J. Elster & J. Roemer (Eds.), *Interpersonal comparisons of well being* (pp. 127–163). New York: Cambridge University Press.

Landman, J. (1993). *Regret: The persistence of the possible*. Oxford, England: Oxford University Press.

Acknowledgments—Support was provided by the National Science Foundation (SBR-94-09819 and SBR-96-15993). We thank Philip Tetlock for comments on an earlier draft.

Notes

1. Address correspondence to Barbara A. Mellers, Department of Psychology, The Ohio State University, Columbus, OH 43210; e-mail: mellers.1@osu.edu; or send e-mail to A. Peter McGraw at mcgraw.27@osu.edu.

2. Pleasure can be derived from acts of virtue, the senses, or relief from pain. Similarly, displeasure can arise from an aggressive impulse, a sense of injustice, or frustration from falling short of a goal. Thus, choices based on pleasure need not imply hedonism.

References

Bell, D.E. (1982). Regret in decision making under uncertainty. *Operations Research, 30,* 961–981.

Bell, D.E. (1985). Disappointment in decision making under uncertainty. *Operations Research, 33,* 1–27.

Gilbert, D.T., Pinel, E.C., Wilson, T.C., Blumberg, S.J., & Wheatley, T.P. (1998). Immune neglect: A source of durability bias in affective forecasting. *Journal of Personality and Social Psychology, 75,* 617–638.

Kahneman, D. (2000). Evaluation by moments: Past and future. In D. Kahneman & A. Tversky (Eds.), *Choices, values, and frames* (pp. 693–708). New York: Cambridge University Press.

Loewenstein, G., & Schkade, D. (1999). Wouldn't be nice? Predicting future feelings. In D. Kahneman, E. Diener, & N. Schwarz (Eds.), *Well-being: The foundations of hedonic psychology* (pp. 85–108). New York: Russell Sage Foundation.

Loomes, G., & Sugden, R. (1982). Regret theory: An alternative of rational choice under uncertainty. *Economic Journal, 92,* 805–824.

Loomes, G., & Sugden, R. (1986). Disappointment and dynamic consistency in choice under uncertainty. *Review of Economic Studies, 53,* 271–282.

Mellers, B.A. (2000). Choice and the relative pleasure of consequences. *Psychological Bulletin.*

Mellers, B.A., & McGraw, A.P. (2001). *Predicting choices from anticipated emotions.* Unpublished manuscript, Ohio State University, Columbus.

Mellers, B.A., Schwartz, A., Ho, K., & Ritov, I. (1997). Decision affect theory: Emotional reactions to the outcomes of risky options. *Psychological Science, 8,* 423–429.

Mellers, B.A., Schwartz, A., & Ritov, I. (1999). Emotion-based choice. *Journal of Experimental Psychology: General, 128,* 332–345.

Savage, L.J. (1954). *The foundations of statistics.* New York: Wiley.

Schkade, D.A., & Kahneman, D. (1998). Does living in California make people happy? *Psychological Science, 9,* 340–346.

Critical Thinking Questions

1. You are at a carnival. You pay one ticket (worth $1) to be able to reach into a dark urn and pull out a chip that indicates whether you have won a prize. You win $3. How happy are you? You say it depends? What does it depend on and why? Is that rational?

   If you have read Spellman & Mandel (from Part 3): How is the answer "it depends" related to issues of counterfactual reasoning and regret?

   (Note: the carnival example reminds us of something the authors tell our students in our statistics classes. Suppose you find out that your friend got a 95 on an exam. You shouldn't offer congratulations until you find out what it was out of—200?—what the mean was—97?—and what the standard deviation was—1? Oops.)

2. Are people good at predicting how they will feel in the future when certain events occur? What kinds of errors do we tend to make? Are we more likely to err in the laboratory or in real life? If you were an insurance salesperson, how could you take advantage of the mispredictions?

3. Emotion is currently a very hot topic in the decision-making literature. One view of emotions is that they lead to mistakes in decision making; another view is that they provide information that can be useful in decision making. Do you think paying attention to emotions when making decisions is "rational"? Why or why not?

The Benefit of Additional Opinions

Ilan Yaniv

Hebrew University of Jerusalem, Jerusalem, Israel

Abstract

In daily decision making, people often solicit one another's opinions in the hope of improving their own judgment. According to both theory and empirical results, integrating even a few opinions is beneficial, with the accuracy gains diminishing as the bias of the judges or the correlation between their opinions increases. Decision makers using intuitive policies for integrating others' opinions rely on a variety of accuracy cues in weighting the opinions they receive. They tend to discount dissenters and to give greater weight to their own opinion than to other people's opinions.

Keywords

judgment and decision making; aggregating opinions; combining information

It is common practice to solicit other people's opinions prior to making a decision. An editor solicits two or three qualified reviewers for their opinions on a manuscript; a patient seeks a second opinion regarding a medical condition; a manager considers several judgmental forecasts of the market before embarking on a new venture. All these situations involve the decision maker in the task of combining other people's opinions, mostly so as to improve the final decision.

People also seek advice when they feel strongly accountable for their decisions. An accountant performing a complex audit might solicit advice to help justify his or her decisions and share the responsibility for the outcome with others. One could justifiably argue, however, that even such reasons for seeking others' opinions are rooted in the belief that this process could improve decision making.

Two main questions arise in the research on combining opinions. One involves the statistical aspects of the combination task: Under what conditions does combining opinions improve decision quality? The other concerns the psychological process of combining judgments: How do judges utilize other people's opinions? These questions, which have been investigated by students of judgment and decision making, statistics, economics, and management, are intertwined, because the quality of the product is related to the way it is produced. In this review, I discuss what researchers have learned about the process and outcomes of combining opinions.

Our focus here is on situations in which a decision maker seeks quantitative estimates, judgments, and forecasts from people possessing the relevant knowledge. The opinions are then combined by the individual decision maker, not by a group (decision making in groups deserves a separate discussion). It is

Address correspondence to Ilan Yaniv, Department of Psychology, Hebrew University, Jerusalem 91905, Israel; e-mail: ilan.yaniv@huji.ac.il.

useful to distinguish between two ways in which expert judgments can be combined: (a) intuitively (subjectively) and (b) mechanically (formally), that is, by using a consistent formula, such as simple or weighted averaging.[1]

ACCURACY GAINS FROM AGGREGATION

Research has demonstrated repeatedly that both mechanical and intuitive methods of combining opinions improve accuracy. For example, in a study of inflation forecasts, the aggregate judgment created by averaging the forecasts of expert economists was more accurate than most of these individual forecasts, though not as good as the best ones (Zarnowitz, 1984). The best forecasts, however, could not be identified before the true value became known. Hence, taking the average was superior to selecting the judgment of any of the individuals. Moreover, a small number of opinions (e.g., three to six) is typically sufficient to realize most of the accuracy gains obtainable by aggregation. These fundamental results have been demonstrated in diverse domains, ranging from perception (line lengths) and general-knowledge tasks (historical dates) to business and economics (sales or inflation forecasts), and are an important reason for the broad interest in research on combining estimates (Johnson, Budescu, & Wallsten, 2001; Sorkin, Hayes, & West, 2001; Yaniv & Kleinberger, 2000).

How Does Combining Opinions Improve Judgment?

The improvement in accuracy is grounded in statistical principles, as well as psychological facts. For quantitative estimates, a common measure of accuracy is the average distance of the prediction from the event predicted. In the special case of judgments made on an arbitrary rating scale (e.g., an interviewer's rating of a job candidate's capability on a 9-point scale), a common measure is the correlation between the judgments and some objective outcome (e.g., the candidate's actual success).

In the case of quantitative estimates, it can be outlined in simple terms why improvement is to be expected from combining estimates. A subjective estimate about an objective event can be viewed as the sum of three components: the "truth," random error (random fluctuations in a judge's performance), and constant bias (a consistent tendency to over- or underestimate the event). Statistical principles guarantee that judgments formed by averaging several sources have lower random error than the individual sources on which the averages are based. Therefore, if the bias is small or zero, the average judgment is expected to converge about the truth (Einhorn, Hogarth, & Klempner, 1977).

The case of categorical, binary judgments (e.g., a physician inspects a picture of a tumor and estimates whether it is benign or malignant) requires a special mention. Suppose a decision maker polls the judgments of N independent expert judges whose individual accuracy levels (chances of choosing the correct

---

[1]More complex methods based on Bayes's theorem are less common in psychological research on combining opinions; hence, they are not treated here.

answer) are greater than 50% and then decides according to the majority. For example, three experts might judge whether or not a witness is lying, and the final decision would be the opinion supported by two or more experts. According to a well known 18th-century theorem (known as Condorcet's jury theorem), the accuracy of the majority increases rapidly toward 100% as N increases (e.g., Sorkin et al., 2001). Thus, the majority outperforms the individual judges. For instance, the majority choice of five independent experts who are each correct 65% of the time is expected to be correct approximately 76% of the time.

Conditions Under Which Accuracy Gains Are Observed

A central condition for obtaining optimal accuracy gains through aggregation is that the experts are independent (e.g., little gain is expected if judge B is essentially a replica of judge A). But gains of appreciable size can be observed even when there are low or moderate positive correlations between the judgments of the experts (Johnson et al., 2001). The gains from aggregating quantitative judgments are also determined by the bias and the random error of the estimates (the lower the better). If judgments are made on rating scales, then the accuracy gains are related directly to the validity of each judge (i.e., how the judge's ratings correlate with the objective value of what is rated) and indirectly to the correlations between different judges' ratings (Einhorn et al., 1977; Hogarth, 1978; Johnson et al., 2001).

Number of Opinions Needed

As already noted, as few as three to six judgments might suffice to achieve most of what can be gained from averaging a larger number of opinions. This puzzling result that adding opinions does not contribute much to accuracy is related to my previous comments. Some level of dependence among experts is present in almost any realistic situation (their opinions tend to have some degree of correlation for a variety of reasons—they may rely on similar information sources or have similar backgrounds, or simply consult one another; cf. Soll, 1999). Therefore, the benefits accrued from polling more experts diminish rapidly, with each additional one amounting to "more of the same." Similarly, bias or low judge validity limits the potential accuracy gains and further diminishes the value of added opinions.

PSYCHOLOGICAL EFFECTS ON THE AGGREGATION OF OPINIONS

Consider generic scenarios involving intuitive methods of combining opinions: A moviegoer receives conflicting reviews about a movie, or an undergraduate student hears mixed evaluations from fellow students about an elective course. Although formal approaches deal with the conflict by assigning explicit weights to the various opinions, people often attempt to resolve the conflict by trying to form well-justified, coherent judgments, assessing the merit of each source and the arguments for or against each opinion and trying to explain away the differences. Specifically, several factors affect the weighting of opinions in intuitive decision making, including (a) cues for accuracy, (b) responses to dissension, and (c) self-versus-other effects.

Cues for Accuracy

A decision maker's trust in a given opinion depends on his or her assessment of the accuracy of the source. How are expectations about this accuracy formed? How does trust develop? Studies suggest that a variety of cues serve as proxy measures of the actual accuracy of sources. These cues include expertise, confidence, and past performance.

First, people are sensitive to the expertise (or credibility) ascribed to various sources and assign weights to sources as a function of such attributions (Birnbaum & Stegner, 1979). Second, a frequent and immediate cue for accuracy is the judge's stated confidence about his or her opinion (Sniezek & Van Swol, 2001). Subjective statements such as "Trust me" or "I am 60% sure" are used as factors in weighting judgments. Such a policy is beneficial to the extent that confidence and accuracy are correlated (Yaniv, 1997). Finally, an expert's past performance serves as a cue to his or her accuracy. In studies in which the same experts give multiple opinions, participants form impressions about the accuracy of each expert and adjust their weights accordingly. Trust in experts is fragile, being "hard to gain, easy to lose," because negative experiences with a source have proportionally greater influence than positive ones (Yaniv & Kleinberger, 2000).

Ignoring Dissenters' Opinions

Certain configurations of opinions present particularly sharp dilemmas as to the appropriate weighting policy. Suppose that three out of four reviewers of a research proposal agree closely (consensus), but the fourth differs widely (dissension). A decision maker attempting to aggregate these opinions might rationalize the disagreement. Indeed, the need to form and maintain consonance, or harmony, is prominent in classical theories of social psychology (e.g., those of Heider and Festinger).

One mental process used to maintain consonance amounts simply to ignoring the dissonant pieces of information. Indeed, early studies of information integration and studies of judgments formed on the basis of numerical inputs of judgment (Slovic, 1966) have shown that people discount inconsistent inputs. Similarly, when intuitively combining a sample of opinions, people discount or completely ignore dissenters and assign greater weight to consensus opinions (Yaniv, 1997). Also, a dissenter's impact on a group's final decision declines as the discrepancy from the consensus increases (Davis, 1996).

On the one hand, decision makers who disregard divergent opinions could be ignoring good data because a dissenting estimate is not necessarily wrong. In general, the tendency to resolve inconsistencies by ignoring outlying views could reduce the quality of decision making. On the other hand, a policy of discounting outlying opinions might be justified if they tend to be wrong more often than consensus opinions. Certain structural aspects of the task might indicate when an outlying opinion is likely to be wrong. For example, suppose the distribution of opinions (in the population) is bell-shaped and thick-tailed. This implies that the prevalence of outlier opinions is larger than would be expected under a standard bell-shaped (normal) distribution. In such cases (assuming the bias is zero or small), an extreme opinion in a set is particularly likely to be wrong (see, e.g., DeGroot, 1986, for a discussion of the advantage of excluding outliers in estimating the center of a thick-tailed distribution). Therefore, discounting dissenters is useful if one suspects that the distribution of opinions is thick-tailed, a situation not uncommon in behavioral studies (Yaniv, 1997).

Discounting dissenters might also be justified in scenarios in which one suspects exaggeration or manipulation. For example, in certain sports competitions, such as diving and gymnastics meets, performance is evaluated by several judges whose evaluations are then combined. Suppose a judge develops a liking for a certain performer and thus, consciously or unconsciously, produces an extreme, exaggerated evaluation that could unduly affect the aggregate opinion. A common practice in combining evaluations in such competitions involves dropping the most extreme evaluations (e.g., one on each end) and averaging the middle ones. Enacting a policy that discounts extreme judgments presumably dissuades judges from acting strategically and attenuates their influence if they do so (Yaniv, 1997).

Updating One's Own Opinion: Self Versus Other

Combining one's own opinion and an advisor's opinion is a special case that requires a separate discussion. Suppose you are responsible for hiring someone to fill a job, and you initially had a strongly favorable opinion about a candidate but are told that a colleague of yours has a lukewarm opinion of the same candidate. How might you revise your opinion in light of this conflict between your own and the other opinion? You could completely ignore the other opinion, make some adjustment of your own opinion toward the other, or completely adhere to the other opinion.

From a formal point of view, other things being equal, the two opinions (own and other) might be equally weighted. However, from your internal point of view, the two opinions are not on a par. Decision makers place more weight on beliefs for which they have more evidence. Because decision makers are privy to their own thoughts, but not to the reasons underlying an advisor's opinion, they place a higher weight on their own opinion than on an advisor's. Indeed, studies show that other things being equal, people discount others' opinions and prefer their own, with the weights split roughly 70% on self and 30% on other; this balance changes when differences in ability or knowledge between self and other are made salient (Harvey & Fischer, 1997; Yaniv, 2004; Yaniv & Kleinberger, 2000). That individuals stick closely to their initial opinions is reminiscent of findings regarding attitude change—people favor their prior opinions even in the presence of contradictory evidence. But, despite the tendency to prefer one's own opinion over another person's opinion and the difficulty of assigning optimal weights to own versus other opinions, the benefit of utilizing others' estimates is appreciable. In one study (Yaniv, 2004), respondents made initial estimates of the dates of historical events and final estimates after seeing other respondents' estimates, selected at random from a pool. Using just one other opinion reduced judgment errors by about 20%.

CONCLUDING COMMENTS

Students of reasoning, judgment, and decision making have traditionally underscored the importance of generating alternatives to one's current thoughts. Other people's opinions direct decision makers to additional alternatives or unintended consequences, as these opinions may provide a different framing of a problem, an alternative explanation, or disconfirming information. Soliciting opinions is therefore an adaptive process that helps improve decisions by compensating for a pervasive weakness of human thinking.

Two theoretical issues deserve attention. First, the view of opinions as alternatives is pertinent to opinions expressed in either numerical or verbal form. Although I have focused here on combining quantitative opinions, similar psychological processes might apply to verbal opinions (advice). Surprisingly, the use of such advice in decision making has received little attention. Future research needs to consider how qualitative advice is elicited and used best.

Second, opinions about matters of fact (estimates or forecasts) differ from opinions about matters of taste (evaluations or attitudes). Theories about the benefit accrued from combining opinions about matters of fact are well developed. In contrast, simple aggregation of tastes (e.g., opinions about resorts or about types of music) for the purpose of individual decision making raises conceptual difficulties, because people are entitled to their different tastes. Nevertheless, other people's opinions about matters of taste could be used advantageously and constructively, challenging the decision maker's established preferences and inducing him or her to consider alternatives. Conceptual and empirical work is needed to clarify these issues.

Recommended Reading

Armstrong, J.S. (2001). Combining forecasts. In J.S. Armstrong (Ed.), *Principles of forecasting: A handbook for researchers and practitioners* (pp. 417–439). Norwell, MA: Kluwer.
Clemen, R.T. (1989). Combining forecasts: A review and annotated bibliography. *International Journal of Forecasting, 5*, 559–583.
Hill, G.W. (1982). Group versus individual performance: Are N+1 heads better than one? *Psychological Bulletin, 91*, 517–539.

Acknowledgments—This research was supported by Grant No. 822 from the Israel Science Foundation.

References

Birnbaum, M.H., & Stegner, S.E. (1979). Source credibility in social judgment: Bias, expertise, and the judge's point of view. *Journal of Personality and Social Psychology, 37*, 48–74.
Davis, J.H. (1996). Group decision making and quantitative judgments: A consensus model. In E. Witte & J.H. Davis (Eds.), *Understanding group behavior: Consensual action by small groups* (pp. 35–59). Hillsdale, NJ: Erlbaum.
DeGroot, M.H. (1986). *Probability and statistics* (2nd ed.). Reading, MA: Addison-Wesley.

Einhorn, H.J., Hogarth, R.M., & Klempner, E. (1977). Quality of group judgment. *Psychological Bulletin, 84*, 158–172.

Harvey, N., & Fischer, I. (1997). Taking advice: Accepting help, improving judgment and sharing responsibility. *Organizational Behavior and Human Decision Processes, 70*, 117–133.

Hogarth, R.M. (1978). A note on aggregating opinions. *Organizational Behavior and Human Performance, 21*, 40–46.

Johnson, T.R., Budescu, D.V., & Wallsten, T.S. (2001). Averaging probability judgments: Monte Carlo analyses of asymptotic diagnostic value. *Journal of Behavioral Decision Making, 14*, 123–140.

Slovic, P. (1966). Cue-consistency and cue-utilization in judgment. *The American Journal of Psychology, 79*, 427–434.

Sniezek, J.A., & Van Swol, L.M. (2001). Trust, confidence, and expertise in a Judge-Advisor System. *Organizational Behavior and Human Decision Processes, 84*, 288–307.

Soll, J.B. (1999). Intuitive theories of information: Beliefs about the value of redundancy. *Cognitive Psychology, 38*, 317–346.

Sorkin, R.D., Hayes, C.J., & West, R. (2001). Signal detection analysis of group decision making. *Psychological Review, 108*, 183–203.

Yaniv, I. (1997). Weighting and trimming: Heuristics for aggregating judgments under uncertainty. *Organizational Behavior and Human Decision Processes, 69*, 237–249.

Yaniv, I. (2004). Receiving other people's advice: Influence and benefit. *Organizational Behavior and Human Decision Processes, 93*, 1–13.

Yaniv, I., & Kleinberger, E. (2000). Advice taking in decision making: Egocentric discounting and reputation formation. *Organizational Behavior and Human Decision Processes, 83*, 260–281.

Zarnowitz, V. (1984). The accuracy of individual and group forecasts from business and outlook surveys. *Journal of Forecasting, 3*, 11–26.

Critical Thinking Questions

1. People tend to rely on their own opinions more than the opinions of others. Why does that happen? Could it ever be rational?

2. In general it is better to rely on more than fewer opinions. How much of the benefit is due to statistical principles and how much to psychology?

3. Suppose you were in charge of evaluating an advertising campaign for a new type of product. You want to select several people from your company to figure out whether it will be successful. Who do you pick? A bunch of people who have worked together many times before and who get along well? People from different previous teams who have not worked together before and don't personally like each other? What are the cognitive and social trade-offs to each of these selections?

    (How and why would your personnel selection for this task differ from your personnel selection if you were in charge of designing an advertising campaign? See Brown in Part 2.)

Language

Language is often characterized as the pinnacle of human cognition and, indeed, understanding its cognitive basis has proven formidable. Language presents two deep problems: how it is acquired, and how it is used once it is learned. In addition to these two problems, language entails multiple levels of analysis, from individual sound units (phonemes), to words, to sentences, to stories or texts. In this section we offer a sample of each type of work.

In "Statistical Language Learning: Mechanisms and Constraints," Saffran offers a solution to a problem that, at first blush, appears virtually impossible to solve. Our perception of others' speech is that there are short pauses or breaks between the words. In fact, there are not such breaks. Speech comes to us in a relatively continuous stream, and when there are brief pauses they might well be in the middle of a word, not between words. We perceive speech as coming to us in distinct words because we are able to segment the continuous stream of sounds into discrete words. Consider, then, a baby trying to learn language. When a parent coos "Do you want the purple pacifier?" the baby hears "Doyouwantthepurplepacifier?" As far as the baby is concerned, "purple" might be a word, but so might "pur" or "plepa" or any other combination of adjacent sounds. Saffran suggests that humans are sensitive to the statistical probability of different sounds appearing next to one another, and that we use this information to segment the continuous stream of sounds into coherent words.

Once one can reliably segment words in spoken language, the challenges of language learning are not over. Estimates vary, but most college students know about sixty or seventy thousand words, most of which they encountered in print. Word learning must therefore occur at a remarkable rate, and most of it must be incidental, meaning that it's not the product of studying vocabulary in school (which researchers estimate adds only about 400 words a year). How are all of these other words learned? Landauer describes his solution in "Learning and Representing Verbal Meaning: The Latent Semantic Analysis Theory." He suggests that we must maintain a mental database that locates word meanings in a multidimensional space, similar to the way locations can be placed on a two-dimensional map. And just as familiar landmarks can help you localize an unfamiliar town without being explicitly told its location, knowledge of the surrounding words and concepts in a paragraph may help you determine the meaning of an unfamiliar word.

From language learning we turn to the use of language by experienced speakers. Obviously, knowledge of word meaning is not sufficient to ensure comprehension because word order is crucial to meaning. For example, "Sue wished she hadn't sweated" and "She wished Sue hadn't sweated" differ only in that two words have switched positions, but their meanings are altogether different. Traditional theories of sentence comprehension have assumed that the mind treats sentences in a rather algorithmic fashion, meaning that words are evaluated for the potential part of speech they might play in the sentence, and the best match is determined by a set of rules for syntactic assignment. In "Good-enough Representations in Language Comprehension," Ferreira, Bailey, and Ferraro review recent research that offers a different interpretation. Sentence syntax is one source of information to help derive meaning from a sentence, but there are others, such as context. People may only process syntax to the point that they are satisfied that they have derived a meaning that is sensible. Surprisingly, when it becomes clear that this meaning is incorrect, the old meaning seems to persist, along with the new, correct meaning.

Even if we fully understood how sentences are comprehended, we still would not have a complete account of language. Research shows that when we read a story, for example, we keep track of what is happening in the story. This representation is called a situation model. For example, if we are told that a character is in her apartment, we note that fact, and if we are later told that she goes down a stairway to the basement of the building, we update the situation model to reflect her new location. In "Situation Models: The Mental Leap into Imagined Worlds," Zwaan reviews the basic components of a story that are maintained in the situation model, and also presents evidence that the reader sees him or herself as being *in* the narrative; when our hero goes downstairs to the basement, we go with her.

The four articles in this section offer the latest perspectives on the cognitive processes that support language; however, they also offer a perspective on a large-scale issue of enduring interest: is language innate or learned? Most any researcher would argue that the answer must be "both"; we learn language, but are biologically prepared to do so, and the relative importance of learning and innateness varies with different aspects of language. Nevertheless, it's worth noting that two of the articles in this section—Saffran and Landauer—show the power of relatively simple learning processes that focus on key information in the environment, and accrue learning over the long term.

Statistical Language Learning: Mechanisms and Constraints

Jenny R. Saffran[1]

Department of Psychology and Waisman Center, University of Wisconsin-Madison, Madison, Wisconsin

Abstract

What types of mechanisms underlie the acquisition of human language? Recent evidence suggests that learners, including infants, can use statistical properties of linguistic input to discover structure, including sound patterns, words, and the beginnings of grammar. These abilities appear to be both powerful and constrained, such that some statistical patterns are more readily detected and used than others. Implications for the structure of human languages are discussed.

Keywords

language acquisition; statistical learning; infants

Imagine that you are faced with the following challenge: You must discover the underlying structure of an immense system that contains tens of thousands of pieces, all generated by combining a small set of elements in various ways. These pieces, in turn, can be combined in an infinite number of ways, although only a subset of those combinations is actually correct. However, the subset that is correct is itself infinite. Somehow you must rapidly figure out the structure of this system so that you can use it appropriately early in your childhood.

This system, of course, is human language. The elements are the sounds of language, and the larger pieces are the words, which in turn combine to form sentences. Given the richness and complexity of language, it seems improbable that children could ever discern its structure. The process of acquiring such a system is likely to be nearly as complex as the system itself, so it is not surprising that the mechanisms underlying language acquisition are a matter of long-standing debate. One of the central focuses of this debate concerns the innate and environmental contributions to the language-acquisition process, and the degree to which these components draw on information and abilities that are also relevant to other domains of learning.

In particular, there is a fundamental tension between theories of language acquisition in which learning plays a central role and theories in which learning is relegated to the sidelines. A strength of learning-oriented theories is that they exploit the growing wealth of evidence suggesting that young humans possess powerful learning mechanisms. For example, infants can rapidly capitalize on the statistical properties of their language environments, including the distributions of sounds in words and the orders of word types in sentences, to discover important components of language structure. Infants can track such statistics, for example, to discover speech categories (e.g., native-language consonants; see, e.g., Maye, Werker, & Gerken, 2002), word boundaries (e.g., Saffran, Aslin, & Newport, 1996), and rudimentary syntax (e.g., Gomez & Gerken, 1999; Saffran & Wilson, 2003).

However, theories of language acquisition in which learning plays a central role are vulnerable to a number of criticisms. One of the most important arguments against learning-oriented theories is that such accounts seem at odds with one of the central observations about human languages. The linguistic systems of the world, despite surface differences, share deep similarities, and vary in nonarbitrary ways. Theories of language acquisition that focus primarily on preexisting knowledge of language do provide an elegant explanation for cross-linguistic similarities. Such theories, which are exemplified by the seminal work of Noam Chomsky, suggest that linguistic universals are prespecified in the child's linguistic endowment, and do not require learning. Such accounts generate predictions about the types of patterns that should be observed cross-linguistically, and lead to important claims regarding the evolution of a language capacity that includes innate knowledge of this kind (e.g., Pinker & Bloom, 1990).

Can learning-oriented theories also account for the existence of language universals? The answer to this question is the object of current research. The *constrained statistical learning framework* suggests that learning is central to language acquisition, and that the specific nature of language learning explains similarities across languages. The crucial point is that learning is constrained; learners are not open-minded, and calculate some statistics more readily than others. Of particular interest are those constraints on learning that correspond to cross-linguistic similarities (e.g., Newport & Aslin, 2000). According to this framework, the similarities across languages are indeed nonaccidental, as suggested by the Chomskian framework—but they are not the result of innate linguistic knowledge. Instead, human languages have been shaped by human learning mechanisms (along with constraints on human perception, processing, and speech production), and aspects of language that enhance learnability are more likely to persist in linguistic structure than those that do not. Thus, according to this view, the similarities across languages are not due to innate knowledge, as is traditionally claimed, but rather are the result of constraints on learning. Further, if human languages were (and continue to be) shaped by constraints on human learning mechanisms, it seems likely that these mechanisms and their constraints were not tailored solely for language acquisition. Instead, learning in nonlinguistic domains should be similarly constrained, as seems to be the case.

A better understanding of these constraints may lead to new connections between theories focused on nature and theories focused on nurture. Constrained learning mechanisms require both particular experiences to drive learning and preexisting structures to capture and manipulate those experiences.

LEARNING THE SOUNDS OF WORDS

In order to investigate the nature of infants' learning mechanisms, my colleagues and I began by studying an aspect of language that we knew must certainly be learned: word segmentation, or the boundaries between words in fluent speech. This is a challenging problem for infants acquiring their first language, for speakers do not mark word boundaries with pauses, as shown in Figure 1. Instead, infants must determine where one word ends and the next begins without access

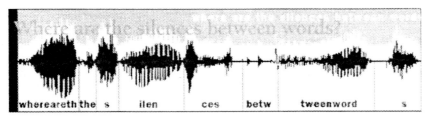

Fig. 1. A speech waveform of the sentence "Where are the silences between words?" The height of the bars indicates loudness, and the *x*-axis is time. This example illustrates the lack of consistent silences between word boundaries in fluent speech. The vertical gray lines represent quiet points in the speech stream, some of which do not correspond to word boundaries. Some sounds are represented twice in the transcription below the waveform because of their continued persistence over time.

to obvious acoustic cues. This process requires learning because children cannot innately know that, for example, *pretty* and *baby* are words, whereas *tyba* (spanning the boundary between *pretty* and *baby*) is not.

One source of information that may contribute to the discovery of word boundaries is the statistical structure of the language in the infant's environment. In English, the syllable *pre* precedes a small set of syllables, including *ty*, *tend*, and *cedes*; in the stream of speech, the probability that *pre* is followed by *ty* is thus quite high (roughly 80% in speech to young infants). However, because the syllable *ty* occurs word finally, it can be followed by any syllable that can begin an English word. Thus, the probability that *ty* is followed by *ba*, as in *pretty baby*, is extremely low (roughly 0.03% in speech to young infants). This difference in sequential probabilities is a clue that *pretty* is a word, and *tyba* is not. More generally, given the statistical properties of the input language, the ability to track sequential probabilities would be an extremely useful tool for infant learners.

To explore whether humans can use statistical learning to segment words, we exposed adults, first graders, and 8-month-olds to spoken nonsense languages in which the only cues to word boundaries were the statistical properties of the syllable sequences (e.g., Saffran et al., 1996). Listeners briefly heard a continuous sequence of syllables containing multisyllabic words from one of the languages (e.g., *golabupabikututibubabupugolabubabupu* . . .). We then tested our participants to determine whether they could discriminate the words from the language from sequences spanning word boundaries. For example, we compared performance on words like *golabu* and *pabiku* with performance on sequences like *bupabi*, which spanned the boundary between words. To succeed at this task, listeners would have had to track the statistical properties of the input. Our results confirmed that human learners, including infants, can indeed use statistics to find word boundaries. Moreover, this ability is not confined to humans: Cotton-top tamarins, a New World monkey species, can also track statistics to discover word boundaries (Hauser, Newport, & Aslin, 2001).

One question immediately raised by these results is the degree to which statistical learning is limited to language-like stimuli. A growing body of results suggests that sequential statistical learning is quite general. For example, infants can track sequences of tones, discovering "tone-word boundaries" via statistical cues (e.g., Saffran, Johnson, Aslin, & Newport, 1999), and can learn statistically defined visual patterns (e.g., Fiser & Aslin, 2002; Kirkham, Slemmer, & Johnson, 2002); work in progress is extending these results to the domain of events in human action sequences.

Given that the ability to discover units via their statistical coherence is not confined to language (or to humans), one might wonder whether the statistical learning results actually pertain to language at all. That is, do infants actually use statistical learning mechanisms in real-world language acquisition? One way to address this question is to ask what infants are actually learning in our segmentation task. Are they learning statistics? Or are they using statistics to learn language? Our results suggest that when infants being raised in English-speaking environments have segmented the sound strings, they treat these nonsensical patterns as English words (Saffran, 2001b). Statistical language learning in the laboratory thus appears to be integrated with other aspects of language acquisition. Related results suggest that 12-month-olds can first segment novel words and then discover syntactic regularities relating the new words—all within the same set of input. This would not be possible if the infants formed mental representations only of the sequential probabilities relating individual syllables, and no word-level representations (Saffran & Wilson, 2003). These findings point to a constraint on statistical language learning: The mental representations produced by this process are not just sets of syllables linked by statistics, but new units that are available to serve as the input to subsequent learning processes.

Similarly, it is possible to examine constraints on learning that might affect the acquisition of the sound structure of human languages. The types of sound patterns that infants learn most readily may be more prevalent in languages than are sound patterns that are not learnable by infants. We tested this hypothesis by asking whether infants find some phonotactic regularities (restrictions on where particular sounds can occur; e.g., /fs/ can occur at the end, but not the beginning, of syllables in English) easier to acquire than others (Saffran & Thiessen, 2003). The results suggest that infants readily acquire novel regularities that are consistent with the types of patterns found in the world's languages, but fail to learn regularities that are inconsistent with natural language structure. For example, infants rapidly learn new phonotactic regularities involving generalizations across sounds that share a phonetic feature, while failing to learn regularities that disregard such features. Thus, it is easier for infants to learn a set of patterns that group together /p/, /t/, and /k/, which are all voiceless, and that group together /b/, /d/, and /g/, which are all voiced, than to learn a pattern that groups together /d/, /p/, and /k/, but does not apply to /t/.[2] Studies of this sort may provide explanations for why languages show the types of sound patterning that they do; sound structures that are hard for infants to learn may be unlikely to recur across the languages of the world.

STATISTICAL LEARNING AND SYNTAX

Issues about learning versus innate knowledge are most prominent in the area of syntax. How could learning-oriented theories account for the acquisition of abstract structure (e.g., phrase boundaries) not obviously mirrored in the surface statistics of the input? Unlike accounts centered on innate linguistic knowledge, most learning-oriented theories do not provide a transparent explanation for the ubiquity of particular structures cross-linguistically. One approach to these issues is to ask whether some nearly universal structural aspects of human languages may result from constraints on human learning (e.g., Morgan, Meier, & Newport, 1987). To test this hypothesis, we asked whether one such aspect of syntax, phrase structure (groupings of types of words together into subunits, such as noun phrases and verb phrases), results from a constraint on learning: Do humans learn sequential structures better when they are organized into subunits such as phrases than when they are not? We identified a statistical cue to phrasal units, predictive dependencies (e.g., the presence of a word like *the* or *a* predicts a noun somewhere downstream; the presence of a preposition predicts a noun phrase somewhere downstream), and determined that learners can use this kind of cue to locate phrase boundaries (Saffran, 2001a).

In a direct test of the theory that predictive dependencies enhance learnability, we compared the acquisition of two nonsense languages, one with predictive dependencies as a cue to phrase structure, and one lacking predictive dependencies (e.g., words like *the* could occur either with or without a noun, and a noun could occur either with or without words like *the*; neither type of word predicted the presence of the other). We found better language learning in listeners exposed to languages containing predictive dependencies than in listeners exposed to languages lacking predictive dependencies (Saffran, 2002). Interestingly, the same constraint on learning emerged in tasks using nonlinguistic materials (e.g., computer alert sounds and simultaneously presented shape arrays). These results support the claim that learning mechanisms not specifically designed for language learning may have shaped the structure of human languages.

DIRECTIONS FOR FUTURE RESEARCH

Results to date demonstrate that human language learners possess powerful statistical learning capacities. These mechanisms are constrained at multiple levels; there are limits on what information serves as input, which computations are performed over that input, and the structure of the representations that emerge as output. To more fully understand the contribution of statistical learning to language acquisition, it is necessary to assess the degree to which statistical learning provides explanatory power given the complexities of the acquisition process.

For example, how does statistical learning interact with other aspects of language acquisition? One way we are addressing this question is by investigating how infants weight statistical cues relative to other cues to word segmentation early in life. The results of such studies provide an important window into the ways in which statistical learning may help infant learners to determine the relevance of the many cues inherent in language input. Similarly, we are studying how statistics meet up with meaning in the world (e.g., are statistically defined "words" easier to learn as labels for novel objects than sound sequences spanning word boundaries?), and how infants in bilingual environments cope with multiple sets of statistics. Studying the intersection between statistical learning and the rest of language learning may provide new insights into how various nonstatistical aspects of language are acquired. Moreover, a clearer picture of the learning mechanisms used successfully by typical language learners may increase researchers' understanding of the types of processes that go awry when children do not acquire language as readily as their peers.

It is also critical to determine which statistics are available to young learners and whether those statistics are actually relevant to natural language structure. Researchers do not agree on the role that statistical learning should play in acquisition theories. For example, they disagree about when learning is best described as statistically based as opposed to rule based (i.e., utilizing mechanisms that operate over algebraic variables to discover abstract knowledge), and about whether learning can still be considered statistical when the input to learning is abstract. Debates regarding the proper place for statistical learning in theories of language acquisition cannot be resolved in advance of the data. For example, although one can distinguish between statistical versus rule-based learning mechanisms, and statistical versus rule-based knowledge, the data are not yet available to determine whether statistical learning itself renders rule-based knowledge structures, and whether abstract knowledge can be probabilistic. Significant empirical advances will be required to disentangle these and other competing theoretical distinctions.

Finally, cross-species investigations may be particularly informative with respect to the relationship between statistical learning and human language. Current research is identifying species differences in the deployment of statistical learning mechanisms (e.g., Newport & Aslin, 2000). To the extent that nonhumans and humans track different statistics, or track statistics over different perceptual units, learning mechanisms that do not initially appear to be human-specific may actually render human-specific outcomes. Alternatively, the overlap between the learning mechanisms available across species may suggest that differences in statistical learning cannot account for cross-species differences in language-learning capacities.

CONCLUSION

It is clear that human language is a system of mind-boggling complexity. At the same time, the use of statistical cues may help learners to discover some of the patterns lurking in language input. To what extent might the kinds of statistical patterns accessible to human learners help in disentangling the complexities of this system? Although the answer to this question remains unknown, it is possible that a combination of inherent constraints on the types of patterns acquired by learners, and the use of output from one level of learning as input to the next, may help to explain why something so complex is mastered readily by the human mind. Human learning mechanisms may themselves have played a prominent role in shaping the structure of human languages.

Recommended Reading

Gómez, R.L., & Gerken, L.A. (2000). Infant artificial language learning and language acquisition. *Trends in Cognitive Sciences, 4*, 178–186.

Hauser, M.D., Chomsky, N., & Fitch, W.T. (2002). The faculty of language: What is it, who has it, and how did it evolve? *Science*, 298, 1569–1579.

Pena, M., Bonatti, L.L., Nespor, M., & Mehler, J. (2002). Signal-driven computations in speech processing. *Science*, 298, 604–607.

Seidenberg, M.S., MacDonald, M.C., & Saffran, J.R. (2002). Does grammar start where statistics stop? *Science*, 298, 553–554.

Acknowledgments—The preparation of this manuscript was supported by grants from the National Institutes of Health (HD37466) and National Science Foundation (BCS-9983630). I thank Martha Alibali, Erin McMullen, Seth Pollak, Erik Thiessen, and Kim Zinski for comments on a previous version of this manuscript.

Notes

1. Address correspondence to Jenny R. Saffran, Department of Psychology, University of Wisconsin-Madison, Madison, WI 53706; e-mail: jsaffran@wisc.edu.

2. Voicing refers to the timing of vibration of the vocal cords. Compared with voiceless consonants, voiced consonants have a shorter lag time between the initial noise burst of the consonant and the subsequent vocal cord vibrations.

References

Fiser, J., & Aslin, R.N. (2002). Statistical learning of new visual feature combinations by infants. *Proceedings of the National Academy of Sciences, USA*, 99, 15822–15826.

Gomez, R.L., & Gerken, L. (1999). Artificial grammar learning by 1-year-olds leads to specific and abstract knowledge. *Cognition*, 70, 109–135.

Hauser, M., Newport, E.L., & Aslin, R.N. (2001). Segmentation of the speech stream in a nonhuman primate: Statistical learning in cotton-top tamarins. *Cognition*, 78, B41–B52.

Kirkham, N.Z., Slemmer, J.A., & Johnson, S.P. (2002). Visual statistical learning in infancy: Evidence of a domain general learning mechanism. *Cognition*, 83, B35–B42.

Maye, J., Werker, J.F., & Gerken, L. (2002). Infant sensitivity to distributional information can affect phonetic discrimination. *Cognition*, 82, B101–B111.

Morgan, J.L., Meier, R.P., & Newport, E.L. (1987). Structural packaging in the input to language learning: Contributions of intonational and morphological marking of phrases to the acquisition of language. *Cognitive Psychology*, 19, 498–550.

Newport, E.L., & Aslin, R.N. (2000). Innately constrained learning: Blending old and new approaches to language acquisition. In S.C. Howell, S.A. Fish, & T. Keith-Lucas (Eds.), *Proceedings of the 24th Boston University Conference on Language Development* (pp. 1–21). Somerville, MA: Cascadilla Press.

Pinker, S., & Bloom, P. (1990). Natural language and natural selection. *Behavioral and Brain Sciences*, 13, 707–784.

Saffran, J.R. (2001a). The use of predictive dependencies in language learning. *Journal of Memory and Language*, 44, 493–515.

Saffran, J.R. (2001b). Words in a sea of sounds: The output of statistical learning. *Cognition*, 81, 149–169.

Saffran, J.R. (2002). Constraints on statistical language learning. *Journal of Memory and Language*, 47, 172–196.

Saffran, J.R., Aslin, R.N., & Newport, E.L. (1996). Statistical learning by 8-month-old infants. *Science*, 274, 1926–1928.

Saffran, J.R., Johnson, E.K., Aslin, R.N., & Newport, E.L. (1999). Statistical learning of tone sequences by human infants and adults. *Cognition*, 70, 27–52.

Saffran, J.R., & Thiessen, E.D. (2003). Pattern induction by infant language learners. *Developmental Psychology*, 39, 484–494.

Saffran, J.R., & Wilson, D.P. (2003). From syllables to syntax: Multi-level statistical learning by 12month-old infants. *Infancy*, 4, 273–284.

Critical Thinking Questions

1. Foreign languages often sound as if they are spoken rapidly, and without pauses between words. Explain why that is so, given Saffran's explanation of statistical language learning.

2. Saffran suggests that the statistical properties of language help us to learn the boundaries of words, and also help us to learn syntax. But we also have knowledge of higher-level structures in language. For example, we know the basic structure that a story follows, commonly called a story grammar: a story has a main character with a goal, but there are obstacles to achieving the goal, and so forth. Could the statistical learning framework be extended to account for high-level language knowledge like that represented in story grammars?

3. The fact that tamarins are sensitive to the statistical properties of linguistic input might make us think that this sensitivity is a general learning ability of primates. In turn, that might make us think that this learning ability is not special to language, but rather, we are, in general, good at learning the statistical properties of sequences that we experience. From your personal experience, do you think that's true?

Learning and Representing Verbal Meaning: The Latent Semantic Analysis Theory

Thomas K. Landauer[1]

Department of Psychology, University of Colorado, Boulder, Colorado

Abstract

Latent semantic analysis (LSA) is a theory of how word meaning—and possibly other knowledge—is derived from statistics of experience, and of how passage meaning is represented by combinations of words. Given a large and representative sample of text, LSA combines the way thousands of words are used in thousands of contexts to map a point for each into a common semantic space. LSA goes beyond pairwise co-occurrence or correlation to find latent dimensions of meaning that best relate every word and passage to every other. After learning from comparable bodies of text, LSA has scored almost as well as humans on vocabulary and subject-matter tests, accurately simulated many aspects of human judgment and behavior based on verbal meaning, and been successfully applied to measure the coherence and conceptual content of text. The surprising success of LSA has implications for the nature of generalization and language.

Keywords

latent semantic analysis; latent semantic indexing; LSA; learning; meaning; lexicon; knowledge; machine learning; simulation

By age 18, you knew the meaning of more than 50,000 words that you had met only in print. How did you do that? My colleagues and I think that we may have cracked this and some other persistent mysteries of verbal meaning. We have been exploring a mathematical computer model and corresponding psychological learning theory called *Latent Semantic Analysis* (LSA). Although far from perfect or complete as a theory of meaning and language, LSA accurately simulates many aspects of human understanding of word and passage meaning and can effectively replace human text comprehension in several educational applications. Among other things, it mimics the rate at which schoolchildren learn recognition vocabulary from text, makes humanlike assessments of semantic relationships between words, passes college multiple-choice exams after "reading" a textbook, and makes it possible to automatically assess the content of factual essays as reliably as expert humans.

THE LATENT SEMANTIC ANALYSIS THEORY

The formal LSA model relies on sophisticated mathematical and computer methods—ones we think may mirror what the brain does (Landauer & Dumais, 1997; Landauer, Foltz, & Laham, 1998). Although a technical discussion of these methods is beyond the scope of this review, a nontechnical description can show what LSA assumes and how it works.

LSA expresses a venerable idea about word meaning, that words occupy positions in a *semantic space* and their meaning is the relation of each to all. Since the arrival of computers, the idea has often taken the form of programs in which words are linked by labels such as *part of*, or through common features like *is living*. Unfortunately, because knowledgeable humans supply the links and labels, these programs beg the question in which we are most interested: how people acquire meaning from experience in the first place.

The psychological theory underlying LSA assumes that people start by associating perceptual objects and experiences, including words, that are met near each other in time. Doing this helps people predict the world in advance and deal with it better as a result. But human cognition (at least) goes far beyond piecewise association. It somehow takes all the billions of local contiguity relations and fits them together into an overall map, a semantic space that represents how each object, event, or word is related to each other.

LSA as mathematical computer model also constructs a semantic space. It gets its experience by being fed a large body of electronic text. It starts by using a computer version of associative learning to establish a link between each unique word type (e.g., the word *psychology* wherever it appears) and every paragraph in which it appears (say, this one). This step is essential, and the particular way it is done matters. However, the real power of LSA comes from a succeeding process, one that combines these billions of little links to form a common semantic space in which each word and any passage in the language has its own place.

LSA does this in a way that is analogous to how cartographers once mapped the Earth. They started with rough estimates of point-to-point distances based on sightings from hilltops, camel-travel days, sailing times, or the likelihood of anyone getting from there to here. They put all of this piecewise data together by placing the points—towns, river junctions, islands, headlands, mythic places—onto a single picture in a way that preserved all the measurements as well as possible. If successful., they not only improved the distance estimates and produced a pleasing graphic, but got an immense added benefit in the ability to read off the infinite number of point-to-point distances that had never been measured. Such maps let people understand their world in an entirely new and better way.

This kind of mapping works because of a simple fact of geometry. A structure of points in which each is connected to many others in properly interlocking triangles is rigid, so any missing paths are strictly defined. However, this works well only if you have assumed the correct shape—the right dimensionality—of the surface that the points are mapped onto. For example, mapping the spherical surface of the Earth onto a flat piece of paper creates major distortions; a globe does much better.

LSA assumes that the mind or brain applies essentially the same method to mapping the meaning of words (and other experiences) into semantic space. It starts with rough estimates of closeness in the form of local temporal associations, then finds a way to fit them together. For LSA's initial estimate of the closeness of two words, it observes how often they occur in the same meaningful contexts (such as sentences or paragraphs) relative to how often they occur in different contexts, and computes a correlation index that approximates Pavlovian conditioning. However, this is just the first step; the correlations themselves are not the answer. Instead, LSA fits all the separate relations into a common space in a manner—and dimensionality—that distorts them as little as possible. It is this second step that constitutes its understanding of meaning.

In the case of physical mapping, two towns may be quite close, but because they are on different sides of a mountain, the distance between them may never have been measured. Nevertheless, it is easily determined from a map. Similarly, two words may be quite alike in meaning but rarely used in the same context, for example, because they are synonyms. However, when LSA places them in semantic space, it can bring them close together (or, if the data so imply, spread them out).

In applying LSA as a simulation model, using the right number of dimensions can make a big difference, and the best number is usually between 100 and 1,000, not the two or three of physical maps. My colleagues and I believe that the high dimensionality needed by LSA is a product of the way the brain is structured combined with the statistical structure of experience.

SIMULATIONS

An invaluable feature of LSA—a result of advances in computer power and the availability of large amounts of electronic text—is that we can often have the model learn from almost the same sources, both in size and in content, as a human does. We can then test LSA in some of the same ways that humans are tested and have it perform some of the same meaning-based tasks that humans perform. We believe that central human cognitive abilities often depend on immense amounts of experience, and that theories that cannot be applied to comparable data may be fundamentally wrong, or at least unprovable. One reason that we have applied LSA to practical problems is to assess how well it captures everything that is important.

The best example of this strategy comes from simulating schoolchildren's learning of vocabulary from reading. Of the roughly 400,000 words that a typical seventh grader could have encountered in print but nowhere else, she knows the meaning of 10 today that she did not know yesterday. Amazingly, she saw only 2 or 3 of these in the interim. Moreover, when psychologists have tried to teach word meanings explicitly, they have not come close to normal rates. Traditional theories of word learning, which are based on specific concrete experiences with individual words, cannot account for these facts. However, when a very large number of other words is known, the indirect effects assumed in LSA offer an explanation: Correctly mapping a few new relations can help to define many more.

Dumais and I had LSA learn from samples of 5 million words of natural text, comparable to the total lifetime reading of a seventh grader, then take standard multiple-choice vocabulary tests (Landauer & Dumais, 1997). From the results, we calculated that if LSA were exposed to the same reading materials as the student, it too would know about 10 new words each day. Moreover, by varying the content of the training text, we deduced that about three fourths of LSA's gain per paragraph was indirect, knowing more about words the model did not encounter because of how the ones it did encounter improved the overall semantic space.

We have also looked in more detail at LSA's knowledge of word meanings, showing, for example, that it usually represents synonym and antonym pairs, singulars and plurals, and members of the same conceptual categories as related to each other in the way people think they are. It also mimics the way that words like ball have more than one meaning, and the way that such words are disambiguated by context, simulating closely the results of classic psycholinguistic laboratory experiments.

We have also trained LSA on more focused bodies of knowledge. Figure 1 shows the results when LSA learned from an electronic version of an introductory psychology textbook, then was tested on the same multiple-choice exams used in large classes at two universities. For each test item, LSA separately represented the question and each alternative as points in its semantic space by averaging points for their contained words. It then chose the alternative most similar to the question. LSA did not do quite as well as the average student (who, unlike LSA, had also attended lectures), but did well enough to pass and showed the same pattern of better performance on easy (for students) factual than difficult conceptual questions.

APPLICATIONS

A set of educational applications of LSA provides additional, and we think important, confirmation of its ability to capture meaning. LSA represents the meaning of a passage just as it does the meaning of a word, as a single point

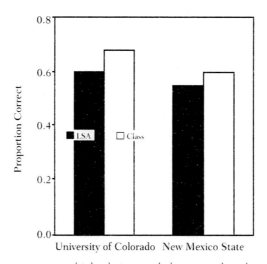

Fig. 1. Performance on a multiple-choice psychology exam by a latent semantic analysis (LSA) simulation and students at two universities. For the classes, mean scores are shown.

in its semantic space. The point for a passage is just the center, or average, of all the words it contains, independent of their order. This means that LSA ignores any meaning that could not be recovered after a passage is scrambled, a possibly serious limitation. However, the single-point representation also implies that the same meaning, as represented in LSA, can be verbally expressed by many different wordings, an important ability if it is to mimic human understanding.

In one application, we accurately predicted how well students would comprehend an expository text about the heart by having LSA compute how similar the meaning of each paragraph was to the one before it. In another, we had students first write short essays on the heart, then correctly predicted that they would learn most from a text that was a little, but not too much, more sophisticated than their essays, measuring the content of the essays, the content of the texts, and their relations with LSA alone.

In the most recent application, we have used LSA to measure the quality and quantity of conceptual content in essay answers to questions about factual subjects. By comparing the LSA representation of a student's essay to a large number of essays on the same topic that have been scored by an expert human, we can accurately estimate the score that the same expert would give a new essay. As shown in Figure 2, over some 1,200 individual essays on seven diverse topics, the correlation between LSA and a human expert was virtually identical to the correlation between one human expert and another.

We have also successfully field-tested a system that lets sixth graders write summaries of chapters, get quality scores and automatically generated commentaries, and then try again until they are ready to turn their summaries in to the teacher. We think such systems can greatly increase the number of opportunities for valuable practice in understanding and composing expository text.

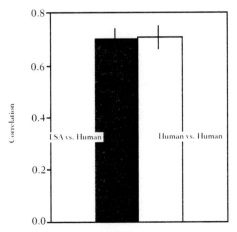

Fig. 2. Average agreement (correlation) between latent semantic analysis (LSA) and human expert scoring of 1,203 essay-exam answers on topics in biology, psychology, history, sociology, and business.

IMPLICATIONS

The rather remarkable successes of the LSA model raise a variety of interesting theoretical questions and suggest a number of potentially practical uses.

- *Generalization.* LSA induces similarity relations among words from contextual experience alone rather than from the sharing of primitive features; LSA knows nothing about what two words have in common except how they occur in the company of other words. Similar mechanisms may help to explain other similarity-based phenomena, such as object recognition.
- *Innateness and experience.* LSA's simulation successes prove that it is possible to induce much more humanlike knowledge from experience than is apparent in surface correlations. This calls into question intuitions about "the poverty of the stimulus," the notion that what people know far exceeds what they could have learned, an apparent paradox that has plagued theorists of meaning from Plato through Wittgenstein and Quine to Chomsky and Pinker.
- *Relative importance of words and syntax.* LSA captures a surprising amount of word and passage meaning without paying attention to word order within sentences and paragraphs. It may be that people, too, represent verbal meaning primarily as word combinations. Perhaps syntax largely serves other functions—such as easing encoding or decoding—and is relatively less important for storing meaning than has usually been assumed in linguistics and psychology.
- *Utility.* Because LSA can mirror human comprehension of language, there are many human cognitive tasks for which it may offer significant help. Comparisons of meaning that are too cumbersome for human processing may become routine. For example, in its oldest application, LSA improved automatic information retrieval by matching queries and documents on meaning rather than words, and in recent pilot experiments, LSA measured the conceptual overlap between 4,000 pairs of military training courses in a few minutes.

In sum, we think LSA offers exciting possibilities on many fronts. There is, of course, much work left to do. We would like to find ways to improve LSA's learning databases, to include syntax-based meaning, to make LSA generate as well as understand discourse, and to test more directly its psychological and neural reality. Meanwhile, to see what it does now, and to use its products for new research and applications, you can visit the World Wide Web: http://LSA.colorado.edu/.

Recommended Reading

Kintsch, W. (1998). *Comprehension: A paradigm for cognition.* Cambridge, England: Cambridge University Press.
Landauer, T.K., & Dumais, S.T. (1997). (See References)
Landauer, T.K., Foltz, P., & Laham, R.D. (1998). (See References)
Miller, G.A. (1996). *The science of words.* New York: W.H. Freeman.

Acknowledgments—I thank many colleagues and collaborators, especially Curt Burgess, Susan Dumais, Peter Foltz, Walter Kintsch, Darrell Laham, Bob Rehder Missy Schreiner, and Michael Wolfe. Financial support was provided by Bellcore, the McDonnell Foundation, and the Defense Advanced Research Project Administration. Stephen Ivens of Touchstone Applied Science Associates and Patrick Kylonnen of the U.S. Air Force Research Laboratory generously shared data.

Note

1. Address correspondence to Thomas K. Landauer, Department of Psychology, Campus Box 344, University of Colorado, Boulder, CO 80309-0344.

References

Landauer, T.K., & Dumais, S.T. (1997). A solution to Plato's Problem: The Latent Semantic Analysis theory of acquisition, induction and representation of knowledge. *Psychological Review*, 104, 211–240.

Landauer, T.K., Foltz, P., & Laham, R.D. (1998). An introduction to latent semantic analysis. *Discourse Processes*, 25, 259–284.

Critical Thinking Questions

1. This article presents a model that attempts to solve two of the most challenging problems in language: first, how meaning is acquired from experience, and second, how we are able to learn word meanings at such a rapid rate through childhood. How does the model solve each of these problems?

2. The model assumes that word meaning is represented in a multidimensional space. To appreciate what is meant by "space" the author draws an analogy to a map. You can locate a town by knowing its location on two dimensions: a north-south dimension and an east-west dimension. Thus a map is a two-dimensional space. Semantic space will have not two dimensions, but 100 or 1000, and they will not all be spatial, but ones that create meaning. But where do these dimensions come from? Words have meaning because they are characterized on these 100 or 1000 dimensions—but how do the dimensions have meaning?

3. The model accounts for how new words and concepts are acquired. If the model is correct, what sort of reading should you do to maximize how much you learn from what you read?

Good-Enough Representations in Language Comprehension

Fernanda Ferreira,[1] Karl G.D. Bailey, and Vittoria Ferraro

Department of Psychology and Cognitive Science Program, Michigan State University, East Lansing, Michigan

Abstract

People comprehend utterances rapidly and without conscious effort. Traditional theories assume that sentence processing is algorithmic and that meaning is derived compositionally. The language processor is believed to generate representations of the linguistic input that are complete, detailed, and accurate. However, recent findings challenge these assumptions. Investigations of the misinterpretation of both garden-path and passive sentences have yielded support for the idea that the meaning people obtain for a sentence is often not a reflection of its true content. Moreover, incorrect interpretations may persist even after syntactic reanalysis has taken place. Our good-enough approach to language comprehension holds that language processing is sometimes only partial and that semantic representations are often incomplete. Future work will elucidate the conditions under which sentence processing is simply good enough.

Keywords

language comprehension; satisficing; syntax; linguistic ambiguity

Over the past three decades, various theories of language comprehension have been developed to explain how people compose the meanings of sentences from individual words. All theories advanced to date assume that the language-processing mechanism applies a set of algorithms to access words from the lexicon, organize them into a syntactic structure through rules of grammar, and derive the meaning of the whole structure based on the meaning of its parts. Furthermore, all theories assume that this process generates complete, detailed, and accurate representations of the linguistic input.

MODELS OF SENTENCE PROCESSING

Two approaches to sentence processing that have been widely contrasted are the *garden-path model* (Ferreira & Clifton, 1986; Frazier, 1978) and the *constraint-satisfaction model* (MacDonald, Pearlmutter, & Seidenberg, 1994; Trueswell, Tanenhaus, & Garnsey, 1994). According to the garden-path account, the language processor initially computes a single syntactic analysis without consideration of context or plausibility. Once an interpretation has been chosen, other information is used to evaluate its appropriateness. For example, a person who heard, "Mary saw the man with the binoculars," would tend to understand the sentence to mean that Mary used the binoculars as an instrument. If it turned out that the man had the binoculars, the initial interpretation would be revised to be compatible with that contextual knowledge.

138

Constraint-satisfaction theorists, in contrast, assume that all possible syntactic analyses are computed at once on the basis of all relevant sources of information. The analysis with the greatest support is chosen over its competitors. The constraint-based approach predicts that people who hear the sentence about Mary, the man, and the binoculars will activate both interpretations and then select the one that is more appropriate in the context. Thus, the two classes of models assume radically different approaches to sentence processing: According to the garden-path model, analyses are proposed serially, and syntactic information is processed entirely separately from real-world knowledge and meaning. According to constraint-based models, analyses are proposed in parallel, and the syntactic processor communicates with any relevant information source. Nevertheless, both models incorporate the assumption that interpretations of utterances are compositionally built up from words clustered into hierarchically organized constituents.

IS THE MEANING OF A SENTENCE ALWAYS THE SUM OF ITS PARTS?

This assumption of compositionality seems eminently plausible, but results in the literature on the psychology of language call it into question. For example, people have been observed to unconsciously normalize strange sentences to make them sensible (Fillenbaum, 1974). The *Moses illusion* (Erickson & Mattson, 1981) is typically viewed as demonstrating the fallibility of memory processes, but it is also relevant to issues of language interpretation and compositionality. When asked, "How many animals of each sort did Moses put on the ark?" people tend to respond "two," instead of objecting to the presupposition behind the question. Similarly, participants often overlook the anomaly in a sentence such as "The authorities had to decide where to bury the survivors" (Barton & Sanford, 1993).

A study conducted to examine whether sentence meaning can prime individual words (i.e., activate them so that they are more accessible to the comprehension system) also demonstrates that language processing is not always compositional, and that the semantic representations that get computed are shallow and incomplete (rather than computing the structure to the fullest degree possible, the comprehension system just does enough to contend with the overall task at hand; Duffy, Henderson, & Morris, 1989). Participants were asked to speak aloud the final word in various sentences after reading the sentences. On average, they took less time to say the word in biased sentences like (1) than in sentences such as (2), indicating that "cocktails" had been activated, or primed, earlier in the sentence. But, unexpectedly, the times were as fast for sentences like (3) as they were for sentences like (1), even though the word "bartender" has no semantic connection to "cocktails" in (3).

(1) The boy watched the bartender serve the cocktails.
(2) The boy saw that the person liked the cocktails.
(3) The boy who watched the bartender served the cocktails.

Clearly, the semantic representation that yielded priming in (1) and (3) was not detailed enough to distinguish the difference in meaning between the two sentences. The representation was "good enough" to provide an interpretation that satisfied the comprehender, but not detailed enough to distinguish the important differences in who was doing what to whom.

RECENT STUDIES OF WHETHER INTERPRETATIONS ARE GOOD ENOUGH

In two series of studies, our lab has been investigating some situations in which good-enough, or noncompositional, processing may occur.

Misinterpretations of Garden-Path Sentences

One series (Christianson, Hollingworth, Halliwell, & Ferreira, 2001) addressed the straightforward question whether people delete from memory their initial misinterpretation of a sentence after reanalysis. When people were visually presented sentence (4), they initially took "the baby" to be the object of "dressed."

> (4) While Anna dressed the baby played in the crib. (presented without commas)

As a result, readers spent a great deal of time processing the disambiguating word "played" and often reread the preceding material. Sentences such as this one are often termed *garden-path sentences*, because the first part of the sentence sends the language comprehension system in an ultimately wrong direction. The comprehender will have no difficulty with (4) if the clauses are separated by a comma or if the main clause is presented before the subordinate clause. In these cases, there is no temptation to take "the baby" to be the object of "dressed," and therefore the reader has no difficulty integrating "played."

It has generally been assumed that if comprehenders restructure their initial interpretation of (4) so as to make "the baby" the subject of the main clause, they will end up with an appropriate representation of the sentence's overall meaning. This assumption was tested by asking participants to respond to questions after reading (at their own pace) garden-path sentences or non-garden-path control versions of the same sentences (Christianson et al., 2001). The questions were of two sorts:

> (5) Did the baby play in the crib?
> (6) Did Anna dress the baby?

Question (5) assessed whether the phrase "the baby" was eventually taken to be the subject of "played." Recall that initially it is not; the syntactic processor makes "the baby" the object of "dressed," and so "played" ends up without a subject. Thus, successful syntactic restructuring requires that "the baby" be removed from that first clause and included in the second, making "yes" the correct answer to (5). Question (6) assessed whether comprehenders then adjusted the meaning of the sentence to correspond to that reanalysis: Under this reinterpretation, "the baby" is no longer the object of "dressed," and so the sentence means that Anna is dressing herself. Therefore, the participants should have said "no" in response to (6).

Participants were virtually 100% correct in responding that the baby played in the crib. Performance was equally good in the garden-path and non-garden-path conditions. Yet when the sentence led the comprehenders down a syntactic garden path, they were inaccurate in answering (6). That is, people initially took "the baby" to be the object of "dressed." Then, they restructured the sentence to make "the baby" the subject of "played," but they persisted in thinking that the baby was being dressed. People who read the non-garden-path control version, however, almost always correctly replied that Anna did not dress the baby. In summary, the initial misinterpretation lingered and caused comprehenders to end up with a representation in which "the baby" was both the subject of "played" and the object of "dressed." This is clear evidence that the meaning people obtain for a sentence is often not a reflection of its true content.

Misinterpretations of Passive Sentences

The other series of experiments (Ferreira & Stacey, 2000) was designed to investigate an even more basic question: Are people ever tricked by simple, but implausible, passive sentences? Consider an active sentence like (7). People have little trouble obtaining its implausible meaning. In contrast, the passive sentence (10) is much more difficult to understand, and one's impression is that it is hard to keep straight whether the dog is the perpetrator or the victim in the scenario.

(7) The man bit the dog.
(8) The man was bitten by the dog.
(9) The dog bit the man.
(10) The dog was bitten by the man.

In one experiment (Ferreira & Stacey, 2000), participants read sentences like (7) through (10) and were instructed to indicate whether the event described in each sentence was plausible. For the active sentences, people were almost always correct. However, they called passive sentences like (10) plausible more than 25% of the time. In another experiment, participants heard one of these four sentences and then identified either the agent or the patient of the action. Again, people were accurate with all sentences except (10). Thus, when people read or hear a passive sentence, they use their knowledge of the world to figure out who is doing what to whom. That interpretation reflects the content words of the sentence more than its compositional, syntactically derived meaning. It is as if people use a semantic heuristic rather than syntactic algorithms to get the meaning of difficult passives.

OUR GOOD-ENOUGH APPROACH

The linguistic system embodies a number of powerful mechanisms designed to enable the comprehender to obtain the meaning of a sentence that was intended by the speaker. The system uses mechanisms such as syntactic analysis to achieve this aim. Syntactic structure allows the comprehender to compute algorithmically who did what to whom, because it allows thematic roles such as agent to be bound to the individual words of the sentence. The challenges in comprehension, however, are twofold. First, as the earliest work in cognitive psychology revealed, the structure built by the language processor is fragile and decays rapidly (Sachs, 1967). The representation needs almost immediate support from context or from schemas (i.e., general frameworks used to organize details on the basis of previous experience). In other words, given (10), syntactic mechanisms deliver the proper interpretation that the dog is the patient and the man is the agent; but the problem is that the delicate syntactic structure needs reinforcement. Schemas in long-term memory cannot provide that support, and so the source of corroboration must be context. Quite likely, then, sentences like this would be correctly understood in normal conversation, because the overall communicative context would support the interpretation. The important concept is that the linguistic representation itself is not robust, so that if it is not reinforced, a merely good-enough interpretation may result.

The second challenge to the linguistic system is that it must cope with potentially interfering information. The garden-path studies show that an initial incorrect representation of a sentence lingers and interferes with obtaining the correct meaning for the sentence. In the case of implausible passive sentences, information from schemas in long-term memory causes interference. As a result, people end up believing that (10) means what their schema tells them rather than what the output of the syntactic algorithms mandates. This interfering information must be inhibited for comprehension to be successful.

FUTURE DIRECTIONS

Experiments are under way to examine the characteristics of the memory representations for garden-path sentences, and to focus on how misinformation is suppressed during successful comprehension. The studies on passives are intriguing because they demonstrate that complex syntactic structures can be misinterpreted, but what makes a structure likely to be misinterpreted? One of the experiments (Ferreira & Stacey, 2000) demonstrated that the surface frequency of the sentence form is not critical to determining difficulty. People were as accurate with sentences such as "It was the man who bit the dog" as they were with common active sentences, even though the former structure is rare. One possible explanation for why the passive structure is difficult to comprehend is that passives require semantic roles to be assigned in an atypical order: patient before agent. This hypothesis can be addressed by examining languages that permit freer word order than does English. We are currently focusing on the aboriginal Native American language Odawa, which orthogonally crosses voice and word order—that is, an active sentence may have the patient either before or after the agent, as may a passive sentence. Thus, Odawa provides a unique opportunity for us to study the factors that cause linguistic representations to be particularly fragile and vulnerable to influence from schemas.

The good-enough approach also leads us in several other less traditional directions. For example, speech disfluencies that occur during conversation include pauses filled with "uh" or "um," repeated words, repairs that modify or replace earlier material, and false starts (utterance fragments that are begun and abandoned). Disfluencies will often yield a string of words that violates gram-matical principles. Nevertheless, comprehenders seem able to process such strings efficiently, and it is not clear how interpretation processes are affected by these disfluencies. Are abandoned fragments incorporated into the semantic representation of a sentence? Our work on misinterpretations of garden-path sentences suggests that the answer could well be yes. In the same way that the incorrect interpretation of a garden-path sentence lingers even though its under-lying structure is ultimately corrected, an interpretation built upon an ultimately abandoned fragment (e.g., "Turn left—I mean right at the stop sign") might per-sist in the comprehender's overall representation.

We are also investigating whether syntactically ambiguous sentences such as (11) and (12) are given incomplete syntactic representations. A recent study found that people were faster at reading sentences like (11), for which the attach-ment of the relative clause is semantically ambiguous, than at reading semanti-cally unambiguous versions like (12) (Traxler, Pickering, & Clifton, 1998).

(11) The son of the driver that had the mustache was pretty cool.
(12) The car of the driver that had the mustache was pretty cool.

One proposed explanation for this finding is that the syntactic representa-tion in the ambiguous case remains underspecified. That is, perhaps the lan-guage processor does not bother to attach the relative clause "that had the mustache" to either "son" or "driver" because it does not have enough informa-tion to support one interpretation over the other.

More generally, the good-enough approach to language comprehension invites a more naturalistic perspective on how people understand utterances than has been adopted in psycholinguistics up to this point. Psycholinguists have focused on people's ability to understand individual sentences (or short texts) in almost ideal circumstances. In laboratories, stimuli are (usually) shown visually in quiet rooms that offer no distractions. The results that have emerged from this work are central to any theory of comprehension, but examination of only those conditions will not yield a complete story. Outside the laboratory, utterances are often difficult to hear because of background noise; dialect and idiolect differences as well as competing sounds can make it difficult for the hearer to extract every word from an utterance; and speakers often produce utterances with disfluencies and outright errors that the processing system must handle somehow. We have shown in our research that, even in the ideal condi-tions of the laboratory, comprehension is more shallow and incomplete than psy-cholinguists might have suspected. In the real world, interpretations are even more likely to be "just good enough."

Perhaps good-enough interpretations help the language system coordinate listening and speaking during conversation. Usually when people talk to one another, turns are not separated by gaps. Therefore, comprehension and pro-duction processes must operate simultaneously. The goal of the comprehension system might be to deliver an interpretation that is just good enough to allow the production system to generate an appropriate response; after all, it is the response that is overt and that determines the success of the participants' joint activity. An adequate theory of how language is understood, then, will ultimately have to take into account the dynamic demands of real-time conversation.

Recommended Reading

Christianson, K., Hollingworth, A., Halliwell, J., & Ferreira, F. (2001). (See References)

Clark, H.H. (1996). *Using language*. Cambridge, England: Cambridge University Press.

Clifton, C., Jr. (2000). Evaluating models of human sentence processing. In M.W. Crocker, M. Pickering, & C. Clifton, Jr. (Eds.), *Architectures and mechanisms for language processing* (pp. 31–55). Cambridge, England: Cambridge University Press.

Tanenhaus, M.K., Spivey-Knowlton, M.J., Eberhard, K.M., & Sedivy, J.E. (1995). Integration of visual and linguistic information in spoken language comprehension. *Science, 268*, 632–634.

Note

1. Address correspondence to Fernanda Ferreira, 129 Psychology Research Building, Michigan State University, East Lansing, MI 48824-1117; e-mail: fernanda@eyelab.msu.edu.

References

Barton, S.B., & Sanford, A.J. (1993). A case study of anomaly detection: Shallow semantic processing and cohesion establishment. *Memory & Cognition, 21*, 477–487.

Christianson, K., Hollingworth, A., Halliwell, J., & Ferreira, F. (2001). Thematic roles assigned along the garden path linger. *Cognitive Psychology, 42*, 368–407.

Duffy, S.A., Henderson, J.M., & Morris, R.K. (1989). Semantic facilitation of lexical access during sentence processing. *Journal of Experimental Psychology: Learning, Memory, and Cognition, 15*, 791–801.

Erickson, T.A., & Mattson, M.E. (1981). From words to meaning: A semantic illusion. *Journal of Verbal Learning and Verbal Behavior, 20*, 540–552.

Ferreira, F., & Clifton, C., Jr. (1986). The independence of syntactic processing. *Journal of Memory and Language, 25*, 348–368.

Ferreira, F., & Stacey, J. (2000). *The misinterpretation of passive sentences*. Manuscript submitted for publication.

Fillenbaum, S. (1974). Pragmatic normalization: Further results for some conjunctive and disjunctive sentences. *Journal of Experimental Psychology, 102*, 574–578.

Frazier, L. (1978). *On comprehending sentences: Syntactic parsing strategies*. Unpublished doctoral dissertation, University of Connecticut, Storrs.

MacDonald, M.C., Pearlmutter, N.J., & Seidenberg, M.S. (1994). The lexical nature of syntactic ambiguity resolution. *Psychological Review, 101*, 676–703.

Sachs, J.S. (1967). Recognition memory for syntactic and semantic aspects of connected discourse. *Perception & Psychophysics, 2*, 437–442.

Traxler, M.J., Pickering, M.J., & Clifton, C., Jr. (1998). Adjunct attachment is not a form of ambiguity resolution. *Journal of Memory and Language, 39*, 558–592.

Trueswell, J., Tanenhaus, M., & Garnsey, S. (1994). Semantic influences on parsing: Use of thematic role information in syntactic disambiguation. *Journal of Memory and Language, 33*, 285–318.

Critical Thinking Questions

1. A consequence of the "good enough" approach is that errors of comprehension *will* occur. If language comprehension is "good enough" these errors should rarely cause significant problems. Think carefully about how often you

misunderstand someone or someone misunderstands you; do you believe that such harmless misunderstandings often occur in your day-to-day life?

2. Do you think that there are other cognitive processes that have this "good enough" quality? For example, is your visual system perfect? Is your motor system perfect? If not, what sorts of errors do they make, and in what sense are these systems (or others) "good enough?"

3. It is surprising that we fail to fully update our knowledge even when we *know* we have made a comprehension error, as in the sample sentence of the baby in the crib being the one getting dressed. Can you think of any reason that one would fail to correct this misinformation? Does this failure mean that incorrect representations are still good enough? If so, what would *not* be good enough?

Situation Models: The Mental Leap Into Imagined Worlds

Rolf A. Zwaan[1]

Department of Psychology, Florida State University, Tallahassee, Florida

Abstract

Situation models are mental representations of the state of affairs described in a text rather than of the text itself. Much of the research on situation models in narrative comprehension suggests that comprehenders behave as though they are in the narrated situation rather than outside of it. This article reviews some of this evidence and provides an outlook on future developments.

Keywords

situation models; language; comprehension

When reading a fictional text, most readers feel they are in the middle of the story, and they eagerly or hesitantly wait to see what will happen next. Readers get inside of stories and vicariously experience them. They feel happy when good things occur, worry when characters are in danger, feel sad, and may even cry when misfortune strikes. While in the middle of a story, they are likely to use past tense verbs for events that have already occurred, and future tense for those that have not. (Segal, 1995, p. 65)

In the 1980s, researchers proposed that understanding a story, or any text for that matter, involves more than merely constructing a mental representation of the text itself. Comprehension is first and foremost the construction of a mental representation of what that text is about: a situation model. Thus, situation models are mental representations of the people, objects, locations, events, and actions described in a text, not of the words, phrases, clauses, sentences, and paragraphs of a text. The situation-model view predicts that comprehenders are influenced by the nature of the situation that is described in a text, rather than merely by the structure of the text itself.

As a first illustration, consider the following sentences: *Mary baked cookies but no cake* versus *Mary baked cookies and cake*. Both sentences mention the word *cake* explicitly, but only the second sentence refers to a situation in which a cake is actually present. If comprehenders construct situation models, the concept of cake should be more available to them when the cake is in the narrated situation than when it is not, despite the fact that the word *cake* appears in both sentences. Consistent with this prediction, students who read (from a computer screen) short narratives containing sentences such as these recognized words (presented immediately after each text) more quickly when the denoted object was actually present in the narrated situation than when it was not (MacDonald & Just, 1989).

G.A. Radvansky and I have recently reviewed the extensive literature on situation models (Zwaan & Radvansky, 1998). Here, I focus specifically on the evidence pertaining to situation models as vicarious experiences in narrative comprehension. When we place ourselves in a situation, we have a certain spatial, temporal, and psychological "vantage" point from which we vicariously experience the events. Such a perspective has been termed a *deictic center*, and the shift to this perspective a *deictic shift* (Duchan, Bruder, & Hewitt, 1995). In everyday life, we are typically aware of our location and time. We are also aware of our current goals. We are aware of people in our environment and their goals and emotions. And we are aware of objects that are relevant to our goals. This is a useful first approximation of what should be relevant to a deictic center.

SPACE

People exist in, move about in, and interact with environments. Situation models should represent relevant aspects of these environments. Very often (but not necessarily), objects that are spatially close to us are more relevant than more distant objects. Therefore, one would expect the same for situation models. Consistent with this idea, comprehenders are slower to recognize words denoting objects distant from a protagonist than those denoting objects close to the protagonist (Glenberg, Meyer, & Lindem, 1987).

When comprehenders have extensive knowledge of the spatial layout of the setting of the story (e.g., a building), they update their representations according to the location and goals of the protagonist. They have the fastest mental access to the room that the protagonist is currently in or is heading to. For example, they can more readily say whether or not two objects are in the same room if the room mentioned is one of these rooms than if it is some other room in the building (e.g., Morrow, Greenspan, & Bower, 1987). This makes perfect sense intuitively; these are the rooms that would be relevant to us if we were in the situation.

People's interpretation of the meaning of a verb denoting movement of people or objects in space, such as *to approach*, depends on their situation models. For example, comprehenders interpret the meaning of *approach* differently in *The tractor is just approaching the fence* than in *The mouse is just approaching the fence*. Specifically, they interpret the distance between the figure and the landmark as being longer when the figure is large (tractor) compared with when it is small (mouse). The comprehenders' interpretation also depends on the size of the landmark and the speed of the figure (Morrow & Clark, 1988). Apparently, comprehenders behave as if they are actually standing in the situation, looking at the tractor or mouse approaching a fence.

TIME

We assume by default that events are narrated in their chronological order, with nothing left out. Presumably this assumption exists because this is how we experience events in everyday life. Events occur to us in a continuous flow, sometimes in close succession, sometimes in parallel, and often partially overlapping. Language allows us to deviate from chronological order, however. For example, we can say, "Before the psychologist submitted the manuscript, the journal changed its policy." The psychologist submitting the manuscript is reported first, even though it was the last of the two events to occur. If people construct a situation model, this sentence should be more difficult to process than its chronological counterpart (the same sentence, but beginning with "After"). Recent neuroscientific evidence supports this prediction. Event-related brain potential (ERP) measurements[2] indicate that "before" sentences elicit, within 300 ms, greater negativity than "after" sentences. This difference in potential is primarily located in the left-anterior part of the brain and is indicative of greater cognitive effort (Münte, Schiltz, & Kutas, 1998).

In real life, events follow each other seamlessly. However, narratives can have temporal discontinuities, when writers omit events not relevant to the plot. Such temporal gaps, typically signaled by phrases such as *a few days later*, are quite common in narratives. Nonetheless, they present a departure from everyday experience. Therefore, time shifts should lead to (minor) disruptions of the comprehension process. And they do. Reading times for sentences that introduce a time shift tend to be longer than those for sentences that do not (Zwaan, 1996).

All other things being equal, events that happened just recently are more accessible to us than events that happened a while ago. Thus, in a situation model, *enter* should be less accessible after *An hour ago, John entered the building* than after *A moment ago, John entered the building*. Recent probe-word recognition experiments support this prediction (e.g., Zwaan, 1996).

GOALS AND CAUSATION

If we have a goal that is currently unsatisfied, it will be more prominent in our minds than a goal that has already been accomplished. For example, my goal to assist my wife in preparing for a party at our house tonight is currently more active in my mind than my goal to write a review of a manuscript if I finished the review this morning. Once a goal has been accomplished, there is no need for me to keep it on my mental desktop. Thus, if a protagonist has a goal that has not yet been accomplished, that goal should be more accessible to the comprehender than a goal that was just accomplished by the protagonist. In line with this prediction, goals yet to be accomplished by the protagonist were recognized more quickly than goals that were just accomplished (Trabasso & Suh, 1993).

We are often able to predict people's future actions by inferring their goals. For example, when we see a man walking over to a chair, we assume that he wants to sit, especially when he has been standing for a long time. Thus, we might generate the inference "He is going to sit." Keefe and McDaniel (1993) presented subjects with sentences like *After standing through the 3-hr debate, the tired speaker walked over to his chair (and sat down)* and then with probe words (e.g., *sat*, in this case). Subjects took about the same amount of time to name *sat* when the clause about the speaker sitting down was omitted and when it was included. Moreover, naming times were significantly faster in both of these conditions than in a control condition in which it was implied that the speaker remained standing.

As we interact with the environment, we have a strong tendency to interpret event sequences as causal sequences. It is important to note that, just as we infer goals, we have to infer causality; we cannot perceive it directly. Singer and his colleagues (e.g., Singer, Halldorson, Lear, & Andrusiak, 1992) have investigated how readers use their world knowledge to validate causal connections between narrated events. Subjects read sentence pairs, such as 1a and then 1b or 1a' and then 1b, and were subsequently presented with a question like 1c:

(1a) Mark poured the bucket of water on the bonfire.

(1a') Mark placed the bucket of water by the bonfire.

(1b) The bonfire went out.

(1c) Does water extinguish fire?

Subjects were faster in responding to 1c after the sequence 1a–1b than after 1a'–1b. According to Singer, the reason for this is that the knowledge that water extinguishes fire was activated to validate the events described in 1a–1b. However, because this knowledge cannot be used to validate 1a'–1b, it was not activated when subjects read that sentence pair.

PEOPLE AND OBJECTS

Comprehenders are quick to make inferences about protagonists, presumably in an attempt to construct a more complete situation model. Consider, for example, what happens after subjects read the sentence *The electrician examined the light fitting*. If the following sentence is *She took out her screwdriver*, their reading speed is slowed down compared with when the second sentence is *He took out his screwdriver*. This happens because *she* provides a mismatch with the stereotypical gender of an electrician, which the subjects apparently infer while reading the first sentence (Carreiras, Garnham, Oakhill, & Cain, 1996).

Comprehenders also make inferences about the emotional states of characters. For example, if we read a story about Paul, who wants his brother Luke to be good in baseball, the concept of "pride" becomes activated in our mind when we read that Luke receives the Most Valuable Player Award (Gernsbacher, Goldsmith, & Robertson, 1992). Thus, just as in real life, we make inferences about people's emotions when we comprehend stories.

Just as we empathize with real people, we seem to empathize with story protagonists. Comprehenders' preferences for a particular outcome of a story interfere with the verification of previously known information about the actual outcome of the story. For example, comprehenders had difficulty verifying that "Margaret made her flight" when they had learned previously that Margaret's plane would plunge into the sea shortly after takeoff (Allbritton & Gerrig, 1991). Allbritton and Gerrig hypothesized that during reading, comprehenders generated *participatory* responses (e.g., "I hope she will miss the flight") that interfered with their verification performance.

THE FUTURE OF SITUATION MODELS

How close are we to a scientific account of the vicarious experiences described in the epigraph to this article? Advances are to be expected on two fronts. On the theoretical front, there will be discussion of the proper representational format for situation models. Researchers, most notably Kintsch (1998), have proposed computer models of how people construct situation models. The question has been raised recently as to whether such computer-based models can account for the full complexity of situation-model construction (and human cognition in general), or whether a biologically oriented approach has more explanatory power (e.g., Barsalou, in press). On the methodological front, the repertoire of cognitive tasks is being supplemented with measures of brain activity. Initial findings provide converging evidence (e.g., Münte et al., 1998).

To summarize, many aspects of narrated situations have already been shown to affect our understanding of stories. However, there is still a great deal that must be learned before we have a good understanding of people's fascinating ability to make a mental leap from their actual situation, reading a book on the couch, to an often fictional situation at a different time and place. Recent theoretical and methodological developments give reason to be optimistic about this endeavor.

Recommended Reading

Duchan, J.F., Bruder, G.A., & Hewitt, L.E. (Eds.). (1995). (See References)
Graesser, A.C., Millis, K.K., & Zwaan, R.A. (1997). Discourse comprehension. *Annual Review of Psychology, 48,* 163–189.
Johnson-Laird, P.N. (1983). *Mental models: Towards a cognitive science of language, inference, and consciousness.* Cambridge, MA: Harvard University Press.
Kintsch, W. (1998). (See References)
Zwaan, R.A., & Radvansky, G.A. (1998). (See References)

Notes

1. Address correspondence to Rolf A. Zwaan, Department of Psychology, Florida State University, Tallahassee, FL 32306-1270; e-mail: zwaan@psy.fsu.edu; World Wide Web: http://freud.psy.fsu.edu:80/~zwaan/.

2. ERPs are modulations of electrical activity in the brain that occur as a result of the processing of external stimuli.

References

Allbritton, D.W., & Gerrig, R.J. (1991). Participatory responses in prose understanding. *Journal of Memory and Language, 30,* 603–626.
Barsalou, L.W. (in press). Perceptual symbol systems. *Behavioral and Brain Sciences.*
Carreiras, M., Garnham, A., Oakhill, J., & Cain, K. (1996). The use of stereotypical gender information in constructing a mental model: Evidence from English and Spanish. *Quarterly Journal of Experimental Psychology, 49A,* 639–663.
Duchan, J.F., Bruder, G.A., & Hewitt, L.E. (Eds.). (1995). *Deixis in narrative: A cognitive science perspective.* Hillsdale, NJ: Erlbaum.
Gernsbacher, M.A., Goldsmith, H.H., & Robertson, R.W. (1992). Do readers mentally represent characters' emotional states? *Cognition and Emotion, 6,* 89–111.
Glenberg, A.M., Meyer, M., & Lindem, K. (1987). Mental models contribute to foregrounding during text comprehension. *Journal of Memory and Language, 26,* 69–83.
Keefe, D.E., & McDaniel, M.A. (1993). The time course and durability of predictive inferences. *Journal of Memory and Language, 32,* 446–463.

Kintsch, W. (1998). *Comprehension: A paradigm for cognition*. Cambridge, England: Cambridge University Press.

MacDonald, M.C., & Just, M.A. (1989). Changes in activation level with negation. *Journal of Experimental Psychology: Learning, Memory, and Cognition, 15*, 633–642.

Morrow, D.G., & Clark, H.H. (1988). Interpreting words in spatial descriptions. *Language and Cognitive Processes, 3*, 275–291.

Morrow, D.G., Greenspan, S.L., & Bower, G.H. (1987). Accessibility and situation models in narrative comprehension. *Journal of Memory and Language, 26*, 165–187.

Münte, T.F., Schiltz, K., & Kutas, M. (1998). When temporal terms belie conceptual order. *Nature, 395*, 71–73.

Segal, L.M. (1995). Cognitive-phenomenological theory of fictional narrative. In J.F. Duchan, G.A. Bruder, & L.E. Hewitt (Eds.), *Deixis in narrative: A cognitive science perspective* (pp. 61–78). Hillsdale, NJ: Erlbaum.

Singer, M., Halldorson, M., Lear, J.C., & Andrusiak, P. (1992). Validation of causal bridging inferences. *Journal of Memory and Language, 31*, 507–524.

Trabasso, T., & Suh, S. (1993). Understanding text: Achieving explanatory coherence through online inferences and mental operations in working memory. *Discourse Processes, 16*, 3–34.

Zwaan, R.A. (1996). Processing narrative time shifts. *Journal of Experimental Psychology: Learning, Memory, and Cognition, 22*, 1196–1207.

Zwaan, R.A., & Radvansky, G.A. (1998). Situation models in language comprehension and memory. *Psychological Bulletin, 123*, 162–185.

Critical Thinking Questions

1. Zwaan discusses situation models for stories. Do you think that we create situation models when we read other sorts of texts, for example, a scientific article, or a historical account of a civil war battle? How about the instructions for constructing a piece of furniture?

2. Some researchers have suggested that the process in our minds that creates situation models when we read actually evolved for dealing with social situations. In other words, we create situation models to keep track of what others are doing in real life, not just when we read, and we often adopt others' point of view when we do so. From your own experience, evaluate this idea, paying special attention to those properties of a situation model that Zwaan says are important.

3. Does our knowledge of situation models help us to predict what people will say is a well-formed story? Does it help us to predict what sort of story people will find surprising, exciting, or interesting?

Minds and Brains

Cognitive psychology is the study of the individual mind: What is its architecture and how does it function? Addressing these questions inevitably leads to closely related questions. What is the relationship of the mind and brain? How do animal minds differ from human minds? Considering such questions may help us to better understand the human mind, or even to reveal problems in our science that had previously gone undetected.

Perhaps the most obvious (and most timely) of these problems is the relationship of mind and brain. Enormous strides have been made in neuroscience in the last twenty years. How may such data be integrated into a study of the mind? More important, do they render study of the mind irrelevant? Perhaps cognitive psychologists are merely playing tic-tac-toe until the neuroscientists get it right. According to Miller and Keller in "Psychology and Neuroscience: Making Peace," this view—that psychology can be reduced to neuroscience—is flawed, although it is a common perception even among psychologists.

Neuroscience will not someday supplant cognitive psychology—but will it help us now? The answer is a clear "yes," as exemplified by Roser and Gazzaniga's article titled "Automatic Brains—Interpretative Minds." The problem they address is this: most cognitive psychologists hold a modular view of the mind. That means they believe that the mind is composed of different processing modules, each performing a different job, and each relatively autonomous of the others (although they do communicate). For example, different processing modules within the visual system appear to process color information, shape, motion, and so on. This view of the mind is supported by data from neuroscience showing different anatomic substrates for these different cognitive modules. Modularity, however, gives rise to a question: If the mind is composed of separate processing modules, why is conscious experience unitary, rather than fragmented? Although the mind appears to tear the world apart into different components that are processed separately, the products of these processes are somehow knit together again. Roser and Gazzaniga argue that a process in the left hemisphere constructs and interprets a personal narrative from these disparate data, affording a single, coherent experience.

In trying to understand the human mind, cognitive psychologists turn not only to the human brain, but also to animal minds. Scientists often begin with simple systems and, once those systems are understood, move onto more complex systems. In studying non-human primates, we might suspect that we are observing simpler versions of ourselves. In "The Mentality of Apes Revisited," Povinelli and Bering argue that this attitude is pervasive in psychology, and that it is dead wrong. There is not an "evolutionary ladder" with humans at the pinnacle and other animals at various rungs below us. Evolution does not create a ladder with a single dimension of human-like intelligence, with different animals endowed with more or less. The product of evolution is diversity of mental function, with each animal possessing different cognitive abilities, suited to its environment. Povinelli and Bering believe that most psychologists have made a fundamental mistake when thinking about the cognition of other species.

Most humans have only a passing interest in what apes know, but we are all forced to make judgments about what other humans know. For example, suppose that I want to tell you about an interesting news story I heard on the radio about current events in Central America. I need to judge how much you likely know about the topic already. If I underestimate, I will provide background that you already know and I will likely bore you. If I overestimate your background knowledge, the story will likely be incomprehensible to you. How do I judge what you know? In "The Projective Way of Knowing: A Useful Heuristic that Sometimes Misleads," Nickerson reviews evidence indicating that we use a very simple rule of thumb in these situations: in the absence of other information we assume that others know what we know. In other words we assume that other brains, or other minds, are like our own.

Psychology and Neuroscience: Making Peace

Gregory A. Miller[1] and Jennifer Keller

Department of Psychology, University of Illinois at Urbana-Champaign, Champaign, Illinois

Abstract

There has been no historically stable consensus about the relationship between psychological and biological concepts and data. A naively reductionist view of this relationship is prevalent in psychology, medicine, and basic and clinical neuroscience. This view undermines the ability of psychology and related sciences to achieve their individual and combined potential. A nondualistic, nonreductionist, noninteractive perspective is recommended, with psychological and biological concepts both having central, distinct roles.

Keywords

psychology; biology; neuroscience; psychopathology

With the Decade of the Brain just ended, it is useful to consider the impact that it has had on psychological research and what should come next. Impressive progress occurred on many fronts, including methodologies used to understand the brain events associated with psychological functions. However, much controversy remains about where biological phenomena fit into psychological science and vice versa. This controversy is especially pronounced in research on psychopathology, a field in which ambitious claims on behalf of narrowly conceived psychological or biological factors often arise, but this fundamental issue applies to the full range of psychological research. Unfortunately, the Decade of the Brain has fostered a naively reductionist view that sets biology and psychology at odds and often casts psychological events as unimportant epiphenomena. We and other researchers have been developing a proposal that rejects this view and provides a different perspective on the relationship between biology and psychology.

A FAILURE OF REDUCTIONISM

A term defined in one domain is characterized as *reduced* to terms in another domain (called the reduction science) when all meaning in the former is captured in the latter. The reduced term thus becomes unnecessary. If, for example, the meaning of the (traditionally psychological) term "fear" is entirely representable in language about a brain region called the amygdala, one does not need the (psychological) term "fear," or one can redefine "fear" to refer merely to a particular biological phenomenon.

Impressive progress in the characterization of neural circuits typically active in (psychologically defined) fear does not justify dismissing the concept or altering the meaning of the term. The phenomena that "fear" typically refers to include a functional state (a way of being or being prepared to act), a cognitive processing bias, and a variety of judgments and associations all of which are conceived psychologically (Miller & Kozak, 1993). Because "fear" means more than a given type of neural activity, the concept of fear is not reducible to neural activity. Researchers are learning a great deal about the biology of fear—and the psychology of fear—from studies of the amygdala (e.g., Lang, Davis, & Öhman, in press), but this does not mean that fear is activity in the amygdala. That is simply not the meaning of the term. "Fear" is not reducible to biology.

This logical fact is widely misunderstood, as evidenced in phrases such as "underlying brain dysfunction" or "neurochemical basis of psychopathology." Most remarkably, major portions of the federal research establishment have recently adopted a distinctly nonmental notion of mental health, referring to "the biobehavioral factors which may underly [sic] mood states" (National Institute of Mental Health, 1999). Similarly, a plan to reorganize grant review committees reflects "the context of the biological question that is being investigated" (National Institutes of Health, 1999, p. 2). Mental health researchers motivated by psychological or sociological questions apparently should take their applications elsewhere.

More subtly problematic than such naive reductionism are terms, such as "biobehavioral marker" or "neurocognitive measure," that appear to cross the boundary between psychological and biological domains. It is not at all apparent what meaning the "bio" or "neuro" prefix adds in these terms, as typically the data referred to are behavioral. Under the political pressures of the Decade of the Brain, psychologists were tempted to repackage their phenomena to sound biological, but the relationship of psychology and biology cannot be addressed by confusing them.

WHOSE WORK IS MORE FUNDAMENTAL?

Such phrases often appear in contexts that assume that biological phenomena are somehow more fundamental than psychological phenomena. Statements that psychological events are nothing more than brain events are clearly logical errors (see the extensive analysis by Marr, 1982). More cautious statements, such as that psychological events "reflect" or "arise from" brain events, are at best incomplete in what they convey about the relationship between psychology and biology. It is not a property of biological data that they "underlie" psychological data. A given theory may explicitly propose such a relationship, but it must be treated as a proposal, not as a fact about the data. Biological data provide valuable information that may not be obtainable with self-report or overt behavioral measures, but biological information is not inherently more fundamental, more accurate, more representative, or even more objective.

The converse problem also arises—psychology allegedly "underlying" or being more fundamental than biology. There is a long tradition of ignoring biological phenomena in clinical psychology. As Zuckerman (1999) noted, "One thing that both behavioral and post-Freudian psychoanalytic theories had in common was the conviction that learning and life experiences alone could account for all disorders" (p. 413). In those traditions, it is psychology that "underlies" biology, not the converse. Biology is seen as merely the implementation of psychology, and psychology is where the intellectually interesting action is. Cognitive theory can thus evolve without the discipline of biological plausibility. As suggested at the midpoint of the Decade of the Brain (Miller, 1995), such a view would justify a Decade of Cognition.

Such a one-sided emphasis would once again be misguided. Anderson and Scott (1999) expressed concern that "the majority of research in the health sciences occurs within a single level of analysis, closely tied to specific disciplines" (p. 5), with most psychologists studying phenomena only in terms of behavior. We advocate not that every study employ both psychological and biological methods, but that researchers not ignore or dismiss relevant literatures, particularly in the conceptualization of their research.

Psychological and biological approaches offer distinct types of data of potentially equal relevance for understanding psychological phenomena. For example, we use magnetoencephalography (MEG) recordings of the magnetic fields generated by neural activity to identify multiple areas of brain tissue that are generating what is typically measured electrically at the scalp (via electroencephalography, or EEG) as the response of the brain associated with cognitive tasks (Cañive, Edgar, Miller, & Weisend, 1999). One of the most firmly established biological findings in schizophrenia is a smaller than normal brain response called the P300 component (Ford, 1999), and there is considerable consensus on the functional significance of P300 in the psychological domain. There is, however, no consensus on what neural generators produce the electrical activity or on what distinct functions those generators serve. Neural sources are often difficult to identify with confidence from EEG alone, whereas for biophysical reasons MEG (which shows brain function) coupled with structural magnetic resonance data (which show brain anatomy) promises localization as good as any other available noninvasive method. If researchers understand the distinct functional significance of various neural generators of P300, and if only some generators are compromised in schizophrenia, this will be informative about the nature of cognitive deficits in schizophrenia. Conversely, what researchers know about cognitive deficits will be informative about the function of the different generators.

MEG and EEG do not "underlie" and are not the "basis" of (the psychological phenomena that define) the functions or mental operations invoked in tasks associated with the P300 response. Neural generators implement functions, but functions do not have locations (Fodor, 1968). For example, a working memory deficit in schizophrenia could not be located in a specific brain region. The psychological and the neuromagnetic are not simply different "levels" of analysis, except in a very loose (and unhelpful) metaphorical sense. Neither underlies the other, neither is more fundamental, and neither explains away the other. There are simply two domains of data, and each can help to explicate the other because of the relationships theories propose.

Psychophysiological research provides many other examples in which the notion of "underlying" is unhelpful. Rather than attributing mood changes to activity in specific brain regions, why not attribute changes in brain activity to changes in mood? In light of EEG (Deldin, Keller, Gergen, & Miller, 2000) or behavioral (Keller et al., 2000) data on regional brain activity in depression, are people depressed because of low activity in left frontal areas of the brain, or do they have low activity in these areas because they are depressed? Under the present view, such a question, trying to establish causal relations *between* psychology and biology, is misguided. These are not empirical issues but logical and theoretical issues. They turn on the kind of relationship that psychological and biological concepts are proposed to have.

CLINICAL IMPLICATIONS

In psychopathology, one of the most unfortunate consequences of the naive competition between psychology and biology is the assumption that dysfunctions conceptualized biologically require biological interventions and that those conceptualized psychologically require psychological interventions. The best way to alter one system may be a direct intervention in another system. Even, for example, if the chemistry of catecholamines (chemicals used for communication to nerve, muscle, and other cells) were the best place to intervene in schizophrenia, it does not follow that a direct biological intervention in that system would be optimal. A variety of experiences that people construe as psychosocial prompt their adrenal glands to flood them with catecholamines. There are psychological interventions associated with this chemistry that can work more effectively or with fewer side effects than medications aimed directly at the chemistry.

Unfortunately, the assumption that disorders construed biologically warrant exclusively biological interventions influences not only theories of psychopathology but also available treatments. For example, major depression is increasingly viewed as a "chemical imbalance." If such (psychological) disorders are assumed to "be" biological, then medical insurers are more likely to fund only biological treatments. Yet Thase et al. (1997) found that medication and psychotherapy were equally effective in treating moderately depressed patients and that the combination of these treatments was more effective than either alone in treating more severely depressed patients. Hollon (1995) discussed how negative life events may alter biological factors that increase risk for depression. Meany (1998) explained how the psychological environment can affect gene activity. The indefensible conceptualization of depression solely as a biological disorder prompts inappropriately narrow (biological) interventions. Thus, treatment as well as theory is hampered by naive reductionism.

WHAT TO DO?

"Underlying" (implying one is more fundamental than the other) is not a satisfactory way to characterize the relationship between biological and psychological concepts. We recommend characterizing the biological as "implementing" the psychology—that is, we see cognition and emotion as implemented in neural systems. Fodor (1968) distinguished between *contingent* and *necessary* identity in the relationship between psychological and biological phenomena. A person in any given psychological state is momentarily in some biological state as well: There is a *contingent* identity between the psychological and the biological at that moment. The psychological phenomenon implemented in a given neural circuit is not the same as, is not accounted for by, and is not reducible to that circuit.

There is an indefinite set of potential neural implementations of a given psychological phenomenon. Conversely, a given neural circuit might implement different psychological functions at different times or in different individuals. Thus, there is no *necessary* identity between psychological states and brain states. Distinct psychological and biological theories are needed to explain their respective domains, and additional theoretical work is needed to relate them.

Nor is it viable (though it is common) to say that psychological and biological phenomena "interact." Such a claim begs the question of how they interact and even what it means to interact. The concept of the experience of "red" does not "interact" with the concept of photon-driven chemical changes in the retina and their neural sequelae. One may propose that those neural sequelae implement the perceptual experience of "red," but "red" *means* not the neural sequelae, but something psychological—a perception.

Biology and psychology often are set up as competitors for public mind-share, research funding, and scientific legitimacy. We are not arguing for a psychological explanation of cognition and emotion *instead of* a biological explanation. Rather, we are arguing against framing biology and psychology in a way that forces a choice between those kinds of explanations. The hyperbiological bias ascendant at the end of the 20th century was no wiser and no more fruitful than the hyperpsychological bias of the behaviorist movement earlier in the 20th century. Scientists can avoid turf battles by approaching the relationship between the psychological and the biological as fundamentally theoretical, not empirical. Working out the biology will not make psychology obsolete, any more than behaviorism rendered biology obsolete. Scientists can avoid dualism by avoiding interactionism (having two distinct domains in a position to interact implies separate realities, hence dualism). Psychological and biological domains can be viewed as logically distinct but not physically distinct, and hence neither dualistic nor interacting. Psychological and biological concepts are not merely different terms for the same phenomena (and thus not reducible in either direction), and psychological and biological explanations are not explanations of the same things. If one views brain tissue as implementing psychological functions, the expertise of cognitive science is needed to characterize those functions, and the expertise of neuroscience is needed to study their implementation. Each of those disciplines will benefit greatly from the other, but neither encompasses, reduces, or underlies the other.

Fundamentally psychological concepts require fundamentally psychological explanations. Stories about biological phenomena can richly inform, but not supplant, those explanations. Yet when psychological events unfold, they are implemented in biology, and those implementations are extremely important to study as well. For example, rather than merely pursuing, in quite separate literatures, anomalies in either expressed emotion or biochemistry, research on schizophrenia should investigate biological mechanisms involved in expressed-emotion phenomena. Similarly, the largely separate literatures on biological and psychosocial mechanisms in emotion should give way to conceptual and methodological collaboration. Research in the next few decades will need not only the improving spatial resolution of newer brain-imaging technologies and the high temporal resolution of established brain-imaging technologies, but also the advancing cognitive resolution of the best psychological science.

Recommended Reading

Anderson, N.B., & Scott, P.A. (1999). (See References)

Cacioppo, J.T., & Berntson, G.G. (1992). Social psychological contributions to the Decade of the Brain. *American Psychologist, 47*, 1019–1028.

Kosslyn, S.M., & Koenig, O. (1992). *Wet mind: The new cognitive neuroscience.* New York: Free Press.

Miller, G.A. (1996). Presidential address: How we think about cognition, emotion, and biology in psychopathology. *Psychophysiology, 33*, 615–628.

Ross, C.A., & Pam, A. (1995). *Pseudoscience in biological psychiatry: Blaming the body.* New York: Wiley.

Acknowledgments—The authors' work has been supported in part by National Institute of Mental Health Grants R01 MH39628, F31 MH11758, and T32 MH19554; by the Department of Psychiatry of Provena Covenant Medical Center; and by the Research Board, the Beckman Institute, and the Departments of Psychology and Psychiatry of the University of Illinois at Urbana-Champaign. The authors appreciate the comments of Howard Berenbaum, Patricia Deldin, Wendy Heller, Karen Rudolph, Judith Ford, Michael Kozak, Sumie Okazaki, and Robert Simons on an earlier draft.

Note

1. Address correspondence to Gregory A. Miller, Departments of Psychology and Psychiatry, University of Illinois, 603 E. Daniel St., Champaign, IL 61820; e-mail: gamiller@uiuc.edu.

References

Anderson, N.B., & Scott, P.A. (1999). Making the case for psychophysiology during the era of molecular biology. *Psychophysiology, 36*, 1–14.

Cañive, J.M., Edgar, J.C., Miller, G.A., & Weisend, M.P. (1999, April). *MEG recordings of M300 in controls and schizophrenics.* Paper presented at the biennial meeting of the International Congress on Schizophrenia Research, Santa Fe, NM.

Deldin, P.J., Keller, J., Gergen, J.A., & Miller, G.A. (2000). Right-posterior N200 anomaly in depression. *Journal of Abnormal Psychology, 109*, 116–121.

Fodor, J.A. (1968). *Psychological explanation.* New York: Random House.

Ford, J.M. (1999). Schizophrenia: The broken P300 and beyond. *Psychophysiology, 36*, 667–682.

Hollon, S.D. (1995). Depression and the behavioral high-risk paradigm. In G.A. Miller (Ed.), *The behavioral high-risk paradigm in psychopathology* (pp. 289–302). New York: Springer-Verlag.

Keller, J., Nitschke, J.B., Bhargava, T., Deldin, P.J., Gergen, J.A., Miller, G.A., & Heller, W. (2000). Neuropsychological differentiation of depression and anxiety. *Journal of Abnormal Psychology, 109*, 3–10.

Lang, P.J., Davis, M., & Öhman, A. (in press). Fear and anxiety: Animal models and human cognitive psychophysiology. *Journal of Affective Disorders.*

Marr, D. (1982). *Vision: A computational investigation into the human representation and processing of visual information.* New York: Freeman.

Meany, M.J. (1998, September). *Variations in maternal care and the development of individual differences in neural systems mediating behavioral and endocrine responses to stress.* Address presented at the annual meeting of the Society for Psychophysiological Research, Denver, CO.

Miller, G.A. (1995, October). *How we think about cognition, emotion, and biology in psychopathology.* Presidential address presented at the annual meeting of the Society for Psychophysiological Research, Toronto, Ontario, Canada.

Miller, G.A., & Kozak, M.J. (1993). A philosophy for the study of emotion: Three-systems theory. In N. Birbaumer & A. Öhman (Eds.), *The structure of emotion: Physiological, cognitive and clinical aspects* (pp. 31–47). Seattle, WA: Hogrefe & Huber.

National Institute of Mental Health. (1999, February 12). [Announcement of NIMH workshop, "Emotion and mood"]. Unpublished e-mail.

National Institutes of Health. (1999, February). *Peer review notes*. Washington, DC: Author.

Thase, M.E., Greenhouse, J.B., Frank, E., Reynold, C.F., Pilkonis, P.A., Hurley, K., Grochocinski, V., & Kupfer, D.J. (1997). Treatment of major depression with psychotherapy or psychotherapy-pharmacotherapy combinations. *Archives of General Psychiatry*, 54, 1009–1015.

Zuckerman, M. (1999). *Vulnerability to psychopathology: A biosocial model*. Washington, DC: American Psychological Association.

Critical Thinking Questions

1. To further appreciate the point of this article, consider the following analogy. Suppose you know exactly what your computer does as your friend types a paper using Microsoft Word. You have complete knowledge of when the hard drive is accessed, what happens in the memory registers, and so on. But you have no knowledge of the functional software commands associated with MS Word. In other words, when your friend types command-U, you know what the physical guts of the computer do, but you don't know why your friend types command-U, nor what that means for the document he is producing. Suppose, also, that your friend has a complete, exhaustive knowledge of how to use MS Word, including all of the esoteric features. But your friend does not have the slightest idea of what is happening inside the computer as he uses it. Who understands MS Word better, you or your friend? Is one understanding more important or more fundamental than the other?

2. What would it mean for a psychological process (e.g., long term memory) to be implemented in something other than the brain? What would it mean for a neural circuit *not* to be identified with a single psychological function, but rather with multiple functions?

3. Most of the examples Miller and Keller provide are clinical. Clinical disorders are arguably more high-level than the sort of behaviors that cognitive psychologists study: for example, clinical depression has embedded in it memory, emotion, decision-making and other processes. If cognitive processes are lower level, does that mean that Miller and Keller's arguments do not apply to cognition and the brain?

Automatic Brains—Interpretive Minds

Matthew Roser and Michael S. Gazzaniga

Dartmouth College

Abstract

The involvement of specific brain areas in carrying out specific tasks has been increasingly well documented over the past decade. Many of these processes are highly automatic and take place outside of conscious awareness. Conscious experience, however, seems unitary and must involve integration between distributed processes. This article presents the argument that this integration occurs in a constructive and interpretive manner and that increasingly complex representations emerge from the integration of modular processes. At the highest levels of consciousness, a personal narrative is constructed. This narrative makes sense of the brain's own behavior and may underlie the sense of a unitary self. The challenge for the future is to identify the relationships between patterns of brain activity and conscious awareness and to delineate the neural mechanisms whereby the underlying distributed processes interact.

Keywords

neural correlates of consciousness; interpreter; integration

Although it has been known for more than a century that particular parts of the brain are important for particular functions, the past decade of functional magnetic resonance imaging (fMRI) research has lead to a huge upsurge in evidence for functional specialization. This work has identified areas of the cortex, the convoluted outer layer of the brain, that are involved in processing particular stimulus attributes, or performing certain tasks. For example, cortical areas especially responsive to faces, movement, and places have been found, and these experimental results have been replicated by many independent observers. Although some of the initial claims for functional specialization have been tempered somewhat in the light of new findings, it is becoming ever more clear that the cortex is not a homogeneous, general-purpose computing device, but rather is a complex of circumscribed, modular processes occupying distinct locations.

Most of the work undertaken by these specialist systems occurs automatically and outside of conscious control. For instance, if certain stimuli trick your visual system into constructing an illusion, knowing that you have been tricked does not mean that the illusion disappears. The part of the visual system that produces the illusion is impervious to correction based on such knowledge. Additionally, a convincing illusion can leave behavior unaffected, as when observers are asked to scale the distance between their fingers to the size of a line presented with an arrowhead attached to each end. Although the arrowheads can alter the perceived size of a line (the Müller-Lyer illusion), observers do not

---

Address correspondence to Matthew Roser, Department of Psychological and Brain Sciences, Moore Hall, Dartmouth College, Hanover, NH 03755; e-mail: matthew.rosser@dartmouth.edu.

162

make a corresponding adjustment in the distance between their fingers, sug-gesting that the processes determining the overt behavior are isolated from those underlying the perception. Thus, a visuo-motor process in response to a stimu-lus can proceed independently of the simultaneous perception of that stimulus (Aglioti, DeSouza, & Goodale, 1995).

Stimuli that are not consciously perceived by subjects can, nonetheless, affect behavior. For example, stimuli that are presented very briefly and followed by a masking stimulus go unperceived by subjects, but still activate response mechanisms and speed the recognition of following stimuli that share their semantic properties (Dehaene et al., 1998). When you add to this the observa-tion that robust perceptual aftereffects can be induced by stimuli that are not consciously perceived (Rees, Kreiman, & Koch, 2002), it becomes evident that a great deal of the brain's work occurs outside of conscious awareness and con-trol. Thus, the systems built into our brains carry out their jobs automatically when presented with stimuli within their domain, often without our knowledge.

The most striking evidence for the isolation of function from consciousness comes from studies of patients showing either neglect or blindsight. Neglect is a condition in which the patient ignores a part of space, usually the left; it is typically found in people with damage to the right parietal area of the brain and is thought to be due to the disruption of the brain's mechanisms for allocating attention. Astonishingly, patients with neglect often deny that they have any such condition. It is as if their consciousness of the deficit is destroyed by the lesion just as their actual awareness of a part of space is, even though early visual areas of the brain (i.e., areas that receive and process incoming visual information) are intact and functioning.

The even more bizarre condition known as blindsight describes the resid-ual visual function shown by some patients following a lesion in early visual areas. Although these patients claim to be completely blind in the side of visual space contralateral to the lesion, they are nonetheless able to discriminate, locate, and guide motion toward a stimulus in that area, all without a conscious percept (Rees et al., 2002).

Together, these syndromes and studies in normal subjects suggest that the activity of the brain is not strictly continuous with our conscious experience. Instead, we are sometimes oblivious to complex processing that occurs in the brain. The question then becomes, what determines whether a process is con-scious or not?

BRAIN ACTIVITY AND CONSCIOUSNESS

The neural correlates of consciousness in the human brain have been investigated using fMRI and a technique known as binocular rivalry. In this kind of study, a different stimulus is presented to each eye, and the conscious percept typically switches back and forth between the two stimuli, each percept lasting for a few seconds. Subjects indicate when their perception changes from one stimulus to the other, and because the stimuli themselves are static, any changes in neural activity that correlate with a change in the reported percept can be ascribed to changes in the contents of awareness (Tong, Nakayama, Vaughan, & Kanwisher,

1998). Brain activations elicited by rivalrous stimuli are very similar in magnitude and location to activations seen in response to separate stimuli that are presented alternately, suggesting that areas involved in processing a type of stimulus are also involved in the conscious perception of that type of stimulus (Zeki, 2003).

fMRI studies have also revealed substantial brain activations in response to stimuli that are not consciously perceived by subjects (Moutoussis & Zeki, 2002). For example, when color-reversed faces (e.g., an outlined red face on a green background and an outlined green face on a red background) are displayed separately to the two eyes, binocular fusion occurs, and subjects report seeing only the color that results from the combination of the two stimulus color (in this case, yellow). The color inputs to the brain are "mixed," like paint on an artist's palette, and the face stimuli become invisible. Despite not being consciously seen, these stimuli typically activate those areas of the brain that are activated by perceived faces. Why then are some seen and some not?

Brain activations correlated with perceived stimuli and those correlated with unseen stimuli show differences in both their intensity and their spatial extent (Dehaene et al., 2001). Dehaene and his colleagues found that although unperceived stimuli and perceived stimuli activated similar locations in the brain, the activations associated with perceived stimuli were many times more intense than those seen with unperceived stimuli and were accompanied by activity at additional sites. Thus, consciousness may have a graded relationship to brain activity, or a threshold may exist, above which activation reaches consciousness (Rees et al., 2002). At present this issue is unresolved, but the development of increasingly sophisticated designs in fMRI may yield progress by allowing the degree of activation to be determined as the availability of a stimulus to awareness is manipulated.

The increased spatial extent of activations elicited by perceived stimuli in the experiment by Dehaene and his colleagues suggests another possible mechanism for determining whether a stimulus reaches consciousness. Processing of a stimulus may reach consciousness if it is integrated into a large-scale system of cortical activity.

CONSCIOUSNESS SEEMS UNITARY

Despite the evidence that processing is distributed around the brain in functionally localized units, and that much processing proceeds outside of awareness, we personally experience consciousness as a unitary whole. How can these observations be resolved?

One possibility is that processes occurring within localized areas and circumscribed domains become available to consciousness only when they are integrated with other domains. Dehaene and Naccache (2001) have hypothesized that there is a *global neuronal work space* in which unconscious modular processes can be integrated in a common network of activation if they receive amplification by an attentional gating system. Attentional amplification leads to increased and prolonged activation and allows processing at one site to affect processing at another. In this way, brain areas involved in perception, action, and emotion can interact with each other and with circuits that can reinstate past states of this work space.

According to this hypothesis, consciousness is the collection of modular processes that are mobilized into a common neuronal work space and integrated in a dynamic fashion. It is a global pattern of activity across the brain, allowing information to be maintained and influence other processes. For instance, consider the task in which subjects are asked to match the distance between their fingers to the size of a Müller-Lyer figure. If a small delay is introduced between the observation of the figure and the reaching response, subjects must rely on their memory of the perceived size when scaling their grip to the size of the figure. Memory involves a consciously maintained representation. In this situation, the illusion does, in fact, affect the subject's motor response (Aglioti et al., 1995).

This model can explain some of the bizarre deficits of consciousness that occur as the result of brain lesions. As processing that does not achieve amplification remains entirely outside of consciousness, a neglect patient may not be aware of his or her deficit because the mechanism linking local processing to global patterns of activation has been disrupted. Thus, a lesion in a specific location may wipe out not merely processing of an attribute, but also the consciousness of the attribute.

Patients with severe cognitive deficits often confabulate wildly in order to produce an explanation of the world that is consistent with their conscious experience. These confabulations include completely denying the existence of a deficit and probably result from interpretations of incomplete information, or a reduced range of conscious experience (Cooney & Gazzaniga, 2003). Wild confabulations that seem untenable to most people, because of conscious access to information that contradicts them, probably seem completely normal to patients to whom only a subset of the elements of consciousness are available for integration.

MIND IS INTERPRETIVE AND CONSTRUCTIVE

The corpus callosum, which connects the two hemispheres, is the largest single fiber tract in the brain. What happens, then, when you cut this pathway for hemispheric communication and isolate the modular systems of the right hemisphere from those in the left? In the so-called split brain, only processes within a hemisphere can be integrated via cortical routes, and only a limited number of processes that can propagate via subcortical routes can be integrated between the hemispheres. Upon introspection, split-brain patients will tell you that they feel pretty normal. And yet, splitting the brain can reveal some of the most striking disconnections between brain processes and awareness. Each hemisphere can be presented with information that remains unknown to the opposite hemisphere.

Experimental designs that exploit this lack of communication have revealed that the left hemisphere tends to interpret what it sees, including the actions of the right hemisphere (Gazzaniga, 2000). For example, suppose two different scenes are presented simultaneously, one to each hemisphere, and the patient is asked to use his or her left hand to choose an appropriate item from an array of pictures of objects that may or may not be typically found within the presented scenes. The left hand is controlled by the right hemisphere, so the patient's left hemisphere, which has no knowledge of what was presented to the right hemisphere, can observe the subsequent actions of the right hemisphere.

If the patient is asked why he or she chose a particular item, the patient's verbal reply will be largely controlled by the left hemisphere, where the brain's primary language centers are located. Studies using this procedure have shown that patients often reply with an interpretation of the action that is congruent with the scene presented to the left hemisphere. Thus, patients resolve one hemisphere's actions with the other hemisphere's perceptions, by producing an explanation that eliminates conflict between the two. Patients' responses in such studies are very similar to the confabulations produced by brain-damaged patients who deny that they have a serious deficit by rationalizing their bizarre behavior (Cooney & Gazzaniga, 2003).

The hypothesis-generating nature of the left hemisphere has also been demonstrated in a nonlinguistic manner. When each hemisphere of a split brain is asked to predict whether a light will appear on the top or the bottom of a computer screen on a series of trials, and to indicate its prediction by pushing one of two buttons with the contralateral hand, the two hemispheres employ radically different strategies. The right hemisphere takes the simple approach and consistently chooses the more probable alternative, thereby maximizing performance. By contrast, the left hemisphere does what neurologically normal subjects do and distributes its responses between the two alternatives according to the probability that each will occur, despite the fact that this is a suboptimal strategy (Wolford, Miller, & Gazzaniga, 2000). It seems that the left hemisphere is driven to hypothesize about the structure of the world even when this is detrimental to performance.

The left-hemisphere interpreter may be responsible for our feeling that our conscious experience is unified. Generation of explanations about our perceptions, memories, and actions, and the relationships among them, leads to the construction of a personal narrative that ties together elements of our conscious experience into a coherent whole. The constructive nature of our consciousness is not apparent to us. The action of an interpretive system becomes observable only when the system can be tricked into making obvious errors by forcing it to work with an impoverished set of inputs, such as in the split brain or in lesion patients. But even in the damaged brain, this system still lets us feel like "us."

CONCLUSIONS AND FUTURE DIRECTIONS

It is becoming increasingly clear that consciousness involves disunited processes that are integrated in a dynamic manner. It is assembled on the fly, as our brains respond to constantly changing inputs, calculate potential courses of action, and execute responses. But it is also constrained by the nature of modular processes that occur without conscious control, and large parts of it can be destroyed, leaving a rump that operates only within its reduced sphere. Progress toward an overarching theory of consciousness will involve putting our picture of the brain back together. Although carving cognition and brain function up at the joints has been a hugely productive approach, future progress must depend on a variety of approaches that integrate disparate and circumscribed processes.

To this end, developing techniques in brain mapping hold much promise. Statistical analysis of fMRI data allows the correlations between activations in different areas to be assessed, yielding maps of cerebral interactivity. The application of these techniques to investigation of the neural correlates of consciousness is extremely relevant, as the activation of large networks is thought to be a necessary condition for consciousness.

A further step involves integrating maps of cerebral interactivity with data about neuroanatomical connections. This technique allows a subset of brain processes to be explicitly modeled as a functional network and yields a map of the strengths of anatomical connections that best fits the imaging data (Horwitz, Tagamets, & McIntosh, 1999). At present, much of the data on neuroanatomical connections comes from postmortem studies in monkeys, but a developing noninvasive MRI technique known as diffusion-tensor imaging (DTI) allows the paths of neurons to be tracked and should provide more accurate data about the human brain. DTI is set to have a huge future impact on this field (Le Bihan et al., 2001).

The brain sciences of the coming years promise to yield great progress in our understanding of integrative processes in the brain. The ultimate aim is to come to a theory of consciousness that, while acknowledging that our brains are elaborate assemblies of myriad processes, explains how it is that we feel so unified.

Recommended Reading

Driver, J., & Mattingley, J.B. (1998). Parietal neglect and visual awareness. *Nature Neuroscience, 1,* 17–22.
Gazzaniga, M.S. (2000). (See References)
Savoy, R.L. (2001). History and future directions of human brain mapping and functional neuroimaging. *Acta Psychologica, 107,* 9–42.

Acknowledgments—Preparation of this article was supported by Grant NS31443 from the National Institutes of Health. We are grateful to Margaret Funnell, Paul Corballis, and Michael Corballis for interesting discussions on the topics covered here.

References

Aglioti, S., DeSouza, J.F., & Goodale, M.A. (1995). Size-contrast illusions deceive the eye but not the hand. *Current Biology, 5,* 679–685.
Cooney, J.W., & Gazzaniga, M. (2003). Neurological disorders and the structure of human consciousness. *Trends in Cognitive Sciences, 7,* 161–165.
Dehaene, S., & Naccache, L. (2001). Towards a cognitive neuroscience of consciousness: Basic evidence and a workspace framework. *Cognition, 79,* 1–37.
Dehaene, S., Naccache, L., Cohen, L., Bihan, D.L., Mangin, J.F., Poline, J.B., & Riviere, D. (2001). Cerebral mechanisms of word masking and unconscious repetition priming. *Nature Neuroscience, 4,* 752–758.
Dehaene, S., Naccache, L., Le Clec, H.G., Koechlin, E., Mueller, M., Dehaene-Lambertz, G., van de Moortele, P.F., & Le Bihan, D. (1998). Imaging unconscious semantic priming. *Nature, 395,* 597–600.
Gazzaniga, M.S. (2000). Cerebral specialization and interhemispheric communication: Does the corpus callosum enable the human condition? *Brain, 123,* 1293–1326.
Horwitz, B., Tagamets, M.-A., & McIntosh, A.R. (1999). Neural modeling, functional brain imaging, and cognition. *Trends in Cognitive Sciences, 3,* 91–98.
Le Bihan, D., Mangin, J.F., Poupon, C., Clark, C.A., Pappata, S., Molko, N., & Chabriat, H. (2001). Diffusion tensor imaging: Concepts and applications. *Journal of Magnetic Resonance Imaging, 13,* 534–546.

Moutoussis, K., & Zeki, S. (2002). The relationship between cortical activation and perception investigated with invisible stimuli. *Proceedings of the National Academy of Sciences, USA, 99,* 9527–9532.

Rees, G., Kreiman, G., & Koch, C. (2002). Neural correlates of consciousness in humans. *Nature Reviews Neuroscience, 3,* 261–270.

Tong, F., Nakayama, K., Vaughan, J.T., & Kanwisher, N. (1998). Binocular rivalry and visual awareness in human extrastriate cortex. *Neuron, 21,* 753–759.

Wolford, G., Miller, M.B., & Gazzaniga, M. (2000). The left hemisphere's role in hypothesis formation. *Journal of Neuroscience, 20,* RC64.

Zeki, S. (2003). The disunity of consciousness. *Trends in Cognitive Sciences, 7,* 214–218.

Critical Thinking Questions

1. Why do you think the human brain is set up to generate hypotheses about the world? Why don't we simply accept the world as it is, and say "I don't know" when there are actions we don't understand?

2. The left hemisphere seems to play a significant role in hypothesis generation. It is tempting to propose that this laterality is related to the left-hemisphere dominance for language. Do you think that babies who don't yet have a command of language also generate hypotheses? Do you think that animals do?

3. What does Roser and Gazzaniga's theory say about the anatomic location of consciousness?

The Mentality of Apes Revisited

Daniel J. Povinelli[1] and Jesse M. Bering

Cognitive Evolution Group, University of Louisiana, Lafayette, Louisiana (D.J.P.), and Department of Psychology, Florida Atlantic University, Boca Raton, Florida (J.M.B.)

Abstract

Although early comparative psychology was seriously marred by claims of our species' supremacy, the residual backlash against these archaic evolutionary views is still being felt, even though our understanding of evolutionary biology is now sufficiently advanced to grapple with possible cognitive specializations that our species does not share with closely related species. The overzealous efforts to dismantle arguments of human uniqueness have only served to show that most comparative psychologists working with apes have yet to set aside the antiquated evolutionary "ladder." Instead, they have only attempted to pull chimpanzees up to the ladder's highest imaginary rung—or perhaps, to pull humans down to an equally imaginary rung at the height of the apes. A true comparative science of animal minds, however, will recognize the complex diversity of the animal kingdom, and will thus view *Homo sapiens* as one more species with a unique set of adaptive skills crying out to be identified and understood.

Keywords

chimpanzees; cognitive evolution; theory of mind; comparative psychology

Five to seven million years ago, a small lineage of anthropoid apes came down from the trees. Within a couple of million years, descendants of this lineage had evolved a new form of locomotion (striding bipedalism), and had resculpted their pelvic girdle, head, hands, and feet. They tripled the size of their brain, and even appeared to have reorganized some of the most fundamental systems within that brain (see Preuss & Coleman, in press). In a world already teeming with biological diversity, the human lineage made its debut.

With the appearance of our species came the ability to ponder those origins, and to pose such questions as, what does it mean to be human? and, more central to this essay, what psychological characteristics appear to be uniquely human? Such questions have challenged generations of inquisitive minds, all the while fueling controversy and divisiveness. Typically, the answers to such questions depended on the profession of the individuals being asked: To the theologian, the uniquely human endowment was the possession of a soul; to the psychologist, it was language; to the anthropologist, it was culture.

DEMOLISHING HUMAN UNIQUENESS

Alas, enter the first comparative psychologist, Darwin, who, running against centuries of religious and philosophical dogma, strategically announced that there is no characteristic truly unique to humans. "There can be no doubt," wrote Darwin in 1871, "that the difference between the mind of the lowest man and that of the highest animal is immense. Nevertheless the difference, great as it is, certainly is one of degree and not of kind" (Darwin, 1871/1982, p. 445). How could Darwin be so sure? To him and his followers, the answer was simple: Just observe other species' natural, spontaneous behaviors, and then use introspection to infer the underlying causes of these behaviors. Although this may seem like a sensible enough approach, consider the full implications of this method: It makes the human mind the standard against which all other minds are judged, installing our mental processes—and only ours—into the minds of other species. Even now, as data calling for a radical departure from this canonical view continue to mount, this most anthropomorphic (and ultimately un-evolutionary) of assumptions continues to live on in the field of comparative psychology.

In no case is this truer than in research into the mental abilities of chimpanzees and other great apes. Here, the classic argument by analogy enjoys the protection of the suspect notion of "evolutionary plausibility." Researchers regularly assert that the parsimonious explanation of behavioral similarity between humans and chimpanzees is the operation of equally similar psychological systems. Against this theoretical backdrop, Savage-Rumbaugh, mentor of the bonobo chimpanzee Kanzi, writes a book whose subtitle proclaims that her chimpanzee is "the ape at the brink of the human mind" (Savage-Rumbaugh & Lewin, 1994). In a recent article, Suddendorf and Whiten (2001) conclude that "the gap between human and animal mind has been narrowed" (p. 644). And de Waal's (1999) take on the same trend is that the chimpanzee is "inching closer to humanity" (p. 635).

For some researchers, continuity extends to identity. Savage-Rumbaugh, for example, declares that she has met the mind of another species and discovered that it is human: "I found out that it was the same as ours," she concludes. "I found out that 'it' was me!" (as quoted by Dreifus, 1999, p. 54). More typically, though, chimpanzees are caricatured as watered-down human beings. Echoing Darwin, Fouts (1997) sees any attempt to demonstrate differences between closely related species as symptomatic of "Cartesian delusions," and proclaims, "The cognitive and emotional lives of animals differ only by degree, from the fishes to the birds to monkeys to humans" (p. 372). Likewise, Goodall (1990) writes of a "succession of experiments that, taken together, clearly prove that many intellectual abilities that had been thought unique to humans were actually present, though in a less highly developed form, in other, non-human beings" (p. 18).

All of this adds up to an agenda for psychological research with chimpanzees: "Just how human are chimpanzees?" We suggest, however, that the obsession with establishing psychological continuity between humans and other apes has cast this area of comparative psychology into a great freeze. It has contributed to marginalizing the discipline's mission by reducing it to a series of demonstrations in which one humanlike ability after another is revealed in non-human animals. It is an objective anchored to the mistaken idea that evolution proceeds linearly and that apes are thus playing catch-up to the human intellect. This objective, however, is fundamentally at odds with the central theme of modern biology: Evolution is real, and it produces diversity.

Indeed, differences are seen as somehow obscuring the true evolutionary relationships among living species: "Researchers are regularly finding heretofore unexpected realms and degrees of similarity," noted Russon and Bard (1996), "and these similarities are particularly useful for evolutionary reconstructions" (p. 14). Not only is this point of view 180° out of phase with modern cladistic approaches to evolutionary reconstruction, but if the dramatic resculpting of the human body and brain that occurred over the past 4 million years or so involved the evolution of some qualitatively new cognitive systems, then this insistence on focusing on similarities will leave comparative psychologists unable to investigate hallmarks of their own species—or chimpanzees, for that matter. It is an agenda that does justice to no one.

AN ALTERNATIVE FRAMEWORK: THE REINTERPRETATION HYPOTHESIS

Perhaps the greatest obstacle to overcoming Darwin's a priori straitjacket of unbroken psychological continuity has been the difficulty of imagining an alternative. After all, if there really were a viable alternative, surely whatever intuitive anthropomorphic biases researchers have could be overcome—much in the way that Newtonian mechanics overcame tenets of Aristotelian physics. But here the challenge may be more substantial, because in this case, the very system that comparative psychologists seek to investigate is the one producing the illusion, compelling these researchers to recreate the psychology of other species in their own image.

In recent publications, we have suggested an alternative to the continuity paradigm, an alternative that we initially applied to the evolution of a *theory of mind*—the ability to reason about mental states in the self and others (for an elaboration of this model, see Povinelli, 2000, chap. 2). Our alternative posits that for dozens of millions of years, the primate order produced numerous social species, each one inheriting a core stock of mammalian social behaviors and then tweaking these behaviors to cope with the peculiar demands of its own circumstances. Although there is debate as to the key factors in this process, there can be no doubt that natural selection for social living was intense during the radiation of the primates, and with this drive to sociality came selection for social behavior to both exploit and cope with group living.

However, rather than positing sociality as the prime mover for the evolution of psychological systems for representing other minds, our alternative holds that the vast array of spontaneous behaviors that humans share with chimpanzees, including deception, gaze following, holding a grudge, tool use, reconciliation, and organized hunting, emerged and were in full operation long before additional systems evolved to interpret these behaviors in terms of underlying mental states. Instead, these behaviors were generated through existing psychological processes, motivated by physiological, attentional, perceptual, and affective mechanisms—mechanisms that continue to guide an enormously complicated assemblage of primate (including human) behavior.

The final part of our claim is that it was not until a particular lineage appeared—the human one—that a new representational system was stamped into the old, so that the observable world, and those things transpiring in it, were "reinterpreted" with hidden meaning, allowing humans to reflect on unobservable causes, such as mental states. Without discarding the ancestral mechanisms it built on, this novel causal explanatory system then generated its own subassemblage of behaviors (e.g., progressive cultural transmission, religious rituals, and explicit pedagogy), all of which hinge on the ability to represent a social and physical world governed by abstract causal forces. Because it envisions that a hallmark of human mental evolution was installing a system or series of systems that interprets ancestral behaviors in new ways, we have labeled our model the *reinterpretation hypothesis.*

The reinterpretation hypothesis calls into question the fairly common assertion that if similar behaviors "are the product of a common history, then it is likely that the underlying psychological processes responsible for the overt behavior are similar, too" (Suddendorf & Whiten, 2001, p. 643). To the contrary, the reinterpretation hypothesis makes clear that no a priori argument from similarities in spontaneous behavior will suffice. Although the recently shared ancestral heritage of chimpanzees and humans virtually guarantees behavioral homologies, the totality of the representational software that rides alongside (or in some cases causes) similar behaviors in the two species is not necessarily the same. With this alternative framework—a theoretical approach that embraces both similarity and difference—it is now possible to return to the investigation of human uniqueness in the way that a biologist would address an investigation of the specializations of any species—open to wherever the empirical facts seem to lead.

DIFFERENCES IN THE MENTALITIES OF APES AND HUMANS

Evidence that human evolution was marked by the emergence of novel mental abilities is beginning to accumulate. There is increasing evidence that, at some point after hominids separated from the line leading to the modern African apes, humans developed a unique capacity to mentally represent a world of hidden causal forces, including mental states. Consider the following:

- Although chimpanzees respond to eye gaze by following the visual trajectory of other individuals, even around barriers, they do not appear to grasp the fact that others' visual behaviors are accompanied by the psychological experience of "seeing" (Fig. 1; see the review by Povinelli, 2000). Recent widely reported claims that chimpanzees may attribute the mental states of seeing and knowing to other chimpanzees (e.g., Hare, Call, Agnetta, & Tomasello, 2000) have not been supported by attempts at replication (Karin-D'Arcy & Povinelli, 2002).
- In carefully controlled studies, great apes have failed to appreciate the underlying referential nature of intentional communication (e.g., Povinelli, Reaux, Bierschwale, Allain, & Simon, 1997). Whether the communicative attempt comes in the form of an extended index finger (i.e., pointing) or an iconic device (e.g., a replica of a box containing a food reward), chimpanzees do not seem to understand that the communicative behaviors of other individuals are driven by a desire to share information.
- A number of nonverbal methodological attempts to parallel research on children's understanding of the mental state of belief have shown that chimpanzees do not distinguish between individuals who are ignorant versus knowledgeable. For example, they respond to observers who have

Fig. 1. Gaze following in a 6-year-old chimpanzee. The chimpanzee makes eye contact with a human caretaker (top left), who turns and looks above the chimpanzee (top right); the chimpanzee then follows the caretaker's gaze (bottom). Although it is tempting to assume in such circumstances that the chimpanzee's interpretation of the situation is similar to our own (i.e., that the animal can infer that the person has "seen" something), the reinterpretation hypothesis (see the text for details) makes clear that in this and many other cases, high-level human cognitive systems may have been grafted into a suite of ancient systems that modulate quite ancient behavioral patterns. Only programmatic experimental approaches will suffice to determine the presence or absence of such systems in other species.

witnessed the hiding of food no differently than they do to those who are oblivious to its actual location—choosing at random who they would like to retrieve the reward for them (Call & Tomasello, 1999).

- "Expert" chimpanzees that have been previously trained on how to perform a cooperative task (e.g., jointly pulling a heavy box by two separate ropes) with a human partner do not spontaneously guide naive chimpanzee partners on relevant dimensions of the task; these experts essentially ignore the fact that their new partners lack the requisite knowledge for success, and fail to instruct them through teaching behaviors (e.g., showing, touching, pointing; see Povinelli & O'Neill, 2000).

Interestingly, this pattern may translate to the nonsocial domain as well. Recently, we completed a project that was designed to map our chimpanzees' understanding of unobservable forces in the physical world (see Povinelli, 2000). The initial round of nearly 30 studies, conducted over a 5-year period, was centered on the widely celebrated ability of chimpanzees to make and use simple tools. However, we were interested not in the level of complexity that such tool use and construction might achieve, but rather whether chimpanzees reason about the hidden properties and functions of tools. In particular, we asked whether chimpanzees' understanding of the physical world is mediated by concepts robustly in place by about 3 years of age in human children—things such as gravity, force, shape, physical connection, and mass.

The results converged upon a finding strikingly analogous to what we have described about chimpanzees' understanding of the social world: Although they are very good at understanding and learning about the observable properties of objects, they appear to have little or no understanding that these observable regularities can be accounted for, or explained, in terms of unobservable causal forces. In short, we have speculated that for every unobservable causal concept that humans may form, the chimpanzee will rely exclusively upon an analogue concept, constructed from the perceptual invariants that are readily detectable by the sensory systems.

CONCLUDING REMARKS

If there is one thing that our species is obviously not very good at, it is imagining ways of understanding the world that differ markedly from our own. The popular press overflows with stories of empathic gorillas rescuing young children, cats scaling burning buildings to bring their kittens to safety, and dogs who think they are human. Whatever the behavioral facts of these cases, one thing is certain: We humans will automatically interpret the psychological facts from the perspective of our evolved, but peculiarly distorted, ways of understanding the world.

Research with chimpanzees and other great apes remains marginalized within the cognitive and biological sciences largely because the field has failed to come to grips with the most important tenet of modern biology: Evolution is real, and it produces diversity. Comparative psychology was founded upon the notion that organisms could be arranged into an evolutionary scale or ladder in which the mental operations of living species were said to differ in degree, not kind. Most psychologists who work with chimpanzees continue to espouse this view. Because of the importance of comparing and contrasting the psychological systems of our own species with those of our nearest living relatives, there is an overwhelming need to train the next generation of these psychologists in the intricacies of the evolutionary biology of the organisms they study. Organismal biology can provide the theoretical motivation to look for, and thus celebrate, the marvelous psychological differences that exist among species.

Recommended Reading

Bering, J.M. (in press). The existential theory of mind. *Review of General Psychology.*
Hodos, W., & Campbell, C.B.G. (1969). *Scala naturae:* Why there is no theory in comparative psychology. *Psychological Review, 76,* 337–350.

Kohler, W. (1927). *The mentality of apes* (2nd ed.). New York: Vintage Books.

Povinelli, D.J. (2000). (See References)

Tomasello, M. (1999). *The cultural origins of human cognition.* Cambridge, MA: Harvard University Press.

Acknowledgments—This work was supported by a Centennial Fellowship Award to D.J.P. from the James S. McDonnell Foundation.

Note

1. Address correspondence to Daniel Povinelli, Cognitive Evolution Group, University of Louisiana, 4401 W. Admiral Doyle Dr., New Iberia, LA 70560; e-mail: ceg@louisiana.edu.

References

Call, J., & Tomasello, M. (1999). A nonverbal theory of mind test: The performance of children and apes. *Child Development, 70,* 381–395.

Darwin, C. (1982). *The descent of man.* New York: Modern Library. (Original work published 1871)

de Waal, F.B.M. (1999). Animal behaviour: Cultural primatology comes of age. *Nature, 399,* 635.

Dreifus, C. (1999, August-September). Going ape. *Ms., 9*(5), 48–54.

Fouts, R. (1997). *Next of kin.* New York: William Morrow and Co.

Goodall, J. (1990). *Through a window.* Boston: Houghton Mifflin.

Hare, B., Call, J., Agnetta, B., & Tomasello, M. (2000). Chimpanzees know what conspecifics do and do not see. *Animal Behaviour, 59,* 771–785.

Karin-D'Arcy, R., & Povinelli, D.J. (2002). *Do chimpanzees know what each other see? A closer look.* Manuscript submitted for publication.

Povinelli, D.J. (2000). *Folk physics for apes.* New York: Oxford University Press.

Povinelli, D.J., & O'Neill, D.K. (2000). Do chimpanzees use their gestures to instruct each other? In S. Baron-Cohen, H. Tager-Flusberg, & D.J. Cohen (Eds.), *Understanding other minds* (pp. 459–487). Oxford, England: Oxford University Press.

Povinelli, D.J., Reaux, J.E., Bierschwale, D.T., Allain, A.D., & Simon, B.B. (1997). Exploitation of pointing as a referential gesture in young children, but not adolescent chimpanzees. *Cognitive Development, 12,* 423–461.

Preuss, T.M., & Coleman, G.Q. (in press). Human-specific organization of primary visual cortex: Alternating compartments of dense Cat-301 and calbindin immuno-reactivity in layer 4A. *Cerebral Cortex.*

Russon, A., & Bard, K. (1996). Exploring the minds of the great apes: Issues and controversies. In A.E. Russon, K.A. Bard, & S.T. Parker (Eds.), *Reaching into thought* (pp. 1–20). Cambridge, England: Cambridge University Press.

Savage-Rumbaugh, S., & Lewin, R. (1994). *Kanzi: The ape at the brink of the human mind.* New York: John Wiley & Sons.

Suddendorf, T., & Whiten, A. (2001). Mental evolution and development: Evidence for secondary representation in children, great apes and other animals. *Psychological Bulletin, 127,* 629–650.

Critical Thinking Questions

1. If you believe that animals and humans are on a continuum of cognitive ability, with different animals varying in how far they are below humans on the "ladder," then it is easy to understand why it would be interesting to study the cognitive ability of animals. Povinelli and Bering specifically say that it is wrongheaded to think of such a continuum. If so, why should we study the cognitive abilities of animals?

2. One reason it is so tempting to think of such a continuum is that animals are so obviously worse than humans in many cognitive abilities that we deem important, for example communication and complex problem-solving. If Povinelli and Bering are correct in that animals are not less cognitively able than we are, but rather just have different cognitive abilities than we do, then some animals should have cognitive abilities that far outstrip ours. Can you think of some examples? Why might examples be difficult to think of, given the world in which we live?

3. Domestic dogs are better than great apes in understanding human communicative signals indicating the location of hidden food (Hare et al., 2002), which is one measure of theory-of-mind. It is thought that this ability is the product of domestication and selective breeding. (Wolves don't show this ability.) Would Povinelli and Bering say that this result supports their thesis? Why or why not?

Hare, B., Brown, M., Williamson, C., & Tomasello, M. (2002). The domestication of social cognition in dogs. *Science, 298,* 1634-1636.

The Projective Way of Knowing:
A Useful Heuristic That Sometimes Misleads

Raymond S. Nickerson[1]

Psychology Department, Tufts University, Medford, Massachusetts

Abstract

For many purposes, people need a reasonably good idea of what other people know. This article presents an argument and considers evidence that people use their own knowledge as a basis for developing models of what specific other people know—in particular, that they tend to assume that other people know what they know. This is a generally useful heuristic, but the assumption is often made uncritically, with the consequence that people end up assuming that others have knowledge that they do not have.

Keywords

knowledge; projection; false consensus; expertise; egocentrism

People's behavior is influenced in many ways by what they know about what other people know. Effective conversation, for example, depends not only on shared knowledge between participants, but also on each person having knowledge, or making reasonably accurate assumptions, about what the other knows.

BUILDING A CONCEPTUAL MODEL OF WHAT ANOTHER PERSON KNOWS

Over time, one can develop a detailed conceptual model of what a specific other person (spouse, sibling, friend, associate) knows, fine-tuning and updating the model with information gleaned from frequent interactions. But what does one use for a model of what a stranger knows? How does one cope with the task of communicating with a collection of people—an audience to whom one has to give a talk, or the readership of a newspaper for which one is writing an article—when one has few specifics about its composition? I assume that the basis for the construction of a default model of what a random other person knows is one's model of what one knows oneself.

What an individual knows changes over time. It follows that if a model of another person's knowledge is to be and remain functionally accurate, it too must change on a continuing basis. Several researchers have noted that refining one's model of another person's knowledge dynamically is important if communication is to be successful.

These ideas are incorporated in Figure 1, a conceptualization of how an individual develops a model of another person's knowledge (from Nickerson, 1999). According to this conceptualization, one's model of one's own knowledge serves as a default model of what a random other person knows. This default model is transformed, as individuating information is acquired, into models of specific other individuals. The models of specific others are continually refined and updated as

177

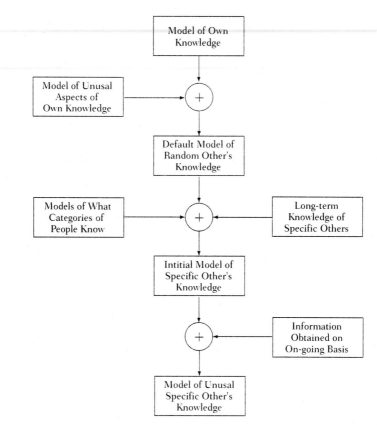

Fig. 1. Illustration of how an individual develops a working model of the knowledge another person has (from Nickerson, 1999).

new information that is relevant to them is acquired. This article focuses on the idea that what serves as the point of departure for developing a model of another person's knowledge is what one knows, or thinks one knows, oneself.

SELF AS A SOURCE OF HYPOTHESES ABOUT WHAT OTHERS KNOW

The notion that a basic source of assumptions or hypotheses regarding what a random other person knows is what one knows oneself has some currency among psychologists. It is closely related to the *simulation* view of how one individual understands another, according to which one imagines oneself in the other's place and discovers how one would think or feel in that situation (Gordon, 1986).

The Projective Way of Knowing

The idea that we understand others by assuming that they are like ourselves is intuitively compelling, and there is much evidence to support the notion that people employ this "projective mode of knowing," as O'Mahony (1984, p. 58)

178

has called it. People who engage in a particular behavior estimate that behavior to be more prevalent than do people who do not engage in that behavior. When attempting to assess the attitudes of specified groups, people tend to project their own attitudes onto those groups. People who are experiencing an experimentally induced emotional state are more likely to project that state to others than are people who are not experiencing it. People generally see their own attitudes and behavior as rational or normative, and they see attitudes and behavior that differ greatly from their own as irrational or deviant. In the political arena, extremists on both ends of the left-right continuum tend to doubt the rationality of those on the opposite end. People sometimes take their own behavior as the norm even in the light of sample-based information to the contrary.

Advantages of Using the Self as a Basis for Developing a Model

Our own knowledge of how we would behave or react in specific situations can be a useful basis, arguably the best basis we have, for anticipating how other people will behave or react in those situations (Hoch, 1987). Projecting our own feelings and reactions to others works because, in fact, people do react similarly in specific situations. The idea that we do well to assume that others are like ourselves is captured in the *principle of humanity*, according to which when trying to understand what someone has said, especially something ambiguous, we should impute to the speaker beliefs and desires similar to our own.

Using our own knowledge as a default model of what a random other person knows simplifies life. If we could not assume, in the absence of contrary evidence, that other people are much like ourselves, the problem of communicating effectively would be overwhelmingly difficult (Davidson, 1982).

THE RISK OF OVERIMPUTATION OF OUR OWN KNOWLEDGE

Although, in the absence of more direct information, one's model of one's own knowledge may be as good a basis as there is for a default assumption about what a random other person knows, evidence suggests that the tendency to impute our own knowledge to others often leads us to assume that others have knowledge they do not actually have, and this can impede communication and mutual understanding in various ways.

Failure to make sufficient allowance for the difference between one's own subjective experiences or perspectives and those of one's hearers or readers has been noted as a source of difficulties in communication in both spoken and written form. Teaching can be adversely affected if teachers underestimate the difference between their own knowledge and that of their students; products of technology may be designed suboptimally if designers underestimate how much difficulty people other than themselves will have in learning to use them. Piaget (1962) remarked on the difficulty that beginning instructors have in placing themselves in the shoes of students who do not know what they themselves do about a course's subject matter, and surmised that they are likely to give incomprehensible lectures for a while as a consequence. Flavell (1977) similarly pointed out that one's own viewpoint can work as an impediment to attaining an accurate appreciation of the viewpoint of another person, and that it may be especially difficult for one to appreciate fully the ignorance of another person with respect to something one understands very well oneself.

The False-Consensus Effect

The *false-consensus effect* refers to a tendency to see oneself as more representative of other people than one really is. The results of numerous studies suggest that the tendency is very common and that it manifests itself in a variety of ways. People are likely to overestimate the amount of general consensus on beliefs and opinions they themselves hold, and to underestimate the degree of agreement on beliefs and opinions that differ from theirs. The effect is illustrated by the finding that U.S. voters typically overestimate the popularity of their favored candidate in a presidential election (Brown, 1982), as well as the extent to which the positions of favored candidates correspond to their own (Page & Jones, 1979).

The Curse of Expertise

People who are experts in specific areas and who recognize themselves as such must realize, almost by definition, that they know more than most other people with respect to their areas of expertise. Nevertheless, the results of several studies suggest that although experts assume that others know less than they about their areas of expertise, they may still overestimate what others know.

The point is illustrated by a study by Hinds (1999), who found that experts in performing a task were more likely than people with only an intermediate level of expertise to underestimate the time novices would take to complete the task. Experts also proved to be resistant to debiasing techniques intended to reduce the tendency to underestimate how difficult novices would find a task to be. In laboratory studies, participants who have been given privileged information for purposes of an experiment may behave as though other participants also have that information even when, if asked, they acknowledge that the other participants do not have the information.

Egocentric Bias in Imputing General Knowledge

The results of many studies suggest that people's estimates of what general knowledge other people have tend to be biased in the direction of the knowledge they themselves have or think they have. When college students were asked to answer general-knowledge questions and to estimate, for each question, the percentage of other college students who would be able to answer that question correctly, they gave higher estimates for questions they thought they knew the answers to (as indicated by confidence ratings), even when their own answers were wrong, than for questions they knew they did not know the answers to, and they were more likely to overestimate the commonality of knowledge if they themselves had it than if they did not.

When students living in New York City rated their familiarity with each of 22 landmarks in the city and estimated the proportions of other city residents who would be able to identify them, students who were highly confident of being able to identify specific landmarks judged those landmarks to be more familiar to others than did students who were not very confident of their own ability to identify them. Students who could identify pictured public figures by name rated the individuals as more recognizable than did students who could not identify them by name. Students who could identify an everyday object estimated the proportion of peers who would be able to identify that object to be higher than did people who could not identify it.

When instructors attempted to answer quiz questions as they expected their students would answer them, they provided about twice as many correct answers as did their students on average. Readers of the account of a conversation attributed to the listener the same understanding of the speaker's utterance as their (the readers') own, even when the utterance was ambiguous and the readers knew that the listener did not have the disambiguating information they had. When observers judged the likelihood that listeners would believe a message that contradicted the listeners' prior belief about a situation, they judged the likelihood to be higher if they (the observers) knew the message to be true than if they knew the message to be false, even though they were aware that the listeners could not have the basis they (the observers) had for judging the message to be true.[2]

A particularly striking example of overimputing one's own knowledge to others comes from an experiment by Newton (1990) in which some participants tapped the rhythms of well-known songs and others attempted to identify the songs on the basis of the tapped rhythms. Tappers estimated the likelihood that listeners would be able to identify the songs to be about .5; the actual probability of correct identification was about .025. Apparently the tappers, who could imagine a musical rendition of a song when they tapped its rhythm, found it overly easy to project their own subjective experience to the listeners, who did not share it.

Illusion of Simplicity

The *illusion of simplicity* refers to the mistaken impression that something is simple just because one is familiar with it. It is illustrated by the findings that people are likely to judge anagrams to be easier to solve if they have been shown the solutions than if they have not and that they are likely to judge sentences to be appropriate for a lower-grade reading level if they have read them before than if they have not (Kelley, 1999). It is a short step from perceiving something to be simpler than it is, only because one is familiar with it, to assuming that someone else, who is not familiar with it, will perceive it the same way.

ANCHORING AND INADEQUATE ADJUSTMENT

This conceptualization of how we build models of what others know can be seen as a case of the general reasoning heuristic of *anchoring and adjustment* (Tversky & Kahneman, 1974), according to which people make judgments by starting with an "anchor" as a point of departure and then make adjustments to it. In this case, the anchor for one's default model of what someone else knows is one's model of what one knows oneself.

Many of the experimental results noted in this article support the idea that, in serving as the anchor for one's model of another person's knowledge, one's model of one's own knowledge is adjusted to take into account differentiating information either about one's own knowledge or about the other person's knowledge. However, as in other documented instances of anchoring and adjustment, the adjustment is often not as great as it should be, and one ends up assuming that another person has knowledge that he or she does not have.

CONCLUDING COMMENTS

In this article, I have emphasized the prevalence of the tendency to overimpute one's own knowledge to other people and the fact that this can be problematic in several respects. Dawes (1989) has made the point that it is possible to err in the other direction as well, and has argued that an uncritical assumption of dissimilarity between oneself and others can also have undesirable consequences. This seems an important caution to bear in mind in interpreting the results noted here. Presumably, relatively accurate models of what others know are more useful than models that are biased either toward or away from what one knows oneself; however, on balance, the literature suggests that biasing one's model of another person's knowledge in the direction of one's own knowledge is a more common problem than biasing it in the opposite direction.

What can be done to improve our conceptions of what specific other people know? I have suggested several possibilities elsewhere (Nickerson, 1999). Here, I mention only the belief that this problem, like many others relating to cognitive or judgmental biases, stems, at least in part, from a failure to be very reflective about assumptions we make—from failing to give much attention to alternative assumptions that could be made. If we generally tend to assume that a random other person knows a fact that we know ourselves, and if we give insufficient consideration to reasons why the other person might not know that fact, we are likely to overimpute our own knowledge to others as a rule.

In many cases, failure to be more critical of our own assumptions may be defended on the grounds that, although the conclusions that we settle on may not be optimal, they are usually close enough for practical purposes and finding better ones would not be worth the effort. However, if judgments of a particular type are relatively consistently biased in a specified way, as judgments of what others know appear to be, search for effective debiasing techniques seems warranted. Simple awareness of a tendency to overimpute one's own knowledge to others may be helpful, but probably not fully corrective. How best to teach people to make more accurate estimates of what other people know, and to counteract the tendency to overimpute their own knowledge to others, remains a challenge to research.

Recommended Reading

Camerer, C., Loewenstein, G., & Weber, M. (1989). The curse of knowledge in economic settings: An experimental analysis. *Journal of Political Economy*, 97, 1232–1254.

Cronbach, L. (1955). Processes affecting scores on "understanding others" and "assumed similarity." *Psychological Bulletin*, 52, 177–193.

Krauss, R., & Fussell, S. (1991). Perspective-taking in communication: Representations of others' knowledge in reference. *Social Cognition*, 9, 2–24.

Mullen, B. (1983). Egocentric bias in estimates of consensus. *Journal of Social Psychology*, 121, 31–38.

Srull, T., & Gaelick, L. (1983). General principles and individual differences in the self as a habitual reference point: An examination of self-other judgments of similarity. *Social Cognition*, 2, 108–121.

Notes

1. Address correspondence to Raymond S. Nickerson, 5 Gleason Rd., Bedford, MA 01730; e-mail: r.nickerson@tufts.edu.

2. References for all the studies alluded to in the preceding three paragraphs are listed in Nickerson (1999).

References

Brown, C. (1982). A false consensus bias in 1980 presidential preferences. *Journal of Social Psychology*, 118, 137–138.

Davidson, D. (1982). Paradoxes of irrationality. In R. Wollheim & J. Hopkins (Eds.), *Philosophical essays on Freud* (pp. 289–305). Cambridge, England: Cambridge University Press.

Dawes, R. (1989). Statistical criteria for establishing a truly false consensus effect. *Journal of Experimental Social Psychology*, 25, 1–17.

Flavell, J. (1977). *Cognitive development*. Englewood Cliffs, NJ: Prentice-Hall.

Gordon, R. (1986). Folk psychology as simulation. *Mind and Language*, 1, 158–171.

Hinds, P. (1999). The curse of expertise: The effects of expertise and debiasing methods on predictions of novice performance. *Journal of Experimental Psychology: Applied*, 5, 205–221.

Hoch, S. (1987). Perceived consensus and predictive accuracy: The pros and cons of projection. *Journal of Personality and Social Psychology*, 53, 221–234.

Kelley, C. (1999). Subjective experience as a basis of "objective" judgments: Effects of past experience on judgments of difficulty. In D. Gopher & A. Koriat (Eds.), *Attention and performance XVII* (pp. 515–536). Cambridge, MA: MIT Press.

Newton, L. (1990). *Overconfidence in the communication of intent: Heard and unheard melodies*. Unpublished doctoral dissertation, Stanford University, Stanford, CA.

Nickerson, R. (1999). How we know—and sometimes misjudge—what others know: Imputing one's own knowledge to others. *Psychological Bulletin*, 125, 737–759.

O'Mahony, J. (1984). Knowing others through the self—Influence of self-perception on perception of others: A review. *Current Psychological Research and Reviews*, 3(4), 48–62.

Page, B., & Jones, C. (1979). Reciprocal effects of policy preferences, party loyalties and the vote. *American Political Science Review*, 73, 1071–1089.

Piaget, J. (1962). Comments [Addendum to L. Vygotsky, *Thought and language* (E. Hanfmann & G. Vakar, Ed. & Trans.)]. Cambridge, MA: MIT Press.

Tversky, A., & Kahneman, D. (1974). Judgment under uncertainty: Heuristics and biases. *Science*, 185, 1124–1131.

Critical Thinking Questions

1. Nickerson suggests that we have a default assumption about what other people know. Wouldn't it make sense in many situations to assume that they know little or nothing, so as to ensure that we don't miscommunicate? What would communication be like if you assumed that the person with whom you spoke had little or no knowledge? In what situations is it *strategically* important to make assumptions about what others know: business? Athletics? Politics?

2. Nickerson assumes that we hypothesize that a "random other person" knows what we know. Do you think that most people would assume that they are like a random other person or that they are a bit smarter than a random other person? If they assume they are a bit smarter than a random other person, why aren't they reducing their estimate of what others know to account for that fact?

3. The author describes a number of disadvantages to assuming that others know what we know; these problems occur because we believe that others have more knowledge than they actually do. What would be an alternative way of assessing how much knowledge people have? Suppose this alternate method led one to assume that people have *less* knowledge than they actually do—would that be better or worse?

THE FOUR ROADS TO HEAVEN

Also by Edwin Mullins

The Pilgrimage to Santiago
In Search of Cluny: God's Lost Empire
Avignon of the Popes
Roman Provence
The Camargue: Portrait of a Wilderness

The
Four Roads
To Heaven

France and the Santiago Pilgrimage

Edwin Mullins

Interlink Books

An imprint of Interlink Publishing Group, Inc.
Northampton, Massachusetts

For Jason

First published in 2018 by
Interlink Books
An imprint of Interlink Publishing Group, Inc.
46 Crosby Street, Northampton, Massachusetts 01060
www.interlinkbooks.com

Published simultaneously in the UK by Signal Books Limited

Library of Congress Cataloging-in-Publication Data

ISBN 978-1-62371-991-3

Printed and bound in the United States of America

1. Paris, No. 1 St. James' Street. The long walk begins. (Adam Woolfitt)

2. Santiago Matamoros— St. Jacques the Moor—slayer and liberator of Spain: an illustration from the *Codex Calixtinus* manuscript— *The Pilgrim's Guide.* (Wikimedia Commons)

3. The porch of the abbey church of Saint-Benoît on the River Loire. Pilgrims came here to revere the relics of St. Benedict, founder of the Benedictine Order, brought here from the abbey he founded, Monte Cassino, in southern Italy.

(Prost.photo/Wikimedia Commons)

4. The ubiquitous scallop-shell, symbol of the Santiago pilgrimage and of St. James, "the saint from the sea": widely carved on churches, on the doors of wayside inns, and worn or woven on pilgrims' clothing. (Adam Woolfitt)

5. Parthenay in south-west France approaching Bordeaux: the medieval bridge and Porte Saint-Jacques leading to the pilgrims' street. (Adam Woolfitt)

6. Musical accompaniment on the pilgrim church of Aulnay in western France. The theme from the Bible of the Old Men of the Apocalypse playing their instruments was a favorite subject carved on churches along the pilgrim roads. (Adam Woolfitt)

7. The village church of Rioux, one of many richly decorated churches in the Saintonge region of southwest France as the pilgrim road approached Bordeaux.
(Adam Woolfitt)

8. Pilgrims would scratch graffiti at the entrance to a hospice as they waited to be admitted at nightfall.
(Adam Woolfitt)

9. St. James represented as a pilgrim in the village chapel at Harambels, near the Pyrenees: one of the earliest pilgrim churches in France, mentioned in the twelfth-century *Pilgrim's Guide*.
(Adam Woolfitt)

10. Meeting point. A traditional Basque headstone now marks the place on a hillside near the Pyrenees where three of the four major pilgrim roads—from Paris, Vézelay and Le Puy—join to become one road leading to Spain. (Adam Woolfitt)

11. A masterpiece of medieval sculpture, the central figure of Christ above the entrance to the former abbey church of St. Mary Magdalene at Vézelay—starting point of the second road described in *The Pilgrim's Guide*. (Vassil/Wikimedia Commons)

12. Mont Saint-Michel, off the Normandy coast: among the most revered centers of pilgrimage in the Middle Ages. English pilgrims bound for Santiago would pause here to offer prayers of thanks for their safe crossing before heading south to join one of the main pilgrim roads.

(Adam Woolfitt)

13. An important diversion for pilgrims: the powerful Burgundian abbey of Cluny, prime sponsor of the Santiago pilgrimage, and probably responsible for *The Pilgrim's Guide.*

(Michal Osmenda/ Wikimedia Commons)

14. A major shrine on the road from Vézelay: the twelfth-century cathedral of Saint-Lazare at Autun. The sculpture throughout the cathedral is the work of one of the few medieval artists we know by name: Gislebertus. (Adam Woolfitt)

15. Autun rivaled Vézelay in popularity with pilgrims for supposedly possessing the relics of Lazarus, brother of Mary Magdalene and Christ's friend whom he brought back to life in Bethany. (Adam Woolfitt)

16. Starting point for the third road: Le Puy-en-Velay—city of rocky hills and pinnacles. To reach the church of Saint-Michel d'Aiguilhe pilgrims needed to climb 267 steps. (Adam Woolfitt)

17. Golden majesty. The statue reliquary of
Sainte-Foy, studded with precious stones:
together with other treasures displayed in
the abbey of Conques regarded as among the
wonders of the Christian world.

(Robert Harding/Alamy)

18. Masterpiece of medieval stone-carving:
a detail of the life-size figure of the prophet
Jeremiah on the central column of the main
entrance to the abbey church of Moissac.

(Adam Woolfitt)

19. Ancient stepping-stones led travelers across the River Bidouze in the Basque Country to where their road from Le Puy links up with the pilgrim roads from Paris and Vézelay.
(Adam Woolfitt)

20. A reminder to pilgrims as they passed by: a simple roadside cross in the Basque Country approaching the Pyrenees.
(Adam Woolfitt)

21. The city of Arles in Provence, starting point for the fourth road. Before setting off, pilgrims would gather for mass in the cathedral of Saint-Trophime, with its rank of carved images of the evangelists on either side of the west door.
(Adam Woolfitt)

22. A major diversion for pilgrims: the fortified church at Les-Saintes-Maries-de-la-Mer in the Camargue, a shrine to the legendary miraculous raft said to have borne Mary Magdalene from the Holy Land.
(Wolfgang Staudt/ Wikimedia Commons)

23. Toulouse: the abbey church of Saint-Sernin had a special appeal for pilgrims heading for the Pyrenees. Its treasures included the so-called Horn of Roland, supposedly sounded to alert the Emperor Charlemagne of the pending attack on his army at Roncesvalles.
(Didier Descouens/ Wikimedia Commons)

24. A familiar musical theme greeting pilgrims on the carved portal of the church of Sainte-Marie at Oloron, near the Pyrenees: the Old Men of the Apocalypse, illustrating the Vision of St. John.
(Adam Woolfitt)

25. Across the Pyrenees into Spanish Navarre: the octagonal chapel of the Knights Templar, Santa Maria de Eunate.
(Adam Woolfitt)

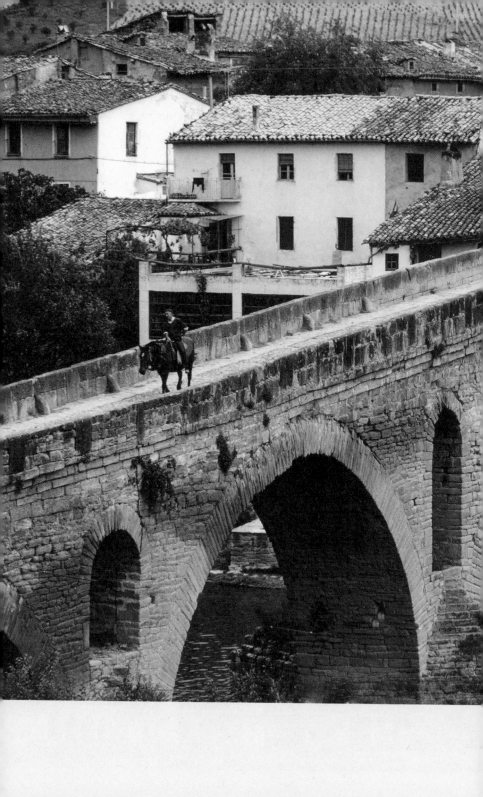

26. The bridge constructed specially for pilgrims at Puente la Reina ("the Queen's bridge"), the Spanish town where all four roads finally become one—the *camino*.

It was the construction of such bridges across northern Spain that made the Santiago pilgrimage possible.

(Adam Woolfitt)

27. At the junction of two rivers, the medieval bridge at Hospital de Órbigo—the longest, with twenty arches, in the 400-mile stretch of the Spanish camino.
(Adam Woolfitt)

28. Sermons in stone on the cathedral at León. Pilgrims were rarely allowed to forget the punishments awaiting sinners at the Last Judgment.
(Adam Woolfitt)

29. Inspired by the cathedrals of the Rhineland, and a major shrine for pilgrims heading for Santiago: the great cathedral at Burgos.
(Adam Woolfitt)

30. Triumphant journey's end: the cathedral of Santiago de Compostela overlooking Praza do Obradoiro. The final bravura touch of twin Baroque towers was added in the eighteenth century. (arousa/Shutterstock)

31. St. James looks down from his cathedral at dusk.

(Adam Woolfitt)

Contents

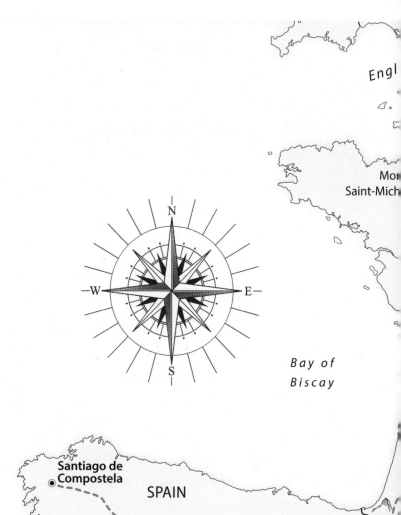

© S.Ballard (2017)

Preface

*"There are four roads leading to Santiago, which converge to form
a single road."*

T HESE ARE THE OPENING WORDS written almost a thousand years
ago in a manuscript which can lay claim to be the world's very
first travelers' guide.

The four roads are all in present-day France, and the manuscript
(in Latin) was written early in the twelfth century for the benefit of
Christian pilgrims preparing to travel to the tomb of the apostle St.
James at Santiago de Compostela in northwest Spain. The *Liber Sancti
Jacobi*, the Book of St. James, is generally known as *The Pilgrim's
Guide*, and was the work of a Benedictine monk—or possibly several
monks—at the instigation of the powerful Burgundian abbey of Cluny.
The abbey possessed strong vested interests in promoting the Santiago
pilgrimage as a vehicle for expanding the power and influence of the
Christian Church in Spain during the turbulent era of the Reconquest.

The Pilgrim's Guide originally existed in a number of copies
which were made available in the libraries of major monasteries
along the pilgrim routes to Santiago. Today only two copies survive,
one in the library of Santiago Cathedral, the other in the monastery
of Ripoll in northeast Spain. Only a tiny literate minority who could
understand and read Latin were able to take advantage of the *Guide's*
words of wisdom and information about where to stay and where to
avoid murderous boatmen and bad wine on their journey. Nonetheless
the message was handed down through the social ranks and widely
preached from the pulpit. And so popular did the pilgrimage become
that by the later Middle Ages as many as half a million travelers are

estimated to have undertaken the journey in a year.

Today even greater numbers, pilgrims and tourists alike, make the same trek across France and into Spain. The historical and cultural significance of the four roads to Santiago has meanwhile been recognized by UNESCO, which added this "serial site" to the World Heritage List in 1993.

The four routes described in *The Pilgrim's Guide* are the principal threads in a huge spider's-web of ancient roads laid across the French landscape. They were distinct from other roads because a high proportion of those using them shared a single objective—the shrine of St. James far away in Spain. They were not only pilgrims on foot: wealthy travelers made the journey on horseback, often accompanied by a small retinue. There would also be merchants, traders, craftsmen, masons, musicians and storytellers, adventurers, fugitives from the law or from their wives, as well as convicted criminals sentenced by magistrates to make the pilgrimage as an alternative to prison or a hefty fine. It would have been a lively mob on the roads to Spain.

These are roads I have traced and have come to know well over the course of many years. In this book I am the invisible traveler on a journey of yesterday and today. I shall be exploring the range of magnificent buildings and works of art which are the legacy of the pilgrimage, as well as trying to capture something of the spirit of hope and endeavor which drove our ancestors to make this same journey all those centuries ago.

*

Many years ago I wrote an account of my first travels along the pilgrim roads in France and Spain. The book was *The Pilgrimage to Santiago*, published in 1974 and reissued by Interlink Books in the millennium year 2000. This was a journey of discovery, and the experience of making that journey has remained vivid in my mind ever since. The first impact of the abbey church of Conques or the cathedral of Santiago can never weaken. And since that time I have seized the chance to make numerous shorter journeys in different parts of France, filling in pieces of the jigsaw I had previously missed. I have also scripted and presented two films on the Santiago pilgrimage for British and French television, which reacquainted me with many of my favorite places on the pilgrim roads. Besides, in recent decades much work has been done on the subject, and on areas closely related to it, including my own book on the history of Cluny, the great Burgundian abbey which was responsible to a large extent for making the pilgrimage possible by creating facilities and protection for medieval travelers in a lawless world frequently torn apart by insurrection and war.

This, then, is a fresh look at an old friend. And it has been a special pleasure that the majority of the illustrations in the book are the work of one of our most distinguished photographers, Adam Woolfitt, who traveled with me on long stretches of the pilgrim roads in both France and Spain, and shared my love of the landscape and the rich variety of historic sites that greeted us.

I. Via Turonensis:
The Road from Paris

1. Setting Out

PILGRIMAGES ARE JOURNEYS OF THE HEART. More than a thousand years separate the modern traveler and pilgrim from those who first trekked across the length and breadth of France, then southwards over the Pyrenees into Spain. Here they followed the pale arm of the Milky Way to guide them to their goal close to the Atlantic Ocean and—for all they knew in those centuries before Christopher Columbus—the very edge of the world. It was an awesome journey, and not one for the faint-hearted.

The goal they sought was a tomb believed to be that of St. James, cousin of Christ and the first of his twelve apostles to be martyred, decapitated by a Roman sword in the year 44 AD. The city that grew up around that tomb became named after him: in Spanish St. Iago, hence Santiago or, in full, Santiago de Compostela, meaning St. James of the Field of the Star, a star having led to the tomb's discovery, so the legend goes.

Since that time so great has been the cult of the saint that over the succeeding centuries the roads leading to his city were adorned with some of the finest monuments of Christian civilization—cathedrals and churches, abbeys and castles, masterpieces of sculpture and painting, bridges and shrines, as well as lined with countless silent witnesses to the slog and hardship of foot travel in medieval Europe—hospices and primitive shelters, stepping-stones across turbulent rivers, stretches of ancient road in the midst of nowhere, a wayside cross marking a death or a thanksgiving, or graffiti of a horseshoe on the wall of an inn scratched by some pilgrim grateful to have made it this far.

For these and so many other reasons, taking the roads across France and into Spain to Santiago can be one of the most uplifting journeys to be undertaken anywhere in the world. It is a journey through surroundings created by some of the most dramatic events of history. The ghosts of past travelers accompany us everywhere we walk, in the places where they worshipped, where they ate and slept, and on the long roads they took. In France it is a journey that can be made along a broad network of pilgrim roads, the *chemins de Saint-Jacques*, all converging into a single principal route across northern Spain known universally as the *camino*, or traditionally as the *camino francés*, the "French road," so-called because pilgrims wherever they came from in northern Europe inevitably had to cross France to reach Spain.

There were also other reasons for the name. Not only did France provide the greatest number of pilgrims to Santiago, but it was in France where the impetus for the pilgrimage arose and gathered strength. The Church in France, supported by its feudal aristocracy, largely facilitated the pilgrimage movement—by building churches, priories and hospices along the way, and by giving support to the beleaguered Spanish rulers in safeguarding the route across northern Spain against the Saracens who controlled much of the country.

And here lies a powerful reason why the Santiago pilgrimage captured the imagination of Christian Europe. There were other important pilgrimages at the time, notably to Jerusalem, and to Rome as the city of the martyrs St. Peter and St. Paul. But Jerusalem was far distant and in the hands of the Saracens. Rome had its obvious appeal, yet to get there involved crossing the Alps: furthermore, it lacked a key element which Santiago possessed—a passionately-held cause. And that cause was the *Reconquista*, the Reconquest of Spain. St. James, who had reappeared in Spain's darkest hour, became a symbol of the fight against the Infidel. Christianity had found a champion. And the road to his shrine was like a journey to the Promised Land. Piety and politics went hand in hand.

*

So, where to begin? One monument has the strongest claim to be the starting point. The city is Paris, and the monument bears the name of the saint himself. This is the Tour Saint-Jacques—the Tower of St. James. It stands in the center of a public garden on the Right Bank, gaunt and lofty, with spiky gargoyles high above craning against the sky. Originally this was the bell tower of a church called Saint-Jacques-de-la-Boucherie, so named because it stood near the Paris meat market. The church itself became a victim of anti-clerical passions during the French Revolution. In the orgy of destruction what the mob also obliterated was a milestone which the original architect, a certain Michel de Felins, had placed meaningfully at the base of his tower. The milestone was inscribed—quite simply—"Zero."

In other words, for pilgrims this was the point of departure: the beginning of the journey of a lifetime. Santiago de Compostela, in northwest Spain, was more than 700 miles away. A few well-to-do pilgrims would be making the pilgrimage on horseback, perhaps with a retinue of servants. But for the vast majority this was a journey on foot—with the aid of a staff, a broad-brimmed hat against the sun and rain, a flask of some sort for water, the stoutest of boots, a few modest worldly possessions, and bountiful hope that charity and the hand of God would guide them safely there and back again six months later. They would depart in the spring, and return—all being well—before winter set in.

The inscription "Zero" was as much a memorial as a milestone, because pilgrims had been setting out from here for at least 400 years before the Tour Saint-Jacques was built early in the sixteenth century. The Church of Saint-Jacques-de-la-Boucherie itself, where they assembled for their last mass before departure, was also a replacement for a much earlier pilgrim church. The site had always been of huge strategic importance, being set at the junction of the two oldest roads in Paris. They were the former Roman roads which sliced right through the city from north to south, forming its principal artery which it continued to be right up until the reshaping of Paris by Baron Haussmann in the mid-nineteenth century, and the creation of the broad boulevards of today.

Here, where the Tour Saint-Jacques now stands, was the great gathering place. Pilgrims from northern France, northern Germany and the Low Countries would make their way to this church or its predecessor, arriving by way of what is now the Rue Saint-Martin, and once here they would join up with travelers from the east and west of the city, all preparing to head south together on the journey to Spain. Before departure, at the climax of the celebratory mass, their pilgrim staffs would be ceremonially blessed by the priest: this would be followed by emotional scenes as families and well-wishers cheered them on their way, maybe accompanying them a short distance as they crossed the River Seine. At this point, in the early Middle Ages, pilgrims would have passed a colossal building site where gantries and wooden scaffolding were enclosing a building far taller than everything around it, topped by twin soaring towers and at ground level a triple portal of huge proportions swarming with builders and stonemasons hauling on ropes to assemble row upon row of carved figures around and above each of the three doors. The Cathedral of Notre Dame was nearing completion.

Then, once on the Left Bank of the river, they would take the straight corridor of a road which is still called the Rue Saint-Jacques[1], making their way through the heart of the old city, past the recently-built Church of Saint-Julien-le-Pauvre, then on up the slow hill past the celebrated university, the Sorbonne, until finally they reached the southern walls of the city and another church bearing the name of their saint, Saint-Jacques-du Haut-Pas, St. James of the High Step.

From here, once through the southern gate of the city, it was the open road—and the beginning of a 700-mile trek into the totally unknown.

Not that they were likely to be on their own. "Thanne longen folk to goon on pilgrimages": so wrote the English poet Geoffrey Chaucer in the fourteenth century in his Prologue to the best-loved of all accounts of a pilgrimage, *The Canterbury Tales*, his collection of stories of a party of pilgrims setting out from London to the shrine of

1. See Plate 1.

the murdered Archbishop of Canterbury Thomas à Becket. As Chaucer makes clear to our delight, medieval pilgrimages could be colorful and convivial occasions. However disparate the company might be— like Chaucer's spicy hotchpotch of humanity—whether for security, for comfort, or simply for companionship, pilgrims generally chose to travel in bands—as early accounts and illuminated manuscripts show. It was the Italian poet Dante, a century before Chaucer, who wrote in *The Divine Comedy* that people would describe the "Way of St. James" as the Milky Way on account of the huge number of people who traveled to the saint's city. In Dante's day, and in the two preceding centuries, more than half a million travelers are said to have used the roads to Santiago every year. And even if not all of them were *bona fide* pilgrims, but merchants and itinerant craftsmen, or merely adventurers and vagabonds, this is still an astonishing volume of humankind to be on the road to a remote corner of Spain, far away under the mists and Atlantic skies where there was little except a tomb, a church and a small community assembled around it. And yet half a million travelers! And this at a time when the total population of the region we now know as France was barely more than twelve million.

What was it that drove them? And what was the story that lay behind so demanding a journey?

*

The pilgrimage to Santiago was one of the most dynamic social movements of the Middle Ages, and it is one that has managed to retain its profound appeal in the radically different social climate of our own day. Even if it can seem sometimes to have become a bauble of the modern tourist industry much of its core of meaning and purpose has survived the long passage of more than ten centuries. In that time the Santiago pilgrimage has been a creative force, a spiritual force and a political force, as well as a triumphant testimony to the enduring power of legend.

And the legend is this.

James the Apostle (sometimes known as James the Greater, not to be confused with James the Lesser who was the half-brother of Jesus) was a fisherman on the Sea of Galilee. His father was Zebedee; his mother was Salome, believed to have been the sister of the Virgin Mary. James therefore seems to have been Christ's first cousin. He was recruited by Jesus as one of his twelve apostles, then after the Crucifixion he responded to Christ's last command that his disciples fan out across the earth to spread the gospel far and wide. Accordingly he voyaged to Spain where he spent a couple of years, until the Virgin Mary (still living at the time) appeared to him in Zaragoza raised on a pillar between twin choirs of angels. Inspired by this miraculous event James proceeded to erect around the pillar the first church ever to be dedicated to the Virgin. Shortly afterwards he returned to Jerusalem where he was beaded by a Roman sword wielded by King Herod Agrippa.

Had the legend ended at this point there would have been no pilgrimage to Santiago, indeed no Santiago at all. Everything rested on the second chapter of the story, treating events that immediately followed James' martyrdom. His body (including the head) was first carried by disciples to Jaffa on the coast of Palestine, then borne by stout ship down the length of the Mediterranean and up the Atlantic coast to the Bay of Padrón in northwest Spain, the voyage being accomplished in a mere week—proof of its miraculous nature. The party arrived thus far only to encounter Roman authorities once again who gaoled them, until in due course they were released by the intercession of an angel. Queen Lupa, who was neither a Christian nor remotely interested in the newcomers' unlikely tale, ordered the body of James to be buried on a remote hillside where a celebrated snake would be sure to polish off the disciples as well as the body of the saint. But the snake it was that died—on seeing the sign of the cross—whereupon Queen Lupa became converted to the Christian faith, and at last the body of the apostle was allowed to be given a decent burial, in a large stone coffin, and at a place where his disciples were themselves later buried.

It is the third chapter of the story which begins to anchor legend to historical fact—and dramatically so. But first there is a gap of six centuries in which there is no evidence that anyone knew of the existence or whereabouts of the saint's grave, or even that James had ever been in Spain at all, dead or alive. Church literature, from the *Acts of the Apostles* onwards, is entirely silent on the subject. In due course the Romans left Spain, the Visigoths took over, and Christianity inched its way into this outpost of Europe near the end of the known world. Then, early in the seventh century, copies began to appear in Western Europe of a Latin translation of a Greek religious text which became known as the *Breviarum Apostolorum*. The text attracted attention because it stated that Christ had awarded his apostles specific "mission fields," whereas in the New Testament Jesus had merely instructed his apostles to make disciples of all nations (*St. Matthew* 28: 19-20). More intriguing still, this Latin version included interpolations, or additions, confirming that James' mission had been to preach in Spain. Here, as far as we know, was the first written evidence of any connection between St. James and Spain, though it may well be that these scribal additions expressed a long-standing oral tradition that the apostle had indeed preached in Spain. On the other hand, it may also be, as skeptics have pointed out, that the translator mistakenly wrote *Hispaniam* (meaning Spain) instead of *Hierusolem* (Jerusalem). Since the original Greek text no longer exists, it is impossible to establish the truth.

However, it was one thing to claim that the apostle had preached the gospel in Spain after the death of Christ; it was quite another to maintain that he was brought back to Spain and buried here after his own martyrdom. Here two historical events proved crucial in creating the framework of the St. James legend as it has remained in the public's imagination ever since. Firstly, in the early eighth century the Saracens invaded from North Africa, and in a short time occupied and controlled most of the Spanish peninsula. Only the northwest corner remained unconquered—the Kingdom of the Asturias as it was known. It was an island of Christianity in a sea of Islam.

The second historical event was the appearance of one of the most influential religious texts of this era and of many centuries to follow. This was the *Commentaries on the Apocalypse*, the work of a Spanish monk and theologian known as St. Beatus of Liébana.

Beatus was born at about the time of the Saracen invasions, and therefore spent his youth and manhood under the shadow of Islam, which gives a special poignancy to his writings. The district of Liébana where he settled lay in the Cantabrian mountains in northern Spain within that Christian pocket unconquered by the Saracens: nonetheless it would have felt an area under siege, and the concept of a Holy War and the dream of Reconquest would have been constantly present in his mind. Christian Spain desperately needed a savior, and it was Beatus who made a key contribution towards finding one. In his celebrated *Commentaries*, which began to be circulated in the third quarter of the eighth century, he became the first to claim in writing that St. James had not only evangelized in Spain but was ultimately buried here. It was a giant step towards establishing the legend surrounding the saint. After all, if St. James was indeed buried here then his grave must surely exist. Somewhere there must be a shrine to be discovered; and the shrines of saints, as was well known, could perform miracles.

We can do little more than speculate, but even in a prevailing climate of religious make-believe and self-deception it seems most unlikely that Beatus would simply have invented the story of the saint's body having been returned to the country where he had evangelized. He is much more likely to have been voicing a conviction rooted in some long-standing local tradition, whatever its origin, that the body of St. James had indeed been brought here. In any event the effect of Beatus' pronouncement was dynamic. Even if a belief in the saint's grave being in Spain may hitherto have been for the most part a wishful tale, now it had acquired the stamp of Church authority. And in the context of a continuing Saracen threat, and the resulting siege mentality of the Church and local rulers in that Christian pocket of northwest Spain, Beatus' statement would have created a mood of high excitement and expectation which was the ideal breeding ground for what was to follow. And it did not take long.

It is tantalizing not to be able to draw back the curtains of history and witness what actually occurred in northwest Spain during these early years of the ninth century. Was there a widespread search for the grave? Were there constant rumours—hopes continually being raised? Was there a suspicion that it was only a fairy tale? How credulous were people? As it is we are restricted to a few surviving records. The first written evidence we have of the saint's grave having been discovered comes from a list of Christian martyrs that was being compiled at about this time by a monk named Florus de Lyon. Referring to this area of northwest Spain, in the year 838 he noted an "extraordinary devotion being paid by the local inhabitants" to the bones of the apostle St. James. Six years later comes the first record of a pilgrim actually visiting the site: ironically he was an Arab traveler, Ibn Dihya, who observed that visitors to the saint's tomb included Normans. How he knew this, and how they managed to communicate, he did not disclose, nor why he chose to be there at all. What is clear, though, is that even before the middle of the ninth century the shrine was well-enough known to be attracting pilgrims from far and wide. The word had got around with remarkable speed.

The story of the actual discovery of the supposed grave has an air of romantic improbability about it, while at the same time containing some intriguing elements of proven fact. There is a further confusion: the earliest written account of the tale of the grave's discovery dates from two hundred years later in a letter purporting—falsely—to be by the hand of Pope Leo II. That aside, the contents of the letter would appear to be a sincere account of events in northwest Spain early in the ninth century as they were now widely believed to be. The story begins with a hermit by the name of Pelagius informing the local bishop, Theodomir, of a vision which had revealed to him the whereabouts of the saint's tomb by means of a guiding star accompanied by celestial music. The seat of Bishop Theodomir was at Iria Flavia (today Padrón), situated at the head of a deep bay opening out on to the Atlantic Ocean. It was in a deserted spot some twelve miles from here that a stone tomb was duly discovered containing the remains of three bodies. The bishop immediately pronounced these to be the bodies of St. James and the two disciples who had brought his body from the Holy Land. The King of the Asturias, Alfonso II, then hastened to the tomb. He promptly declared that St. James was henceforward to be worshipped as the protector and patron saint of Spain. And he had a small church erected over the tomb, and a monastery close by. It was the beginning of what was to become the city and cathedral of Santiago. A town grew up round the new church and monastery during the course of the ninth century, and this became known as Campo de la Estrella, or Campus Stellae, "the Field of the Star," later to become shortened to Compostela.

Here, in the darkness of the ninth century, lie the foundations of the whole St. James legend. And from those foundations grew one of the phenomena of the Middle Ages.

History and legend part company over several aspects of this story. Yet, setting aside the celestial music and guiding star, there is no doubt that a grave, or several graves, came to light about this time. Recent archaeological excavations beneath the present cathedral revealed an ancient tomb inscribed with the name of the bishop mentioned as the one led to the site of the grave by the hermit Pelagius. The long-buried tomb beneath the cathedral reads—very clearly—"Theodomir." Traces of further graves were detected, believed to date from the Roman or early-Christian period, and the prevailing view among scholars today is that here was an ancient cemetery, and the word Compostela is therefore likely to derive not from the Latin noun for a star but from the verb *componere*, "to bury."

The wide discrepancy between fact and fairy tale in the whole St. James story is of trivial importance compared to its contribution in so many areas of life in the era following what has been described by Hugh Trevor-Roper as "the darkest age of Europe."[2] At some indefinable moment early in the ninth century it became widely believed among Christians that in their hour of need one of Christ's closest disciples was present in their midst. From that moment it was as though a climactic change took place in the hearts and minds of European Christians. The contribution of the St. James story

2. Hugh Trevor-Roper, *The Rise of Christian Europe* (London: 1966)

began to make itself felt in so many areas of human activities—in an outpouring of religious fervor, in a widespread resurgence of the human spirit, in artistic achievements and church building on an unprecedented scale, in the field of politics, in the expansion of trade and commerce and, by no means least, on the field of battle.

And it was here on the battlefield—not surprisingly—that St. James first of all made his mark, taking up arms in the cause of beleaguered Christianity. In the year 844, barely a few decades after the discovery of his tomb, the saint is said to have made his first appearance as a knight in shining armor at the side of a Christian army led by Ramiro I, King of the Asturias, who was facing an overwhelmingly superior Saracen force. The battle supposedly took place at a mountainous site known as Clavijo overlooking the fertile plains of northern Castile; and here St. James is credited with riding to the king's rescue on horseback and personally slaughtering 70,000 of the enemy. Thereafter the saint became popularly known as *Santiago Matamoros*—St. James the Moor-Slayer. And it is in this guise (as we shall see later) that he appears on so many churches along the Spanish *camino*, sword raised, banner aloft, crushing the forces of Islam beneath his stallion's feet.[3] After Clavijo the triumphant campaign continued: at Simancas in the tenth century, at Coimbra by the side of El Cid in the eleventh, at Las Navas de Tolosa in the breakthrough to Andalucia in the thirteenth—in all at least forty appearances in battle by the seventeenth century, according to the Spanish historian of the Santiago legend, Don Antonio Calderón. No good Spaniard in the Middle Ages, and for many centuries later, would have believed that his country could have been liberated from the Saracens without the flashing sword of their patron saint.

It was a whirlwind of events to have been generated by a pious claim first put about by a Spanish monk in the eighth century: St. Beatus of Liébana deserves to be regarded as the true godfather of the Santiago legend. Nor was that influence short-lived: it extends far beyond the era when the shrine of St. James was being established

---

3. See Plate 2

and the first pilgrims were flocking to Santiago. The long-term importance of Beatus' *Commentaries on the Apocalypse* lay in the fact that the work became so popular among Church leaders that in the centuries to follow artists in monastery workshops across Europe were busy illustrating it with fantastic and lurid imagery drawn from descriptions in the Book of Revelations. These illustrated copies of Beatus' text were widely distributed during the Middle Ages and became a primary source of many of the most powerful and disturbing carvings and mural paintings of the Romanesque period. Nothing expresses the doomsday vision of the medieval Church more eloquently than these apocalyptic images which snarl and leer out at you from church after church, assaulting the eyes of generations of pilgrims as they made their way to Santiago, just as they do today on the long roads across France and Spain.

*

Nowadays a brace of motorways heads south from Paris. The traditional road bisects the two, carrying quieter traffic, much of it local, as well as a regular procession of summer pilgrims on the first leg of their journey. For the medieval traveler this would have been a gentle introduction to life on the road, since this first stretch out of Paris is as flat as the famous local cheese, which is Brie. For at least thirty miles it is hard to imagine any distraction, either now or a thousand years ago. But then suddenly the old road passes the battered remains of an ancient fort before leading into a narrow strip of a town, Étampes, and here the modern traveler and the medieval pilgrim are on common ground. At either end of the town rise two churches, Notre-Dame-du-Fort and Saint-Martin, the former with its handsome tower and spire, both of them dominating the town only a little less than they would have done in the twelfth century when pilgrims paused here to refresh themselves from the local well and pray in one or other of these churches, both still in the process of completion.

Heading on south, less than a day's journey brought pilgrims to another narrow passageway of a town, then a mere hamlet. This was Toury, today somewhat larger than in the Middle Ages, but still centered on a chunk of medieval church with a stone colonnade extending from end to end, facing the town square. Then, from here it was only a short distance to the next objective—or, rather, a double objective. Principally this was Orléans, already a focus of pilgrimage more than 300 years before Joan of Arc brought glory to the city by driving out the English army, so launching the revival of the fortunes of France in the Hundred Years' War. A more symbolic objective was the great river that runs past Orléans. The Loire is the longest river in France, 634 miles of it, and from here it flows westwards across the lowlands of central France towards the Atlantic, guiding Santiago pilgrims—as it has done for a thousand years—along the next stretch of their journey.

2. Along the Loire

LIKE A LONG STRING OF PEARLS the medieval pilgrim's journey was strung with sacred places which had to be visited—not unlike the modern tourist ticking off a list of historic sites "not to be missed." The motive may seem to have changed, though less than you might think. The Middle Ages had a hunger for religious relics: we have a hunger for historical relics—cathedrals, castles, ancient monuments, ancient cities. The Middle Ages themselves have become part of our own store of relics.

The pilgrim road from Paris has now reached the River Loire. The numerous shrines "not to be missed" lay along the river, east and west. But first there was Orléans itself. The town was a major stopping place for pilgrims because it boasted two important relics. The lesser of the two was a chalice that had belonged to an early bishop of the town, St. Euverte, about whom little is known except that a nearby abbey was dedicated to him. But the second relic was an object of the deepest veneration, a supposed fragment of the Holy Cross. It was the first of many such fragments which the medieval pilgrim would soon encounter on his travels, and one cannot help wondering if, even in an era of the most naive credulity, doubts would at some stage begin to creep in as to the likelihood of the Cross having been splintered into quite so many pieces and distributed across quite so many far-flung places. Nonetheless, at this early stage of the journey no doubt wonder and awe were uppermost; and in the Middle Ages this particular fragment used to be displayed in the church that was built in its honor. This is the Church of Sainte-Croix, now on the edge of the old town. Today only traces of the medieval crypt survive of the

early church, concealed beneath a grandiose Gothic-style cathedral—
but with no trace of the Holy Cross. Skepticism may in the end have
triumphed.

The main pilgrim road as we know it heads west from here,
following the flow of the Rhône. Yet in the opposite direction,
twenty miles upriver from Orléans, stands a place which had been
a focus of pilgrimage several centuries before the cult of St. James,
and it was where a great many Santiago pilgrims would have made a
respectful detour. This is the ancient abbey Church of Saint-Benoît-
sur-Loire, one of the most striking Romanesque churches in France.
Approached from the west it looks like a massive stone box punched
through with gaping holes. This is merely the porch: the present
church itself, and the former abbey to which it was attached, came
into existence because it came to possess the body of St. Benedict,
founder of the Benedictine Order. And quite apart from the historical
importance of the saint, it was the Benedictines who were prominent
in promoting and aiding the pilgrimage movement. It would have
been widely known that they were the pilgrims' friend.

St. Benedict of Nursia was an Italian monk in the early six
century who founded the great abbey of Monte Cassino in southern
Italy. Here he composed the famous Rule of St. Benedict, a humane
and perceptive handbook setting out how monks should lead their
daily lives. Accordingly, he is generally regarded as the father of
western monasticism.

Then, a century after St. Benedict's death, Monte Cassino was
sacked by marauding Lombards. News of the abbey's destruction
reached the small monastery of Fleury, on the Loire. From here a party
of monks set out for southern Italy in the brave hope of retrieving
the body of the saint. On arrival they found that St. Benedict's grave
had miraculously escaped destruction; so in about the year 672 AD
they brought the body back to their monastery in northern France.
From this moment the abbey of Fleury acquired fame and wealth
as a popular center of pilgrimage, duly changing its name to the
Abbey of Saint-Benoît, and acquiring a new abbey church whose
west front is the imposing porch which confronts travelers today as

they approach from the direction of Orléans. The porch originally supported a belfry, and it is the oldest part of the church, dating from the eleventh century. Now, as in the Middle Ages, it serves as an ideal assembly area for those attending or leaving mass, conveniently sheltered from the sun and from the winter gales hurtling up the Loire from the distant Atlantic.[4]

Travelers and pilgrims waiting in the porch found themselves surrounded by a forest of stone columns whose capitals are carved with images that will be repeated again and again on the pilgrim roads through France and into Spain. They are sermons in stone, designed to match the doomsday sermons delivered every Sunday from the pulpit. There are familiar bible stories—among them the Annunciation and the Flight into Egypt—together with inevitable reminders of the perils of sexual temptation and sin, all embellished by lurid scenes from the Book of Revelations, inspired by those illuminated versions of St. Beatus' *Commentaries on the Apocalypse* described in Chapter 1.

What is particularly unusual at Saint-Benoît is that we know who carved these images. One of the Corinthian capitals in the porch bears the inscription UMBERTUS ME FECIT—"Humbertus made me." We know nothing else about him, but he must have been someone to be reckoned with. The vast majority of medieval church carving are anonymous, and the very few examples we have of signed work suggest that the sculptor in question was held in especially high regard—the outstanding example being that of Gislebertus at Autun, among the greatest of all medieval church sculptors (as we shall see in a later chapter).

The size of the nave of the abbey church itself is an indication of just how many worshippers flocked here to venerate the relics of the founder of the Benedictine Order. Traditionally his remains would probably have been contained in some form of gold or silver reliquary, handsomely engraved, and placed on the high altar during mass in full view of the congregation. Today those relics are kept

---

4. See Plate 3.

down below in the labyrinthine crypt, encased within the massive central pillar. Something of the drama of the original display on the high altar has inevitably been lost in the process.

Forms of worship have inevitably changed over the centuries. To Christians in medieval Europe the value of shrines and the relics they contained needs to be understood in the context of the religious climate of the times. In the Middle Ages the prospect of eternal damnation was preached relentlessly by a Church obsessed with the promise of the Second Coming and the Day of Judgment. The world as people knew it was believed to be coming to an end. This dark prophesy was preached just as vividly by the sculptors and painters who decorated the churches in which those hell-fire sermons were delivered. For the simple illiterate peasant, who knew nothing of the world beyond the fields he tilled and the church he attended on Sundays, this vision of the future was all he was offered. He had no means of knowing otherwise. From the day he learned to use his eyes and ears to the day he died he was indoctrinated with the urgency of obtaining divine forgiveness for sins which he probably had no idea he had committed. It was in the very fabric of human nature to be sinful, and that was it. Doom sounded like a gong in his ears all his life. There was only one way out—to seek forgiveness—*remissio peccatorum*. And the surest way such forgiveness could be obtained was by making contact with the saints, who alone could intercede with God on his or her behalf. The surest way of getting God to listen was to go to them and ask.

The most powerful ingredient in medieval religious life lay in the cult of such relics. They were the essential go-betweens, and visiting the shrines where they were displayed was what pilgrimages were fundamentally all about. Even the great thirteenth-century theologian and philosopher St. Thomas Aquinas, who might be expected to have held a more rational view of such things, was an unequivocal advocate: "We ought to hold them in the deepest possible veneration as limbs of God, children and friends of God, and as intercessors on our behalf," he wrote.[5]

---

5. From the *Collected Works of St. Thomas Aquinas*, c. 397 AD (Oxford: 1999).

Much of the power of holy relics lay in the fact that they were universally believed to perform miracles, as an apparently limitless number of worshippers were eager to testify. Nothing could enhance the prestige and popularity of a shrine more than a reputation for healing the sick, curing blindness, or madness, or whatever human condition was associated with a particular saint. No one in medieval France would seriously have doubted that the relics of a saint could perform such miraculous feats. So closely was religious worship bound up with a belief in these powers that it even became a ruling by Church authorities that no new church could be consecrated unless it was in possession of a holy relic of some description. Nor was it confined to the Church. Monarchs and rulers also formed personal collections of relics. The Emperor Charlemagne early in the ninth century was among the first, having obtained them in quantity from Constantinople. The Byzantine capital, named after the first Roman emperor to convert to Christianity, Constantine I, became something of a marketplace for biblical relics, being relatively close to the Holy Land, which was of course their principal source. The later Byzantine emperors also acquired hoards of such items, and it was after the disgraceful sacking and looting of Constantinople by the Fourth Crusade in 1204 that the flood-gates were truly opened, and many an adventurer masquerading as a crusader returned home with sack-loads of stolen relics, most of them bogus.

Yet, in spite of the flood there were never enough relics to go around. This was the great age of church building and of new monastic foundations all over Western Europe. With demand exceeding supply, fraudsters and dubious merchants were very soon doing a roaring trade in fakes. And in a credulous world, and where a relic might be nothing more than a fragment of bone or a splinter of wood, their job was an easy one: hence many of the—to our eyes—ludicrous examples of relics displayed for the benefit of the trusting faithful. Monks at the twelfth-century Abbey of Saint-Médard de Soissons, in northern France, claimed to possess a tooth of Jesus. Solemn theological objections were raised on the grounds that Christ had been resurrected, presumably with his teeth intact, to which the monks retorted artfully that this must be a milk tooth. That not everyone was taken in by such outlandish claims is made clear by Chaucer in *The Canterbury Tales*, scornfully mocking his Pardoner for carrying in his wallet a fragment of cloth which he swore was from the sail used by St. Peter when he was a fisherman on the Sea of Galilee.

Genuine or make-believe, relics of the saints were enshrined in places it became obligatory for pilgrims to visit. Santiago de Compostela might be the distant goal, but almost every day there would be a shrine of some sort possessing relics to be revered. A pilgrim's journey became a continuous prayer that all would be well.

But there is another side to this journey from shrine to shrine. Whether by chance or good planning it was often the case that a summer pilgrimage through France would find travelers reaching a shrine around the time of a saint's day—with luck the very saint revered at this particular shrine. And these would turn into festive occasions. They might begin with moments of the deepest piety: there are accounts of crowds of pilgrims on the eve of a saint's day keeping vigil all night in the church where the saint's relics were displayed. And on the following morning a solemn mass for a packed congregation would be held. But then, duty done, the atmosphere would change. Secular life took over, and the saint's day became a feast day. The pilgrimage became a party, with music and dancing, acrobats and jugglers performing their acts, hawkers and food vendors plying their trades. Musicians would mingle with the crowds in the square outside the church even while a service was still being conducted inside. There is an account of a saint's day in Conques, in south-central France, when bawdy songs were sung so loud outside the great abbey church that they drowned the words of the litany. Nor was there anything new about such behavior: as early as the fourth century St. Augustine wrote of religious occasions being accompanied by "licentious revels."[6]

In all, what we see in descriptions of saints' days are the beginnings of carnivals, fairs and fiestas. Important Church occasions were the focus of public celebration as much as ceremonial piety. The

---

6. St. Augustine of Hippo, *Confessions*, c. AD 397 (Oxford: 2009).

bucolic scenes of village celebrations immortalized in the paintings of Peter Brueghel in the sixteenth century owe their origin to those annual events when pilgrims, fellow travelers, merchants and local folk, rogues and hangers-on came together to honor some long-dead saint for the good of their souls, then having done so would then step aside from the drudgery of daily life for just a single day, and cast caution to the wind.

The setting is often still there. If we stand by the great gaunt porch of Saint-Benoît's abbey church, it is not hard to imagine the scene as it might have been on a saint's day all those centuries ago. On the stone pillars above the crowd are those apocalyptic images of hellfire and damnation which pilgrims could never get away from. Yet down below in the square there is quite a different world: there is jollity, music and laughter, the singing of ribald songs, all washed down by gallons of sparkling Loire wine. And the party will go on deep into the night. Pilgrimages could be about enjoyment as well as piety.

Then in the morning pilgrims would make their way from Saint-Benoît, perhaps a little worse for wear, along the high bank of the Loire towards Orléans, where a few days earlier they had venerated a supposed fragment of the Holy Cross in the Church of Sainte-Croix. Before continuing westward to the next major stopping place, the city of Tours, they would have joined up with local pilgrims preparing to make the same journey. Orléans was an important town, and we know that a considerable number of Santiago pilgrims came from here. Today a Renaissance mansion in a side street close to the river offers an insight into the medieval pilgrim's world. It is called La Maison de la Coquille. The name has survived from an older building on the site, which was probably a pilgrims' hospice, or simple inn, attached to a religious house close by. *Coquille* means "scallop shell," and most pilgrim hospices came to be identified by this insignia from the time when it became universally recognized as the emblem of the Santiago pilgrimage.[7]

7. See Plate 4.

How this came about is a complex tale. At least by the twelfth century scallop shells from the waters of the Atlantic near Santiago were being sold outside the newly-built cathedral to pilgrims who would then proudly take them home as mementos and as proof of having successfully completed the journey. Later the image of the *coquille* became a badge, often cast in metal and worn on a cord around the neck, or else stitched on to clothing, so distinguishing pilgrims from other travelers who might have less honorable reasons for taking to the road. Later still the French genius for gastronomy led to the creation of a delicious dish of scallops cooked in wine, cream and cheese, for which they appropriately invoked the name of the apostle by calling the dish *Coquilles Saint-Jacques*.

The Orléans hospice on the site of the Maison de la Coquille would have been one of many in this part of the town. A nearby street, which once led to the only bridge over the Loire, still bears the name Rue des Hôteliers. Here too stood a late medieval pilgrims' chapel, the Chapelle Saint-Jacques, which remains largely in the memory today as a place where Joan of Arc prayed after her successful siege of the city. This was a chapel attached to a pilgrims' guild, known as a *confrérie*, or "confraternity," which had been in existence since the thirteenth century. Anyone could join a confraternity—men, women, priests—provided they could provide proof that they had actually been to Santiago de Compostela. In the early days the word of the local priest would have been proof enough, perhaps supported by producing the scallop shell which the pilgrim had brought back with him. Then from the fourteenth century the canons of the cathedral in Santiago took to issuing a certificate to arrivals. These were beneficial to the returning pilgrim, but also served as good publicity for the city of Santiago whose cathedral authorities were always keen to explore fresh avenues of fundraising.

In France alone there were more than 200 confraternities of St. James in virtually all the larger towns. Collectively their members formed a genuine society of pilgrims: they were a distinct and certainly rather special body, with their own customs, folklore, dress, songs and poems, and of course possessing the powerful bond of a shared experience—the fact that they had all of them made that long journey on foot to Santiago and back, in a small tightly-bound community in which few others had ever traveled further than the next market town.

Confraternities were secular bodies run on charitable lines (to some extent at least), with obligations to help the poor and to sponsor those in their town who wished to undertake the pilgrimage themselves. Naturally they were also social clubs, with their own constitution and rules, as well as possessing a great fondness for banquets, especially on every July 25 which was St. James' Day. The largest of these *confréries* was the one in Paris, founded in the thirteenth century, then greatly expanded early in the following century by the French King Philip IV (the Fair), under whose patronage the Paris *confrérie* built a new church, a chapel and a hospice near the Porte Saint-Denis at the northern edge of the city. The French monarch had recently succeeded in suppressing the Knights Templar and burning the order's senior members at the stake, so securing much of the knights' vast wealth—all of which made the king's lavish generosity towards the confraternity possible, bloodstained though it may have been.

As it happens records of the Paris confraternity have survived for the year 1340, just a quarter of a century later. These include the accounts for the July 25 banquet of that year. From these we learn that over one thousand members attended, each paying two sous. Whatever the chefs may have produced on this occasion, it accounted for five cows, twenty pigs, three hundred eggs, two barrels of white wine and three of red. Even so, at some stage a large body of members processed through the streets of Paris carrying banners depicting St. James.

It was a far cry from the bread and mug of rough wine which a pilgrim might feel glad to receive in some wayside hospice after a 25-mile slog through forest and swamp. A pilgrim's life had many complexions.

*

No stretch of the pilgrim road has changed more dramatically since the Middle Ages than the section skirting the Loire. Within a few centuries it was to become the "Royal Loire," studded with spectacular *châteaux* which we marvel at today—Blois, Amboise, Villandry, Langeais, Saumur, as well as those close by: Azay-le-Rideau, Chenonceaux, Chinon, Loches, Chambord. The modern traveler following in the footsteps of the medieval pilgrim needs to be blind to these glories of Renaissance architecture and instead to go in search of more humble places.

Heading west along the river from Orléans we are almost immediately reminded that it was not only the large towns which possessed a community of former pilgrims. Cléry-Saint-André is scarcely more than a village today and can never have been any larger, yet we know from a notary's document that as late as the early seventeenth century a pilgrims' confraternity here boasted a membership of thirty-two men and one woman. (Who was she, one wonders?) Whether they had all of them been to Santiago is not recorded. What seems clear is that *confréries* such as this one continued to play a social role even in small communities over a period of many centuries. The local church here also offers a reminder that pilgrimages often overlapped one another. Cléry, on the road to Santiago, also attracted pilgrims to a shrine of its own. This was a statue of the Virgin which had been discovered in the thirteenth century by a local ploughman in a nearby thicket. The statue was held to be a miraculous arrival in their midst, and a cult grew up which persisted for centuries. Hence when the church was rebuilt in the fifteenth century by King Louis XI he had it dedicated to the Virgin: and so strong was his attachment to the place that he instructed that he should be buried here. And yet his tomb rests in a chapel dedicated not to the Virgin but to St. James. Saints could sometimes be interchangeable.

From Cléry the road continues southwest along the river. Pilgrims passed through Blois, already a city of some substance in the Middle Ages, whose ruler in the eleventh century had married the daughter of William the Conqueror, who bore him a son, Stephen, later to become King of England. Then onwards again to the most important of the cities on the Loire, and one of the oldest and most revered places of pilgrimage in France. This was Tours.

3. A Brave New World

A<small>T THE CITY OF</small> T<small>OURS</small>, the pilgrim road finally ceases to hug the River Loire, and heads south. But before continuing their journey pilgrims would linger here a while: there was no more sacred a place between here and Spain than Tours. Its importance may be measured by the fact that this road from Paris was actually known as the *Via Turonensis*—the Road from Tours. Furthermore, of the four main pilgrim roads to Spain in the Middle Ages this was held to be the principal one: it was the *magnum iter Sancti Jacobi*.

The first account of this network of four roads laid across the landscape of France comes from that celebrated twelfth-century document called the *Liber Sancti Jacobi*, the Book of St. James, the fifth volume of which is more generally known simply as *The Pilgrim's Guide to Santiago de Compostela* (*Iter pro peregrinis ad Compostellam*). It forms part of a larger manuscript known as the *Codex Calixtinus* (named after Pope Callixtus II). *The Pilgrim's Guide* itself is a work with an undisguised agenda, which is to promote the Santiago pilgrimage, and it does so with the aid of barefaced deceit. The *Guide* is presented as having been written by the highest authority in the Church, Calixtus II, who almost certainly had nothing whatever to do with it since he had been dead for at least ten years before the manuscript was composed. (The reasons for this blatant pretence may become clearer later in this book when the political issues surrounding the pilgrimage come into focus.)

The Pilgrim's Guide makes clear the attraction of the city of Tours. "One must visit on this route, on the banks of the Loire, the venerable remains of the Blessed Martin, bishop and confessor"[8]: so begins an account of one of the most revered saints in France, St. Martin of Tours, whose name is invoked again and again in churches and abbeys along the pilgrim roads in France and in Spain. St. Martin had been a soldier in the Roman army in the fourth century, then rose to become the leading bishop in what was still Roman Gaul, recently converted to Christianity. A rich corpus of legends and miracles are attached to his life and career, and a widespread cult grew up around him in the succeeding centuries, making Tours a center of pilgrimage. With the emergence of the cult of St. James the Tours pilgrimage, like so many other local pilgrimages throughout France, became attached to the greater adventure of the Santiago trail. Tours became one of the many pearls strung along the thread of that great journey. The first sanctuary was built over St. Martin's tomb in the fifth century, only to be destroyed in the ninth century by the Vikings—"the Norman fury"—at a time when they laid waste to much of the Loire Valley. The church which replaced it was then destroyed by fire in the year 997. A far larger church was constructed in the eleventh century, further expanded early in the twelfth century; and it is this great new basilica which *The Pilgrim's Guide* urges Santiago pilgrims to visit: "The sarcophagus ... glitters with an immense display of silver and gold as well as precious stones. ... Above this a huge and venerable basilica has been erected in his honor, similar to the church of the Blessed James, and executed in admirable workmanship."

Here the author touches on one of the most distinctive features of the Santiago pilgrimage and the architecture which it inspired. The great Basilica of Saint-Martin-de-Tours was one of a group of five huge churches all of which conform to the same architectural specification. They were designed specifically to accommodate huge numbers of pilgrims, one church located on each of the four French roads leading

---

8. The passages from *The Pilgrim's Guide* which I have quoted in English throughout this book are based on the translation I made for my earlier study of this subject back in 1974. My source was an excellent version in French made by Jeanne Vieillard in 1938, at that time the only translation from the original Latin available to me (see Selected Further Reading).

to Spain as described in the *Guide*. Besides Tours on the Paris road, there was Limoges on the road from Vézelay, Conques on the route from Le Puy, and Toulouse on the southern route from Arles; the fifth being in Spain, the Cathedral of Santiago de Compostela itself. The *Guide*'s author, dedicated as he was to promoting the glories of Santiago, seems to have fallen victim to his own propaganda in suggesting that St. Martin's church was modelled on that of Santiago (possibly in response to pressure from the cathedral authorities in Santiago, as may become clear later in this book). In fact, it was probably the other way round: the Church of Saint-Martin-de-Tours is likely to have been the model and prototype for the four others. Tours was the mother-church of the great pilgrim routes. Revisionist scholars have recently struggled to downplay the importance of this grouping of five archetypal pilgrim churches: nonetheless, in whatever order they were built, together they made up some of the most remarkable ecclesiastical buildings ever created.

All five churches conformed to a common plan because they were designed for the same purpose: this was to accommodate huge congregations on special occasions when pilgrims could be expected in exceptional numbers not only to attend mass but to venerate a saint's relics which would be displayed on the High Altar. These requirements necessitated a vast aisle of double the normal width to ensure that all could participate in the elaborate services held on saints' days and at other special festivals. Then, since honoring the saint's relics was the climax of a pilgrim's day, there needed to be a broad ambulatory around and behind the High Altar so that the congregation could circulate freely. A further feature of these churches was a semicircle of five chapels which were set in the apse beyond the ambulatory so that pilgrims could offer prayers in relative privacy.

Such was the celebrated Basilica of Saint-Martin, where "the sick come and are healed"—*The Pilgrim's Guide* assures us "the possessed are delivered, the blind are restored to their vision, the lame rise, all sorts of illnesses are cured, and upon all those who ask for it a complete relief is conferred."

Alas, no longer. Only three of the five great pilgrim churches survive, and Saint-Martin-de-Tours is not one of them. The history of destruction which preceded it was to continue. It was vandalized by a Protestant mob during the French religious wars in the sixteenth century; then most of what remained was finally demolished in the nineteenth century to make way for a market. Today, following in the footsteps of yesterday's pilgrims, we can walk down the market street, the Rue des Halles, reflecting that this was formerly the nave of the great church. As a reminder, a single broken arch has somehow survived, snapped off above our heads. Further down the street on the right stands the Tour de l'Horloge, once the southwest tower of the church. Further on still, the Tour Charlemagne originally rose above the north transept: now it overshadows the little shops, market stalls and half-timbered houses of the old town. The mother church of the Santiago pilgrimage has become a venue for cabbages and cheese. As for St. Martin's tomb and shrine which once drew worshippers in their thousands, it is—surprisingly—still there, deep in the ancient crypt hidden beneath a grandiose nineteenth-century basilica built in what is politely described as "the neo-Byzantine style."

And so, after a respectful bow to the ghost of St. Martin's basilica, today's travelers follow the trail of their medieval predecessors, and move on—southwards.

For the pilgrim there was always the problem of rivers. Today we cross them without thinking, pausing perhaps to gaze down at the gentle flow of water combing the thickets of bulrushes and water irises. But for the pilgrim in the Middle Ages each river was a potential barrier. There were few bridges, and unless there was a convenient ferry each crossing meant negotiating with a kindly boatman who—as *The Pilgrim's Guide* ghoulishly explains later—sometimes proved to be more murderous than kind. This area of central France was particularly hazardous because the Loire is fed by numerous tributaries splayed like the fingers of a hand across this fertile plain, not joined up into a single arm until further downstream.

On one of these tributaries, the Vienne, is the tiny village of Tavant. It lies a short distance to the west of the pilgrim road, yet like so many isolated settlements in this region it preserves its own record of the great pilgrimage which flowed close by for so long. The squat twelfth-century church is remarkable in possessing frescoes of the same period, including lively figures clearly drawn from life, among them—on the wall of the crypt—a study of what is apparently a pilgrim: and if he is indeed a pilgrim bound for Santiago, then he is among the very first of whom we have a picture. There are anomalies about his appearance. He has the usual leather pouch for carrying food, as well as the long pilgrim's staff. Yet in his hand he also carries a palm branch, the traditional emblem of a pilgrim to Jerusalem (even though it was sometimes adopted by Santiago pilgrims). Then his headgear bears no resemblance to the wide-brimmed hat turned up at the rim generally believed to be part of a Santiago pilgrim's uniform: rather it resembles a turban of some sort, again suggesting a connection with the East. So, perhaps he was a palmer recently returned from the Holy Land in the wake of the First Crusade, maybe wearing a trophy on his head to impress his fellow villagers; and the artist busy painting religious scenes in the church at the time decided to include a portrait of so decorative a local celebrity. Or there again, maybe he was a pilgrim about to set out for Spain wearing what he considered to be an appropriate outfit. After all, most images of Santiago pilgrims in their recognizable gear of hat, jacket, staff, wallet and scallop shell are of a later date; and in these early centuries who knows what pilgrims normally wore?

Then, as if to demonstrate how the conventions of a pilgrim's dress became stereotyped to the point of caricature, a little further south along the pilgrim road, at Châtellerault, the Church of Saint-Jacques displays a seventeenth-century polychrome statue of the apostle whose hat and cloak are smothered with enough scallop shells to have made a feast of *Coquilles Saint-Jacques*. This is pilgrim dress turned caricature.

A mile or so further still, close to the pilgrim road, lies a place so sleepily anonymous that only its name betrays it as the site of one of the most momentous events in the early history of Europe, with a strong indirect bearing on the Santiago pilgrimage. It is a village

with the name of Moussais-la-Bataille. In other words it is the site of
a battle. Somewhere in the lush fields around this village was where,
in the year 732, the general of the Frankish army routed a Saracen
force which was threatening to plunder its way through France at
least as far as Paris. This was the first significant check to the repeated
Moorish invasions from Spain which were threatening to overwhelm
much of Christian Europe. It was the beginning of a tide being turned,
and the first fragile pointer to a counter-offensive which was to
culminate in the crusades and in the Reconquest of Spain. The general
of the conquering Frankish army was Charles Martel, Charles "the
Hammer," also distinguished as being the grandfather of the Emperor
Charlemagne, Charles "the Great," the first Holy Roman Emperor,
who will feature prominently later in the Santiago story.

The road south now enters a region which brings *The Pilgrim's
Guide* colorfully to life, in the process providing a clue as to who
may have been responsible for writing it—or at least writing this
section of it. The region is Poitou: that is, the area centered on the
city of Poitiers. Here, we read, is "a land well-managed, excellent
and full of all blessings. The inhabitants of Poitou are vigorous and
warlike, extraordinarily able users of bows, arrows and lances in
times of war; they are daring on the battle-front, fast in running,
comely in dress, of noble features, clever of language, generous in
the rewards they bestow, and prodigal in the hospitality they offer."

This effusive account of Poitou and its people is in startling
contrast to the sketchy and impersonal nature of what precedes it in
the *Guide*. The chapter continues at unusual length with a description
of the places and people on the route ahead—some of it furiously
hostile—which makes it clear that this section of the book at least
was written by someone who had actually traveled the pilgrim road
and is relating his experiences of it. Furthermore, his unashamed
bias in favor of Poitou suggests that the author is likely to have come
from here. His identity is something of a puzzle, though there are
intriguing clues, as well as any number of false trails. In the latter
category, besides the spurious attribution to Pope Calixtus II, there
is a second name given as author and co-author of two of the other

chapters. He is called Aimery the Chancellor, who is known to have been a cardinal and an adviser to several successive popes as well as a friend of the first Archbishop of Santiago, Diego Gelmirez.

There are other clues besides, and they point to another candidate who happens to share the same name—Aimery. A surviving letter purporting to be from Pope Innocent II states that the *Codex Calixtinus,* the manuscript which includes *The Pilgrim's Guide,* was donated to the Cathedral of Santiago de Compostela by a certain Aimery Picaud who made the pilgrimage in the company of a man named Oliverus and a woman called Ginberga who may have been Oliverus' wife. The same Aimery also appears as the author of a hymn written down immediately after the last words of the *Guide.* Furthermore this Aimery is known to have been a monk in the priory of Parthenay-le-Vieux, which lies in the heart of the Poitou region of France—the area effusively eulogized in *The Pilgrim's Guide.*

Suddenly, from these fragments of circumstantial evidence pieces of the puzzle seem to be falling into place. We have two principal authors of the *Guide,* both called Aimery. One is a high-ranking dignitary of the Church, friend of popes and of Santiago's archbishop, acknowledged as being responsible for a long description of the cathedral and city of Santiago towards the end of the *Guide.* The second Aimery is a lowly monk from the Poitou region, who remains anonymous in the text, yet his presence can be felt everywhere. The account of the landscape and people of his native Poitou, as well as of neighboring regions, is so vivid and opinionated that it suggests he is most likely to have been responsible for at least the chapter entitled "The Quality of the Lands and the People along the Road," and very possibly other chapters as well. And this same Aimery, as we are informed, was the man who later carried the completed *Codex* manuscript all the way to Santiago. He may even have undertaken the pilgrimage twice. In any case he was someone who knew sections of the different pilgrim roads extremely well.

Aimery Picaud's church at Parthenay-le-Vieux still exists, a short distance to the west of the main pilgrim road running south towards Poitiers, capital of the Poitou region. The church itself was built early in the twelfth century, at a time when the young Aimery may well have been a novice at the priory. Today the church stands solidly among trees and rough ground, its handsome octagonal bell tower a prominent landmark in this flat countryside. A rounded apse fans out from the eastern end, while on the west side three sculpted portals greet the visitor with the customary admonitory figures of good and evil stretched along semi-circular bands of carving overhead—pairs of dogs representing Hell, and birds standing for Paradise. Then, in the center of the lefthand portal is a powerful life-size mounted figure wearing a crown and trampling a foe beneath his stallion's hooves. Like the messages of good and evil, this would soon become a familiar figure to pilgrims traveling through this region. The crowned warrior was a symbol of the Church Triumphant, designed to represent the first Christian Roman Emperor, Constantine I, crushing the forces of paganism—a further reminder of the crusading aspect of the Santiago pilgrimage. As for the priory itself, all that remain are retaining walls that enclose fruit trees and vegetable plots, and a fragment of the former cloister where Picaud would so often have strolled, discovered as recently as the 1920s embedded in the wall of a farmhouse close to the church.

In fact there are two Parthenays, a couple of miles apart. And it is this second Parthenay which remains one of the "showcase" pilgrim towns of France, even though it stands only on a tributary pilgrim road. In this region there are so many threads of roads conveying pilgrims, including those from England, who had traveled directly from Normandy and Brittany, avoiding Paris and Tours, and gradually converging as they progressed south. Parthenay today is like a full-scale demonstration model of what a medieval town was like: indeed there is no town along any of the pilgrim roads, even in Spain, which more clearly shows how the Santiago pilgrimage made its mark on the local life and architecture in the Middle Ages. Travelers would approach the town from the north, and characteristically the northern outskirts of Parthenay became the Faubourg Saint-Jacques, the suburb for Santiago pilgrims, because here was generally where they found lodgings for the night. The *faubourg* was deliberately set outside the walls of the town, so enabling travelers to come and go after the gates of the town were locked at night. To this day hundreds of French towns, from Paris down to places smaller than Parthenay, retain an area still named Faubourg Saint-Jacques.

Here, in the heart of the *faubourg* of Parthenay, there survives the shell of a Chapelle Saint-Jacques, long ago incorporated into a domestic building and a neighboring barn. Here pilgrims would offer their prayers on arrival before finding a hospice or lodgings of some sort close by. Then, barely a hundred yards further, the road opens up to reveal a prospect which has remained unchanged for more than three quarters of a millennium. This is the thirteenth-century stone bridge—naturally called the Pont Saint-Jacques[9]—which leads across the gentle River Thouet directly to the principal town gate, the enormous fortified Porte Saint-Jacques; and from here into the main street of the town, called (not surprisingly) the Rue Saint-Jacques. On either side stand half-timbered houses formerly occupied by merchants and tradesmen who had grown prosperous supplying provisions and other services to pilgrims passing through the town in their thousands on their way south. The merchants gave their houses fine stone lintels with the appropriate year engraved on them, and overhanging gables in wood carved with corn motifs and human heads. Today's traveler then climbs up towards the remains of the town castle and, next to it, the Church of Notre-Dame de la Couldre. Only a particularly fine Romanesque west door survives as a witness to a historic occasion which took place here in 1135 in the lifetime of Aimery Picaud (who may well have been present). One of the most forceful and charismatic preachers of the Middle Ages came to this pilgrimage town and delivered a sermon of such hypnotic power that the local ruler, Duke William of Aquitaine, who was present, is said to have become a devotee on the spot.[10] The preacher was none other

---

9. See Plate 5.

10. Duke William's most significant contribution to European history is that he was the father of Eleanor of Aquitaine, whose inheritance of Aquitaine enabled her husband, the English King Henry II, to claim this large area of "France" as belonging to England, a claim that led ultimately to the Hundred Years' War between England and France.

than St. Bernard of Clairvaux, founder of the Cistercian Order and a guiding spirit of resurgent Christianity. Here, as elsewhere along the roads to Santiago, those passionate issues of the day, the Holy War and the Reconquest of Spain, were woven into the very fabric of the pilgrimage movement.

How much or how little of The Pilgrim's Guide was actually composed by a monk from Parthenay-le-Vieux is of far lesser importance than why it was put together and where the inspiration for it lay. Here a strong clue lies in its final words. The author concludes with a list of places where he claims it has been written. These include Italy, Germany, "the lands of Jerusalem" (which sounds unlikely), then significantly "mainly in Cluny." So often in The Pilgrim's Guide claims of authorship and geographical location suggest propaganda rather than actuality. Attributions tend to act as smoke screens. Yet "mainly Cluny" has an altogether more meaningful ring. Whatever the extent of Picaud's contribution to the book, his priory of Parthernay-le-Vieux was a Benedictine house which had strong links to Cluny. The priory's own mother-abbey in the Auvergne, La Chaise-Dieu, had been founded in the mid-eleventh century by a monk of Cluny who had been an apprentice to one of its most forceful abbots, St. Odilo, who had directed much of his abundant energy into supporting the Christian rulers of northern Spain in their campaign against the Saracens, and in doing so safeguarding the pilgrim route to Santiago.

As in so many twists and turns of the pilgrimage story the name of the great abbey is never far away. Cluny was the political powerhouse of the medieval Church, the most influential ecclesiastical body of its day, and for several centuries the spiritual heart of Christendom, shining "as another sun over the earth," in the words of Pope Urban II. With its massive investment in Spain and in the Santiago pilgrimage, what more likely a guiding spirit could there be in the compiling of a book urging pilgrims to go there? Whether it was actually written or merely masterminded there, the Guide has the shadow of Cluny all over it—as we may see more clearly when this journey takes us there.

*

The first modern scholar of the Santiago pilgrimage, the American Arthur Kingsley Porter, wrote in the 1920s that "the pilgrimage roads may be compared to a great river emptying into the sea at Santiago." The image is a memorable one. The continuous flow of a river evokes the unbroken passage of pilgrims over the centuries, as well as the essentially peaceful and restorative nature of a pilgrim's journey. However hazardous, the roads to Santiago were felt to be the roads to heaven, the way to personal salvation. And from our perspective in the twenty-first century the art and architecture of those roads are the signposts which guided people there. Nowadays we travel to admire those achievements, much as medieval pilgrims traveled to venerate the shrines contained within them.

Less identifiable than buildings and works of art is the human element, which is what the pilgrimage is all about. What kind of people were they who undertook such a journey at a time when travel was liable to be dangerous at the best of times, and for most pilgrims a step into the unknown? *The Pilgrim's Guide*, popular though it became, copied and circulated widely over the following centuries, would still have been for a small minority of travelers: the vast majority of pilgrims would have been illiterate. Latin was not their native tongue even if they could manage to read it.

Above all this was the pilgrimage of Everyman. Santiago was the most popular goal for pilgrims throughout much of the Middle Ages, engendering an intensity of devotion and affection which under very different circumstances has been maintained to this day. Largely due to this universal appeal across a broad social spectrum, the route to Santiago in its heyday became a high-road of Christian teaching, Christian institutions and Christian art. And on that road all humanity was there. Chaucer's pilgrims were heading for Canterbury, not Santiago; yet even though they were a more varied and colorful bunch of travelers than one might have encountered on our road to Parthenay and Poitiers, they nonetheless illustrate a general truth about pilgrim bands, particularly in the later Middle Ages, which is that their journey could be a social occasion, pleasure-loving and sometimes raucous, as they whiled away the long hours between shrine and shrine, hospice and hospice.

Chaucer's Squire, "a lusty bachelor ... fresh as the month of May" and "syngynge ... al the day," may not have been the typical pilgrim; yet we know that singing was a popular feature of the Santiago pilgrimage. Special pilgrim songs have survived, and so have manuscripts depicting pilgrims on the road carrying musical instruments. Horns, tambourines and the zither were common, though the classic pilgrimage instrument was the hurdy-gurdy. This would have been far too cumbersome for the ordinary pilgrim to carry all the way to Santiago; but we know that there were poet-minstrels, *jongleurs,* who would entertain pilgrims on their travels, breaking off to sing at fairs along the route, or outside churches on special feast days, just as itinerant musicians today will earning an extra buck serenading tourists on market-days or doing the round of cafés and restaurants.

The pilgrimage was a world on the move: a floating population made up of a cross-section of medieval society. At the top end of the social scale there were aristocrats. In Jonathan Sumption's words, "The annual or biennial pilgrimage became a recognized mark of aristocratic piety."[11] And one suspects that the piety may have been largely for show. At the opposite end of the scale were the criminals sentenced to undertake a pilgrimage by the civil courts as punishment for a variety of offences, among the most serious being arson. Such sentences were relatively few, partly because offenders could be given the choice of paying a fine instead. This was calculated on a sliding scale depending on the severity of the sentence: the fine imposed as the alternative to going to Santiago was roughly the same as that for Rome, but only a quarter as much if the sentence was to go no further than to Saint-Martin-de-Tours. Then, in the middle of the social scale, was another category of traveler for whom piety was also not the primary incentive for setting out on pilgrimage. These were itinerant merchants and salesmen who posed as pilgrims as a

---

11 Jonathan Sumption, *The Age of Pilgrimage* (New York: 2003).

way of plying their trade *en route* while at the same time enjoying the exemption from taxes and tolls which was a pilgrim's coveted privilege.

Yet the overwhelming majority of pilgrims to Santiago in the Middle Ages would never voluntarily have undertaken so long and hazardous a journey far from home unless driven by a spirit of genuine piety. This was a Christian world obsessed with the expectation of a Second Coming and the imminence of the Day of Judgment. Accordingly it was a society plagued by a sense of sin, and by a desperate need to seek redemption if hell was to be avoided. To go on pilgrimage was one means of attaining such a goal. In our own day of rapid travel and easy communications it is hard to grasp just how isolated Christians in Western Europe could feel from the roots of their faith, especially since the Holy Land itself was now in the hands of the Infidel. By claiming to possess the body of one of Christ's apostles Santiago offered a bridge to the birthplace of Christianity; and by traveling there a pilgrim could feel his own footsteps touching that bridge.

At the same time the incentives to set out on the long road to Santiago cannot have been only the fear of hellfire. Pilgrims were not entirely blindfolded against the pleasures of daily life. Again the lesson of Chaucer is a salutary one: a pilgrimage could be fun. It offered possibly the one and only chance to escape the smallness and monotony of village life—and most people still lived in villages, and rarely moved far away from them. Piety and the freedom of the open road could be happy companions on a pilgrim's journey. Freedom of the road may also have brought a welcome escape from the authority of the parish priest. In the Middle Ages people belonged to their local church in a more total sense than parishioners may feel today. The church might offer spiritual guidance, but it also controlled people's thoughts and daily lives—just as the sheer size of a church with its tower or steeple dominated the mean cottages where most of the congregation lived. Churches delivered a double message: a demand to be obeyed on the one hand—but then, high above as if radiating from heaven itself, the sound of bells. In the words of that eminent scholar of the medieval world, Johan Huizinga, church bells were "like good spirits lifting people's spirits into a sphere of order and serenity."

Together with this sense of liberation from the daily grind, a pilgrimage to a distant land would also offer a unique glimpse of a broader life—a life full of strange sights and sounds. A modern tourist setting eyes for the first time on the Taj Mahal or the Egyptian pyramids would have some idea of the impact of a great abbey or cathedral on a French peasant who had never before left his native village. The shock of so many new experiences would have been accompanied by an exhilarating sense of romance which may have lifted the pilgrim's spirits even more than the sound of church bells. Here was an exciting and brave new world—besides which a short journey beyond the city of Santiago took the pilgrim to the end of the known world: Cape Finisterre, *finis terra*. It is difficult to imagine in our more pragmatic world a comparable feeling of spiritual achievement and sheer wonder. In so many ways a pilgrimage was a journey like no other.

In the words of one of the early abbots of Cluny, Abbot Mayeul, aristocrat and scholar, "life itself is but a pilgrimage, and man lives as a fleeting guest on this earth." Not for the first time the voice of Cluny echoes down the pilgrim roads. And we shall hear it again and again.

4. Churches "more beautiful than before"

THE PILGRIM ROAD NOW CROSSES A REGION that is "well-managed, excellent and full of all blessings"; so *The Pilgrim's Guide* assured the medieval traveler. Such praise is not surprising: this was the author's homeland of Poitou, and not a word was to be said against it, least of all against its inhabitants who were so "prodigal in the hospitality they offer." Aimery Picaud, if he was indeed the author, was never to be so kind again when describing those who lived in other regions. In his eyes the people of Poitou enjoyed a monopoly of all human virtues.

Today a motorway runs parallel to the old Roman road which the early pilgrims would have taken. The tributary road running east from Picaud's Parthenay joins it a little to the north of the regional capital, the city of Poitiers. To the medieval traveler this was the city of St. Hilary, Saint-Hilaire, and his shrine was the most important for the pilgrim to venerate after that of St. Martin of Tours on the banks of the Loire. Hilaire had in fact been St. Martin's mentor, a bishop and doctor in Roman Gaul during the fourth century. So the road taking pilgrims from Tours to Poitiers would certainly have been known to him. The saint's principal achievement, according to *The Pilgrim's Guide*, was to have "vanquished the Arian heresy," that famous theological quibble over the precise meaning of the Holy Trinity. One dissident by the name of Leo, it goes on, was most horribly punished for the error of his ways, "dying abominably in the latrines, afflicted by a corruption of the belly." The saint himself enjoyed an altogether more peaceful end, after which his tomb was "decorated with abundant gold, silver, and extraordinarily precious stones, and its large and splendid basilica is venerated by virtue of its many miracles."

The "large and splendid" Church of Saint-Hilaire-le-Grand had been built in the mid-eleventh century, apparently by an English architect by the name of Coorland—a reminder of England's powerful influence in this whole area of western France known as Aquitaine, long before it finally came under English rule in the mid-twelfth century. This was to take place with the marriage of Eleanor, Duchess of Aquitaine, to Henry Plantagenet, Duke of Normandy and Count of Anjou, who soon became King Henry II of England, ruling a vast territory that stretched from Scotland to the Pyrenees.

The church still stands, more or less in its original form, in a southern suburb of Poitiers on the site of two former churches likewise dedicated to the saint which had been destroyed respectively by the Vandals in the fifth century and by the Normans in the ninth. Grand in name and grand in concept, Saint-Hilaire-le-Grand is vast. Without entirely conforming to the pattern of pilgrim churches that were modelled on Saint-Martin-de-Tours, it was nonetheless designed to accommodate pilgrims in extremely large numbers, and there is no church along the entire length of the road from Paris to the Pyrenees which gives so clear an idea of how a massed congregation of pilgrims could be contained for a special service held on the saint's day—which in this case, so the *Guide* informs us, was celebrated on January 13.

Medieval Poitiers enjoyed an explosion of church building during the years of religious fervor that followed the peaceful passing of the Millennium, when people began to believe that the world was not about to come to end after all—at least not just yet. Most of these medieval churches in Poitiers have survived. For pilgrims there was even a second saint to venerate. She was a woman by the name of Radegonde who founded an abbey here in the sixth century. The eleventh- and twelfth-century church dedicated to her stands tucked among old houses at the eastern edge of the city close to the river. Its splendid Gothic portal, intricately carved, evokes a later era when fears of the imminent ending of the world were long past: God was in his heaven and all was well down on earth. The pall of darkness hanging over people's lives had lifted; and with it some of the urge to go on pilgrimage to save their souls had by now weakened.

In contrast to the flamboyant portal of Sainte-Radegonde is a modest little box of a building a mere hundred yards away. This is a baptistery dating from the very first century of Christianity in Roman Gaul, the fourth; and it bears the special credential of being the earliest example of Christian architecture in the whole of France. It is a somber thought that this place of worship was already eight centuries old when Aimery Picaud was setting off on his pilgrimage from here: and in truth he may not even have been aware of its existence. A more poignant question is—would he have known the greatest of all the medieval churches of Poitiers, Notre-Dame-la-Grande, which was almost certainly being built during his lifetime? This former collegiate church had no specific connection with the Santiago pilgrimage; yet no pilgrim passing through the city from north to south could have failed to notice it, and to wonder at it. Today's traveler may feel the same, because here is one of the glories of church building in France.

Notre-Dame-la-Grande stands in the very center of Poitiers, on the site of what was once the Roman forum. As a fervent advocate of Poitou Aimery Picaud would have been especially proud of the twin pine-cone turrets which so handsomely flank the west front of the building, and were a particular feature of medieval churches in this region. But it is the west front itself which remains among the highest achievements of Romanesque architecture. If there are more than a few square inches of stone left uncarved it would be a surprise. Here is a sermon in stone, spelled out in images across the entire face of the church. Arch above arch, frieze above frieze, are decorated with fantastic creatures inspired as ever by the biblical Apocalypse. They lash and snap at each another, devour their own tails, while in between these zoomorphic nightmares are rows of gentle saints and apostles, all bound together by entwining images of flora and fauna, and by intricate geometrical patterns that lead the eye a dance across the face of this extraordinary church. We are gazing at a carved morality play. And high above this tussle for people's souls on the facade of Notre-Dame-la-Grande stands the figure of Christ in Majesty in his oval-shaped mandorla like a seal of office. Here was the promise of ultimate authority, and the reassurance that the struggle with the forces of evil was all worthwhile. Then, in the midst of all this heavy moralizing we come across an endearing touch—a panel depicting the child Jesus being given his first bath, with Joseph looking awkwardly out of place and keeping a safe distance. Ordinary human sentiments did sometimes find their place amid the rhetoric.

The term "Romanesque" was only coined in the nineteenth century. But it is used to describe a style of architecture which came into being in Western Europe from about the middle of the tenth century in response to an upsurge of religious fervor and self-confidence following centuries of anarchy and violence at the hands of Vikings, Saracens and Magyars—triple invasions from north, south and east which devastated much of the continent. Now the demand was for more spacious churches to accommodate swelling congregations. To satisfy these needs architects turned for inspiration to the most substantial buildings to have survived the long centuries of chaos and disruption. These were the monumental civic buildings which had been created during the Roman Empire as long as a thousand years earlier—temples, amphitheaters, bridges, aqueducts. Nothing so impressive or so enduring had been created in the dark centuries since. So the new churches were constructed in emulation of them, with architects and stonemasons learning afresh those engineering skills at which the Romans had excelled. Doors and windows were no longer narrow slots but broad rounded arches supported by solid columns on either side, while roofs were vaulted so that the weight of a stone roof could be borne on massive piers set on either side of a wide central nave. The result was more space, more light and, in particular, more opportunity for decoration.

For these reasons, especially the last, the Romanesque style of church building became ideally suited to the kind of pictorial sermonizing which is such a dominant feature of Notre-Dame-la-Grande, as it is of hundreds of other churches in this region and along the pilgrim roads in general. Every arch and every space offered invitations for stone-carvers to create a corpus of imagery designed to deliver the Christian message. The facades of these churches are like large-scale illustrated books, each chapter telling some aspect of the bible story for the benefit of a largely illiterate population, each story pressing home the perils of a sinful life and the rewards of a godly one.

It is these elaborately decorated churches which seem to set the tone of daily life at a time in European history when no other sources of teaching were available to most people. Churches held the key to all knowledge and all wisdom.

No modern historian has offered a sharper insight into what it was like to be an ordinary citizen in medieval Europe than the Dutch scholar Johan Huizinga. In his *The Waning of the Middle Ages* he wrote: "So violent and motley was life, that it bore the mixed smell of blood and of roses. The men of that time always oscillate between the fear of hell and the most naive joy, between cruelty and tenderness, between harsh asceticism and insane attachment to the delights of this world, between hatred and goodness, always running to extremes."[12] With Huizinga's words in mind, as we gaze up at the facade of one of these Romanesque churches along the pilgrim roads, it is as if we are reading an open book of the medieval mind.

These new churches, more spacious and elegant than the cramped little structures which preceded them, owed their design and craftsmanship to northern Italy. And as so often in these early years of revitalized Christendom the incentive for establishing them here in France came from the monasteries—chiefly from leaders of the Benedictine Order, and in particular from those formidable aristocrats who were the successive abbots of Cluny, in southern

---

12 Johan Huizinga, *The Waning of the Middle Ages* (London: 1967).

Burgundy. It was in the year 987 that Abbot Mayeul of Cluny persuaded one of the most remarkable churchmen of his day to come to Burgundy. He was William (or Guglielmo) of Volpiano, at that time a monk in the Piedmont region of northwest Italy. William was not only a monk: he was an aristocrat (like Mayeul) with social connections to just about everybody who mattered in Europe. In addition he was a highly-gifted designer and architect—which was principally why Abbot Mayeul invited him.

It was an invitation which proved to be a historic moment in the history of church building in Europe. William accepted, and duly arrived in the Duchy of Burgundy, never returning to live in his native Italy. In the course of his time here, and elsewhere in France, William became to a large extent responsible for establishing the first truly international style of church architecture in Western Europe. He was the father—or more appropriately the godfather—of Romanesque, which was to remain the predominant style of church building for more than two hundred years until the evolution of Gothic late in the twelfth century.

But churches on the scale which William planned required builders and craftsmen with skills far beyond those that existed anywhere in Burgundy or in the whole of France. William was aware, however, that those skills did exist in the region of northern Italy where his own powerful family held sway. This was in Lombardy, the area immediately to the south of the Alps and the Italian lakes. No doubt using the influence of his social position, William succeeded in bringing a body of these Lombard craftsmen to Burgundy for the express purpose of carrying out the work of church building which Abbot Mayeul was proposing. Two immensely powerful men were now working together.

As a result Burgundy became the cradle of what has become known as "the First Romanesque style" of church architecture. By the end of the eleventh century a number of monasteries, priories and parish churches, most of them within a short radius of Cluny, all bore the unmistakable hallmark of these Lombard craftsmen: tall slender bell towers, decorative blind arches set into the walls, saw-edged moulding around the roofline of the apse. The Lombard masons and craftsmen operated within a system of itinerant workshops which were part of an extended mobile community that would have included wives, families, servants and domestic animals. The community established itself "on site" wherever the work was, remaining there until the particular job was completed, before moving on to the next. They were like a nomadic tribe with no permanent home, generation after generation. The Lombards possessed a store of specialist skills which kept them in constant demand over a long period, so that within the next hundred years their influence had spread far beyond Burgundy. Today the churches which bear their unmistakeable signature stand proudly in the European landscape as far afield as Germany, Sweden, the Pyrenees, Dalmatia and Hungary. They remain among the unsung heroes and pioneers of European architecture.

The Lombards were in the forefront of what has been described as a spiritual Renaissance in Europe, of which the pilgrimage movement was a vital manifestation. Right across the continent there arose an urge to build, to give thanks, and to celebrate. People had come out of the darkness to seek a new light. A monk from Cluny by the name of Raoul (Rodulfus) Glaber was keeping a chronicle at exactly the time when the Lombards were building their tall handsome churches right across Burgundy. And he made a memorable observation about the prevailing mood of the time: "A little after the year one thousand there was a sudden rush to rebuild churches all over the world ... One would have said that the world itself was casting aside its old age and clothing itself anew in a white mantle of churches ... A veritable contest drove each Christian community to build a more sumptuous church than its neighbors .... Even the little churches in the villages were reconstructed by the faithful more beautiful than before."[13]

*

13. Raoul Glaber, *The Five Books of The Histories*, c. 1026- (Oxford: 1989).

The old Roman road continues southwest across the low-lying pastures and vineyards of Poitou towards the Atlantic and the broad estuary of the Gironde, and ultimately to Bordeaux. If the medieval pilgrim imagined that after the churches of Poitiers he might be free of didactic sermons in stone, he was in for a shock: because within little more than a day's journey he was confronted—right beside the road—by the Church of Saint-Pierre at Aulnay.

The village of Aulnay itself is out of sight, well to the east. But then this was not intended as a village church. It was always for pilgrims. And if the Poitiers churches had been an illustrated book, Aulnay was an illustrated encyclopaedia. It is hard to imagine what the foot-weary traveler would have made of it. Everywhere we look a face looks back at us—a face designed to inspire terror or hope: birds, horses, mermaids, four-legged creatures with wings and beaks, couples entwined by their tails, beasts bound by vine tendrils, elephants, grimacing giants, moustached owls; and then saints, apostles, angels, knights, maidens. The grotesque and the serene live side by side: the light and the dark. The south portal of the church is most intensely carved—one of the most glorious ensembles of Romanesque sculpture in Europe, with four semi-circular bands of massed figures curved above the door, one above the other. Four knights crush demons beneath their shields and spears. Below the knights, along the inner band, is a frieze of beasts caught within the tendrils of a climbing plant. The next tier depicts the twelve apostles and the twelve prophets of the Apocalypse bearing crowns of gold and musical instruments, and vials filled with odors, after the account given in the fourth chapter of the Book of Revelations:

> ...and, behold, a throne was set in heaven, and *one* sat on the throne. And he that sat was to look upon like a jasper and a sardine stone: and there was a rainbow round about the throne, in sight like unto an emerald. And round about the throne were four and twenty seats: and upon the seats I saw four and twenty elders sitting, clothed in white raiment; and they had on their heads crowns of gold.

Only here the sculptors, having already carved twenty-four figures on the band below, took the liberty of raising the number of elders to thirty-one. The precise meaning of these figures, with their crowns and their musical instruments, has remained a topic of debate and dispute for centuries, as to whether they are angels, saints in heaven or simply elders of the Church. But if the answer has been lost on the tides of history what remains is an endearing theme that is echoed over and over again on Romanesque churches in France, like a chorus line forever coming forward to sing and play for us as if to offer a welcome relief from so much persistent gloom.[14]

But it is the outer band of carving on the Aulnay south portal which is the most extraordinary. Here is an illustration of the medieval mind at its most nightmarish: an evocation of chaos, of life without God, a world without love, made up of hideous creatures who do nothing but howl and grimace and chatter like the occupants of an imaginary madhouse. It is yet another dark and visual sermon, designed as ever to assault weary pilgrims making their way into the church to celebrate mass, only to endure an actual sermon delivered most likely on the very same theme. One begins to feel sorry for them. As if months of foot-slogging across unknown lands were not testing enough, to be reminded of hell fire at every place of prayer would seem an undue extra burden. Today's traveler may enjoy the privilege of appreciating these apocalyptic nightmares as works of art. The medieval pilgrim would have possessed no such sophisticated detachment: the very concept of "art" as we understand it would have been quite unknown to him. It is we who, in a sense, have become "illiterate": ignorant of the meaning of these images, insensitive to medieval pilgrims' terrors and anxieties, and by and large unlikely to be sharing the naked simplicity of their faith.

The south portal at Aulnay was carved about the year 1130— around the time St. Bernard was drawing packed congregations to his sermons and converting the elderly Duke of Aquitaine at Parthenay; perhaps, too, at about the time when Aimery Picaud

---

14. See Plate 6.

was gathering material for *The Pilgrim's Guide* in his monastery at Parthenay-le-Vieux, as well as preparing to set off on his own pilgrimage to Santiago. It was a time of fervent energy in all matters of faith, especially in church building. Within the next fifty years there would scarcely be a village within a short radius of Aulnay that would not have a new church decorated, to a great or lesser extent, by the same didactic school of stone carvers.

It was almost two centuries since William of Volpiano had brought his troupe of skilled masons and sculptors from Lombardy to Burgundy to build the first Romanesque churches in France. Since then other workshops had sprung up all over the country: at least six different "schools" of church architecture were operating by the time *The Pilgrim's Guide* was being compiled, each with its distinctive regional style of building and decoration. Today, if we follow those four principal roads listed in the twelfth-century *Guide*, one of the enduring pleasures of the journey is noticing how the appearance and the mood of those churches alters as we move from region to region.

A short distance south from Aulnay pilgrims once again found themselves being given firm guidance by the *Guide* as to where they should go next. "One must visit the venerable head of the Blessed John the Baptist, brought by ecclesiastics from the land of Jerusalem to a place called Angély, for safekeeping in the land of Poitou." This instruction comes from the chapter devoted to saints who should be revered along the route, and it immediately follows Picaud's chapter, already quoted, on "The Quality of the Lands and the People along the Road." While there is no proof that Picaud may have been responsible for this chapter as well, the pride with which the writer claims the head of John the Baptist to be in the safekeeping of Poitou suggests the two authors may be one and the same. Where else, he seems to be suggesting, would so holy a relic have found so safe a haven?

Today the small town of Saint-Jean-d'Angély retains its connection with John the Baptist, though only in name. The head is actually believed to have been brought not from Jerusalem but from Alexandria, where it was given to the Benedictines in the ninth century. They proceeded to build a great abbey to receive it, and where it was placed on display for flocks of pilgrims to venerate, "worshipped day and night by a choir of one hundred monks," so the *Guide* continues. Even at the time there was clearly some question as to its authenticity since the author felt the need to claim numerous miracles as "the reason why one believes this one to be the true head of the saint." His implied doubts were well-founded. Several other claimants to ownership of the Baptist's head began to dent the abbey's special prestige. Worse was to follow. Saint-Jean-d'Angély became a Protestant town during the sixteenth-century religious wars, and the Reformers duly burnt down the abbey along with the abbey church and—we musts assume—the sacred head.

There were saints galore to venerate in this region. A further short journey south along the Roman road took pilgrims to a town actually called Saintes. It took its name in honor of a third-century Christian martyr, St. Eutrope, about whom *The Pilgrim's Guide* had an enormous amount to say. He is described as coming from a noble Persian family, converted to Christianity and then traveled widely preaching the gospel, arriving eventually at "a city called Saintes" (presumably still known by some other name), which was "prosperous in all goods and provisions, abounding in excellent meadows and clear springs ... provided with handsome squares and streets, and beautiful in many ways." Again we can detect the familiar note of pride in Picaud's native region. Yet Eutrope and his Christianity were not welcomed here. He was driven out, and returned to Rome to lick his wounds. Later he returned, and settled himself in a hut outside the town walls, where he was befriended by a beautiful girl who became his devoted disciple, until a jealous father recruited a hostile mob who came to the hut and "stoned the saintly man of God" and finally "slew him by beating his head to pieces with hatchets and axes."

The richly flavored account of St. Eutrope's life and death occupies more than twice the space of any other entry in the chapter on saints to be venerated along the pilgrim roads. Being a local saint

Eutrope clearly meant much to the author. But his heroic tale also tells us a great deal about the nature of religious faith at this period of resurgent Christianity in Western Europe, and why it is that the pilgrimage movement epitomizes the very spirit of that faith. Besides the story's central theme of a spiritual crusade in the face of barbarism and ignorance, a number of other threads are intertwined—a taste for adventure and exotic lands, the importance of myth and legend, belief in the power of miracles, and nostalgia for home and home comforts. Along with these threads is a strong sense of romance, even of romantic and impossible love, reminding us that this was the era of the courtly troubadour poets, and that the local duke at this time, William IX of Aquitaine, has been described as the first troubadour poet.[15]

As we look back at the social phenomenon which has been the Santiago pilgrimage, from the Middle Ages onwards, all these diverse elements would seem to have been present to some degree: and maybe it is the interweaving of these varied threads which accounts for the impact the pilgrimage made on the medieval imagination— and, in our very different world, does so still. A pilgrim's costume has been a coat of many colors.

*

St. Eutrope's tomb still rests in the crypt of the priory church in Saintes which the monks of Cluny built in his honor during the eleventh century. The crypt is all that remains of the original church, which must have been enormous, certainly as large as the Saint-Hilaire pilgrim church in Poitiers, and was designed for precisely the same purpose, in order to accommodate huge congregations, many of them pilgrims on their way to Spain and Santiago de Compostela.

As the *Guide* makes clear the fame and popularity of the saint's shrine was to a great extent due to the healing powers it was believed to possess. "The lame stand up, the blind recover their sight, hearing

---

15. Successive Dukes of Aquitaine were all called William. The troubadour poet, William IX, was the father of Duke William X, the Santiago pilgrim who was also the father of Eleanor of Aquitaine, who married Henry II of England (see Chapter 3).

is returned to the deaf, the possessed are freed, and all those who with a sincere heart ask for it are accorded salutary help." A belief in miracles was central to medieval religious thought. Priests who were attached to a church fortunate enough to have obtained a relic with such powers would compile lists and detailed accounts of the miracles performed, which would then be circulated, thereby recruiting more sufferers eager to be cured, along with all manner of other hangers-on. As Jonathan Sumption has put it in *The Age of Pilgrimage*, "Every major shrine was perpetually besieged by a motley crowd of pilgrims, hawkers, musicians, beggars and idlers whose appetite for new wonders was insatiable." Shrines, and the miracles performed there, could promote a veritable industry of healing, and a highly profitable one, most of all for the church or abbey concerned, for whom donations made in gratitude for miraculous cures became its principal source of income. A holy relic that was stolen, lost or somehow discredited could bring financial ruin to a church or abbey.

The old town of Saintes remains overloaded with medieval churches. But across the River Charente lies a particular treasure of Romanesque architecture. This is among the earliest and most splendid in the region, the Eglise Sainte-Marie: it was attached to an abbey which would certainly have been visited by passing pilgrims, though only by a special minority of them, namely women. This is the Abbaye aux Dames, an immaculately blue-blooded institution headed by an abbess who was always chosen from one of the noblest families in France, and who also supervised a ladies' college that was likewise for the daughters of the rich. What the young ladies actually learned is not on record.

The abbey church itself, Sainte-Marie, was built early in the twelfth century, and it offers one of the earliest examples of those sculpted bands of figures, arranged in tight ranks above the principal door in a series of semi-circles—six of them—all densely and intricately carved, the figures radiating outwards from the hand of God in the very center. Apart from images of angels, suffering martyrs and symbols of the four evangelists the number of crowned elders from the Apocalypse has now grown far beyond the thirty-one at Aulnay to a veritable orchestra of fifty-four, all of them arranged intimately in pairs who face one another with an air of cheerful piety as they play their musical instruments. It is a delightful human touch. The skilled craftsmanship in stone, which is such a feature of the churches along this stretch of the pilgrim road, was clearly in high demand, particularly at this time when it was thought vital by religious leaders that the message of Christianity be broadcast on the face of every new place of worship. The Church was expanding greatly. A missionary spirit was abroad, and sculptors and stonemasons had become essential agents in a vast proselytizing campaign. They became part of another traveling workshop in the manner of the Lombards in Burgundy a century earlier. Along with pilgrims they traveled the same roads. Hence the design of church facades in this region of France, typified by that of Aulnay and Saintes, became carried by craftsmen over the Pyrenees and into Spain, where a number of churches along the *camino*, even as far as Santiago itself, display the unmistakeable "signature" of these skilled carvers and masons. And as a result many of these itinerant French craftsmen working on the churches of northern Spain built in the wake of the Reconquest then chose to settle in their new lands along the pilgrim road, as Spanish towns with names like Villafranca bear witness today.

"Then comes the country of Saintonge," explains *The Pilgrim's Guide*. This was the region close to the Atlantic, with Saintes as its capital and administrative center. The pilgrim road runs through the very center of it, beckoning the traveler to larger objectives— Bordeaux, the Pyrenees and Spain. Yet in this rural backwater it is as though the dynamic of the Santiago pilgrimage spread out across the area like a ripple of the nearby ocean, depositing churches as exquisite as jewels in every village. The Saintonge remains one of the great treasuries of Romanesque architecture in France, with more than one hundred churches dating at least in part from the twelfth century, a great many of them magnificently decorated.

Making a selection is inevitably personal, so wide is the choice. One gem is the church at Corme-Royal, a short distance off the main road from Saintes towards the oyster-beds of Marennes. Here the west door of the Church of Saint-Nazaire, protected from the eroding sea winds by a high wall, is a scaled down version of the portals at Aulnay and Saintes. The carved images tell familiar stories, the parable of the Wise and Foolish Virgins, figures representing Virtue and Vice, various saints and apostles including one who can only be St. James, represented as a pilgrim and carrying a staff—as he frequently is in church paintings and illuminated manuscripts: the apostle as a pilgrim traveling to his own shrine.

A few miles away is Notre-Dame in Rioux, one of the most perfect churches in the Saintonge. Now the decoration across the entire face of the church is geometrical: diamond patterns, shell shapes, floral motifs, zigzags, criss-crosses, curlicues, chevrons, dogs-teeth, stylized foliage twisting and entwining on and on, as well as numerous other motifs spread everywhere. Scarcely a block of stone is left bare. If a church could be embroidered, it is Rioux.[16] Then, unexpectedly amid all this geometry, set into the central arch above the west door, is a carving of a smiling Madonna and Child so touching it would grace any museum in the world.

Then, after Rioux—Marignac, and Saint-Sulpice, its facade weathered by the Atlantic gales, its three-lobed apse commanding a landscape of vines and crumbling stone barns; further east, Saint-Martin at Chadenac, with its elegant figures of archangels whose robes ripple and billow like those of a Botticelli Venus, and which the great American scholar of the pilgrim roads, Arthur Kingsley Porter, considered to be finer even than the figures on the west door of Chartres; a short distance further, Échebrune, where the Church of Saint-Pierre is tucked into the woods with another facade of geometrical tracery, and whose lower columns display crude graffiti of a procession of pilgrims recognizable by their hats and staves[17]; and Perignac, where another Church of Saint-Pierre is fringed today by Cognac vineyards, with its wonderful mandorla, or oval-shaped

16. See Plate 7.

17. See Plate 8.

panel, of Christ flanked by two archangels with flowing robes, and an extraordinary band of horses' heads high above the west window.

The Saintonge offers a banquet of the Romanesque. And Romanesque is quintessentially the architecture of the pilgrim roads. When the monk of Cluny, Raoul Glaber, wrote that "even the little churches in the villages were reconstructed by the faithful more beautiful than before" he would have been thinking mostly of his native Burgundy, and at a time at least a century before these little churches of the Saintonge were created. Had he lived in the twelfth rather than the eleventh century Glaber would certainly have warmed just as fulsomely to these little jewel-boxes on the far side of France. And his celebrated observation about the world "casting aside its old age and clothing itself anew in a whiter mantle of churches" would have applied just as vividly here near the windswept shores of the Atlantic as in the sheltered plains of Burgundy.

5. Battles Long Ago

A GREAT DEAL MORE EVIDENCE SURVIVES of where pilgrims prayed than where they slept. Occasionally, though, the pilgrim roads offer a surprise. Half a day's journey on foot from Saintes took travelers to Pons. Today a motorway bypasses the town, and so does the local main road which preceded it, the N137. As a result, what was once the Bordeaux road out of town appears abandoned in the past, which has happily enabled a unique treasure to be preserved. This is the remains of a twelfth-century pilgrims' hospice, described ironically as the *Hôpital Neuf,* the New Hospital. The main building was long ago converted to domestic and other uses; but what has survived is its stone entrance by the side of the old road, and that of the chapel facing it, as well as more than twenty yards of stone arches which linked the two and gave pilgrims protection from the elements while they waited to be admitted at nightfall. And here is where they sat, resting after the day's long slog in the heat, whiling away the time.

We can still feel their presence, because scattered around these handsome Romanesque arches and columns are the marks which those pilgrims left while they waited on the broad stone benches provided. They are travelers' graffiti scratched deep in the stone. Some are little stylized human figures, perhaps semi-comic self-portraits. There are a number of crosses of different kinds, and at least one heraldic shield, evidently of some passing nobleman not so proud as to spurn a humble hospice. But for the most part the marks are inverted horseshoes, many of them with a cross set within, and which would have been instantly recognizable to everyone as a pilgrim's symbol—a traveler with a mission. They are strangely moving, these personal autographs: they remind us how little there is along the pilgrim roads which speaks of individual people undertaking this journey. Gazing at this graffiti at Pons we are suddenly made conscious that pilgrimages were about human traffic—made up of vulnerable, hopeful people whose homes were far away, seeking a place to rest and feed, and the warm company of fellow travelers. To be a pilgrim was to be a member of a rather special club. And these marks scratched into the wall were a kind of membership badge.

Most pilgrims would have been strangers—even foreigners—in every place they stopped. Their looks, and certainly their language, would have been locally unfamiliar. So, like their customary outfit of staff, wallet, hat and scallop-shell, pictorial symbols like these would help identify a stranger's purpose for being on the road, or asking for food and shelter for the night. They would also have contributed to the protection which the pilgrim was entitled to enjoy. An eleventh-century Archbishop of Lyon made this pronouncement: "All those who travel to the shrines of the saints are safeguarded against attack at all time." Early in the following century the Lateran Council in Rome decreed that anyone who robbed a pilgrim would be punished by excommunication. This decree was to prove of considerable benefit to Santiago pilgrims, and it can hardly be a coincidence that the pope who issued it was none other than Calixtus II, the very pontiff credited (albeit falsely) with being the author of part of *The Pilgrim's Guide*. Maybe it was a form of posthumous thanks by those who compiled the text to a Church leader who had done so much in his lifetime for the wellbeing of pilgrims.

Hence the pilgrim came to be protected by a form of international law: and the same legal protection also absolved him from paying tolls and tariffs on his travels, as well as entitling him to charity and to safe conduct. Anyone robbing or attempting to kill him was now liable to the severest of all punishments in a Christian world, namely the punishment of God who was held to be the pilgrim's divine guardian.

A further protection consisted of the credentials he carried with him. These would have been provided probably by his local priest; or possibly by the local lord to whom he was a vassal and who under the terms of feudal law effectively owned him, and whose permission to depart on pilgrimage would have been essential. Besides such documentation the pilgrim also benefited from the medieval code of chivalry as practiced above all by the various religious orders of knighthood which were springing up at this time, in particular the Knights Templars and the Knights Hospitallers. The Templars, the "Poor Knights of Christ and of the Temple of Solomon," had been established in the wake of the First Crusade in order to protect holy places in Palestine which had been "liberated" from the Infidel; while the role of the Hospitallers, who were the Knights of St. John of Jerusalem, was to protect Christian pilgrims visiting those sites. Both knightly orders were keen to extend their traditional duties to safeguarding the roads to Spain on the grounds that pilgrims to Santiago were seen to be engaged in the same Holy War as the crusaders by contributing indirectly to the Reconquest of Spain. Accordingly both the Templars and Hospitallers became as prominent as the Church itself in providing facilities for the welfare of pilgrims in the form of hospices and fortified garrisons— if necessary supplying armed protection against bandits and wayside robbers. They acted as benevolent hosts and a police force at the same time.

Hospices, like the one at Pons, were generally attached to some religious house, either an abbey, a priory or a fortified *commanderie* of the knights. They varied considerably in the services they provided. Most were fairly basic, offering the simplest of meals with maybe some sour wine if the pilgrim was lucky; otherwise it was just bread and water, followed by a bare room with a large straw mattress on the floor on which a varying number of people would be required to sleep, accompanied by an even larger population of lice and fleas. But there were also more comfortable establishments, particularly where the presiding monastery had grown wealthy through donations, or where the Christian ethic of charity towards the poor prevailed. And pilgrims were of course supposed to be poor, even if there were noblemen who made the journey accompanied by a retinue of servants. Besides the rudimentary hospices there were also inns, for which the pilgrim needed to pay. These were often identified by scallop-shell images fixed to the wall, or by some other traveler's symbol. What pilgrims received for their money evidently varied: if *The Pilgrim's Guide* is to be believed, medieval innkeepers, along with ferrymen and toll-gatherers, tended to deserve all those punishments of Hell so gruesomely described in church carvings all along the pilgrim roads.

*

Those jewel-box churches of the Saintonge are never far from the salt spray of the Atlantic; but one of them is perched on the very edge of the ocean. This is the Church of Sainte-Radegonde at Talmont, high on a cliff-edge close to the broad mouth of the Gironde estuary. All buildings and walls that once surrounded it have long ago fallen into the water, leaving the church perilously alone, battered by the winds and tides. In the Middle Ages pilgrims coming to Talmont were those who had skirted the coast from Brittany, and now needed to cross the estuary before continuing their journey south across what today are some of the rich Bordeaux vineyards of the Médoc. But their immediate destination was a deep natural harbor on the far shore of the estuary. This was Soulac. And here they were liable to find themselves in unexpected company. Anchored in that deep harbor, long since swallowed up by the Atlantic, there would often be one or more trading vessels which had just reached these shores having survived a stormy voyage across the Bay of Biscay and, before that, the English Channel. The cargo they would soon be loading for the return voyage consisted of oak kegs of Bordeaux wine destined for the dinner tables of the well-to-do in London, Winchester and Canterbury. The ships would be capable of storing huge quantities since they were now emptied of the cargo they had brought here from England. That cargo was: pilgrims.

It was a profitable two-way traffic, particularly once England came to own this whole region of France in the mid-twelfth century.

The only losers in this busy trade were the English wine growers who—largely for reasons of climate—could never hope to compete with those of Bordeaux: hence the long decline of the English wine industry. So the English gentry developed a taste for French wines, while English pilgrims took advantage of a short-cut on their way to Santiago.

The most popular sea route to and from England was from the port of Bristol, in the west of the country and therefore conveniently placed for access to the Atlantic. The number of English travelers undertaking the pilgrimage to Santiago is unknown, though some idea of its popularity may be gauged by the fact that there could be as many as six English vessels moored in the harbor at Soulac at any one time. Another clue to numbers is the fact that more than four hundred English churches are known to have been dedicated to St. James during the Middle Ages. One of these churches, in the village of Stoke Orchard in Gloucestershire, even possesses one of the earliest known cycles of wall paintings on the life and legend of St. James: these date from the early thirteenth century and are more or less contemporaneous with the famous stained-glass windows on the same theme in the French cathedrals of Chartres and Bourges.

For English pilgrims to Santiago the golden age was the twelfth century, as it was for pilgrims in France and other countries. Numerous conditions made this possible. This was an era of relative peace in Western Europe. The institutions of the Church had grown stronger and more influential, creating a stabilizing effect on society as a whole, and accordingly making people's lives generally more secure. Much of this sense of security was the product of the new feudalism which now controlled European society. Bred out of the chaos of the later Dark Ages, feudalism was a finely balanced ladder of legal obligations whose rungs stretched across every layer of society from monarch down to serf. Everyone was subject to the authority of someone above them, from the lowly to the highest. Feudalism had emerged in response to a desperate need to deal with foreign invasions that had repeatedly ravaged the continent of Europe since the eighth century. The lynchpin of the whole system was the local lord, to whom everyone on his lands owed allegiance as a vassal. He ruled and effectively owned them, and in return he was obliged to protect them. His castle was his fortress, paid for by the taxes he exacted. Yet the local lord had his obligations too: he was the vassal of some overlord. Only the king, or the pope, at the top of the ladder, were exempt—except that they needed to regard themselves as vassals too... of God.

Feudalism had been brought to England as a result of the Norman Conquest in 1066. Over the following centuries, and in France and elsewhere in Europe, it resulted in the kind of well-ordered society in which pilgrims in ever greater numbers could feel secure enough to embark on a pilgrimage for six months without the fear of their legal rights or their property being removed in their absence. Pilgrims were now protected by royal decree as well as by papal decree.

This was the social climate which enabled Chaucer's motley bunch of pilgrims to undertake their cheerful journey to Canterbury, telling stories as they went. One robust member of that company was the formidable Wife of Bath, who had been married in church to five husbands, not to mention enjoying "oother compaignye in youthe." She was a lady who had found the time and the means to travel to Jerusalem no fewer than three times, in addition to having made her own pilgrimage to Santiago, as Chaucer respectfully explains— though in what spirit Chaucer spares us the details.

The fact that England owned and ruled this region of France greatly facilitated the movement of English pilgrims embarking near Bordeaux before making their way south towards Spain. In any case, the Kings of England were at this time considerably more French than English: the ruling Plantagenet dynasty originated from the Counts of Anjou, on the Loire. The English monarchs spoke French as their first language and generally spent considerably more time in France than in England. And since they also came to own— through marriage—vast chunks of what is now France there was no meaningful distinction between "home" and "abroad."

Equally important for the pilgrimage movement were the strong links which the Church in England had established with the monastic orders in France. These were orders deeply committed to sponsoring the Santiago pilgrimage, to a large extent for political reasons relating to the Reconquest of Spain. Pilgrimage and Reconquest were virtually synonymous in the minds of Church leaders. At the gateway of Spain the Augustinian monastery of Roncesvalles, high in the Pyrenees, kept a hospice in London, at Charing Cross. This London hospice also acted as an information center for prospective pilgrims to Santiago, supplying names of places along the route in France and Spain where pilgrims could safely lodge. But chief among these monastic orders were the immensely powerful Benedictines of Cluny. In the years immediately following the Norman Conquest the great Burgundian abbey of Cluny supplied monks to a number of new monastic foundations in England, principally Lewes Priory (founded by William the Conqueror in 1077), Faversham (founded by King Stephen), and the royal abbey of Reading, founded by the Conqueror's son Henry I, whose daughter Matilda made the pilgrimage to Santiago in about the year 1125. Here she was presented with no less a holy relic than a hand of St. James, which on her return she presented to her father's abbey, immediately making Reading the focus of the St. James cult in England.

*

Such was the background to those English vessels moored in the harbor of Soulac, having shed their cargo of pilgrims who would now be heading south across the Médoc lowlands before joining up with the principal body of French, German and other pilgrims as they approached Bordeaux.

A short distance inland the main pilgrim road continued south from Saintes and Pons, keeping the broad estuary of the Gironde just out of sight until eventually it reached the water's edge at the old Roman town of Blaye. Here, a few miles north of Bordeaux, medieval pilgrims made their first encounter with an enduring legend which lies at the very root of the St. James story. In a church on the seashore at Blaye, so *The Pilgrim's Guide* explains, "the remains of the Blessed Roland the martyr rest. Belonging to the noble family circle of Charlemagne, he entered Spain in order to expel the Infidel." The legend relates to the emperor's campaign in Spain in the year 778 in which Roland, Count of Brittany and nephew of Charlemagne, was killed in an ambush on the emperor's rearguard in the Pyrenees above Roncesvalles, having famously blown his ivory horn too late to alert the emperor. Accordingly a grief-stricken Charlemagne then built a church at Blaye in which to bury his beloved nephew and military commander.

The key to the legend is the first French epic poem, *La Chanson de Roland*, The Song of Roland. This survives in various versions first written down (in Old French) around the middle of the twelfth century, but was certainly recited by traveling poet-minstrels along the pilgrim roads long before that time. The *Chanson* is a highly romanticized account of Charlemagne's Spanish campaign culminating in Roland's three mighty blasts on his horn before he dies.

How the Charlemagne legend became attached to that of St. James and to the Santiago pilgrimage has long been a subject of heated debate among scholars, kept permanently on the boil since there can never be a definitive answer. However, most eyes have been pointed in one direction as the place most likely to have been the key player in creating the link: and as so often in the story of the pilgrimage that place is the great Burgundian abbey of Cluny, which had a profound commitment to the Santiago pilgrimage as well as to the Reconquest of Spain. We know that passages of *La Chanson de Roland* were regularly recited, or sung, along the pilgrim roads by poet-minstrels for the entertainment of travelers. But they also sang *Chansons de Saint-Jacques*, Songs of St. James, which told of the many miracles which the saint had performed, and his numerous heroic deeds against the Saracens in the cause of the Reconquest. A number of these pilgrims' songs are recorded in the *Codex Calixtinus,* the manuscript which includes *The Pilgrim's Guide*, one of these songs appearing in the name of the *Guide*'s likely author, Aimery Picaud. And since its final words are that it was written "mainly in Cluny," it is hard not to see the Song of Roland and the Songs of St.

James as having been promoted, at least, by the abbey in order to add heroism and prestige to the Santiago pilgrimage, coloring it with the authority of history.

But the strongest link between the Charlemagne legend and the pilgrimage is again provided by the *Codex*. One section of the manuscript is the so-called chronicle of Archbishop Turpin, the *Historia Caroli Magna*, the History of Charlemagne. Turpin was an eighth-century archbishop who also features in the Song of Roland as one of Charlemagne's trustiest warriors in his Spanish campaign. And there is no doubt that he existed. Yet his chronicle, as it appears in the *Codex*, is generally described as the *Pseudo-Turpin* on the grounds that Turpin himself had nothing whatsoever to do with it, since it was written at least three centuries later. As with the attribution of *The Pilgrim's Guide* to Pope Calixtus II, this is another case of imaginative name-dropping. The preface to the chronicle presents Turpin as "archbishop of Reims and faithful companion of Charlemagne in Spain." In this latter role Turpin becomes the eye-witness to Charlemagne's campaigns, during which the emperor has a mysterious dream of the Milky Way as a kind of celestial signpost directing him westwards. Its significance is then explained to him by the appearance in his dream of the apostle St. James. The stars, he is told, are beckoning him to the saint's tomb (as yet undiscovered). But at present the road to the tomb is overrun by the Infidel, and urgently needs to be opened up and safeguarded in order that pilgrims may travel there and venerate the saint.

Such is the clear message of the chronicle. As it happens *La Chanson de Roland* also contains the story of Charlemagne's dream: but the account in the chronicle of Turpin differs from it in one crucial respect. In the Song of Roland the emperor is visited by the archangel Gabriel. But in the Turpin chronicle the visitor has become the apostle St. James, so giving the story an entirely new slant. Now Charlemagne is not only the champion of Christianity and of the Reconquest of Spain, but has become specifically the champion of St. James, the first man to be made aware of the apostle's tomb, and therefore destined to become the very first pilgrim to Santiago.

Whereas the Song of Roland hailed the triumph of Christianity over Islam, the chronicle of Turpin hailed the triumph of St. James. The shift of emphasis is so deliberate that it can only have been written in order to boost the cult of St. James and the Santiago pilgrimage. And in doing so it was hugely successful. More than two hundred manuscripts of Archbishop Turpin's chronicle have survived, either in Latin or in Old French, so matching the distribution of *The Pilgrim's Guide*, and offering a testimony to its extraordinary popularity throughout twelfth-century France. It became the basis of the widespread belief, held by churchmen and pilgrims alike, that the Emperor Charlemagne was indeed the founding father of the Santiago pilgrimage.

From now onwards all humble souls who trudged wearily across France and northern Spain in order to pray at the apostle's shrine did so in the footsteps of the very first Holy Roman Emperor. It was a psychological masterstroke. More even than the Song of Roland it gave the journey to Santiago impeccable historical credentials, and a welcome touch of glamor.

The chronicle of Archbishop Turpin was of course a fabrication. Yet it seems to have had its roots in popular legend. Its most likely source was those fanciful pilgrims' songs, *chansons de geste*, which had been recited along the pilgrim roads probably for centuries, passed orally from singer to singer, with appropriate elaborations along the way. Then at some stage, in the eleventh or twelfth century, these tales had been cobbled together and written down as a coherent story. But by whom? Theories abound. The cathedral authorities in Santiago have been named, as have various monasteries in France with vested interests in the pilgrimage, including of course Cluny. Copies were subsequently made by scribes in any number of abbeys which possessed a *scriptorium*; and we need to imagine that passing pilgrims, the majority of them illiterate, would have passages read out to them to cheer them on their way.

To what extent all this was a deliberate exercise in propaganda, designed to promote the Santiago pilgrimage, is peculiarly hard to judge. Seen through the lens of our own times it is tempting to envisage some proto-Machiavellian spirit at work conducting a skilful operation exploiting the susceptibilities of ignorant people. There have been scholarly voices claiming that the Turpin chronicle, along with the pilgrims' songs, even the Song of Roland itself, were all purposely sponsored by Church authorities with vested interests in the opening-up of Spain and the expulsion of the Muslims. But to adopt such a modern viewpoint may be a failure to comprehend the workings of the medieval mind, and in particular the essentially mystic nature of the medieval religious experience. When considering these distant legends, and the social context in which they were recounted from generation to generation, we are reaching back into a time when in the minds of ordinary people distinctions between fact and fiction, history and myth, truths and make-believe, were largely blurred. In the words of Johan Huizinga, "in more than one respect life had the colors of a fairy-tale."

The essence of all fairy tales, as of legends, is that they are not presented as historical fact, but as a kind of metaphor. Part of their appeal is that they cast a certain light on the human condition—which in the Middle Ages was perceived by and large to be bleak. Fear of damnation was rife, and those fears were stimulated by sermons from the pulpit and by sermons in stone carved on church portals all along the pilgrim roads. Beyond the Christian world the human lot was perceived to be even worse. The Roland legend and the legend of St. James tell of people in Spain under the yoke of Islam and therefore living in darkness. These legends, relating the rich exploits of Charlemagne, Roland and St. James offered the promise of release, both for the pilgrim personally and for those living under the shadow of Islam. The triumph of Charlemagne represented the triumph of Christianity, of light over darkness. His legend charted a path of bright stars, the Milky Way, by which people could move out of the darkness of their own lives into the enduring light cast by the Christian God.

These stories were metaphors for human salvation, and as such they answered a vast need: hence their popularity. The fact that the legends were—to our modern eyes—simply fiction, not historical fact, is beside the point. The poets who sang them, and those who wrote them down, copied and translated them, were not passing these stories off as literally true. This was not what concerned them. They were like parables, or morality plays. The truth in them lay in what these legends revealed, that God is the path to a better life, and that through worship and love of God human beings can lift themselves out of the miseries of this world into the glories of the next.

In other words, "truth" is what lies embedded in these legends, not in their historical accuracy. This is a concept that modern consciousness may find particularly hard to grasp. Yet without doing so one of the most inspired social movements in history, the Santiago pilgrimage, runs the risk of being seen as merely the product of an unprincipled and exploitative Church manipulating its gullible congregation like sheep to further its own ambitions—which would be a sad and misguided distortion. This book, at least, would not be written if that was believed to be so.

The question remains: where should we look for the source of energy and inspiration behind the dissemination of these vivid tales which fed the pilgrimage movement with poetry, song and romance? Where was the fountain head? Inevitably, we look again towards the heartland of Christian pilgrimage, whether in the form of the crusades, or the cult of Rome, or that of Santiago. And that place is Burgundy, and in particular the formidable abbey of Cluny—wealthy, immensely influential and politically minded, courted by popes and emperors, and patrons of both. The bejewelled hand of Cluny, and of its succession of gifted abbots, touches so many aspects of the Santiago story, as we have already seen. A driving ambition of those abbots, in the eleventh and twelfth centuries, was to establish the power and influence of the French Church throughout northern Spain. Support for the Santiago pilgrimage was a key element in that ambition, and it led to Cluny helping to establish hospices and priories along the Spanish pilgrim road, as we shall see in the final section of this book. But in order to safeguard the road, and the institutions that were strung along it, it was vitally necessary to provide support for the beleaguered Christian rulers in Spain. Cluny itself could not supply military aid; yet it can be no coincidence that the most successful campaigns against the Saracens were those led by knights from Burgundy, many of them no doubt related to the lordly monks of Cluny who would also have been their spiritual mentors.

This was feudal France: and the top echelon of that feudal society was like an elite club whose members all shared the wealth and privilege of high birth, and were made up of Church leaders, senior monks in the local monasteries and priories, and the various lords and landowners of the area. In this latter category were these Burgundian knights who raised local armies, consisting largely of their own serfs, then crossed the Pyrenees in order to drive the Moors from Spain—all of them riding in the wake of Charlemagne, with *La Chanson de Roland* echoing in their ears, and brandishing the avenging sword of St. James in the spirit of Santiago Matamoros.

It was heady stuff. This was a multifaceted feudal world in which spiritual values and a life of violent action were inextricably bound together. And if in this confusing theater of war and peace the abbey of Cluny was indeed the command post, it is impossible to imagine those wise, and worldly-wise abbots experiencing even a shadow of doubt over the rightness of their vast enterprises. The pilgrimage and the Reconquest—they were twin paths in a single journey into the light.

Meanwhile, in Blaye, on the broad estuary of the Gironde opening on to the Atlantic Ocean, today's traveler may well feel that the story of Roland sounding his horn three times before dying in battle and being buried here by Charlemagne is yet another of those fairy tales; because nothing remains here which speaks of that tale. Charlemagne's church and Roland's body have long gone, victims of the English army during the Hundred Years' War in the fifteenth century. On another of the four pilgrim's roads the traveler will later be shown what is claimed to be the very horn Roland blew near Roncesvalles—his famous oliphant. But for the time being the traveler on this pilgrim road can do little more than sit on the shore and reflect on one of the most beguiling legends of the Middle Ages—one that lies at the very core of the Santiago pilgrimage.

6. To the Bad Lands...
and the Good Lands

BORDEAUX WAS THE CAPITAL AND CHIEF PORT of this extensive region of France which belonged to England at that time, known as Aquitaine. In fact, one of the main attractions for Santiago pilgrims as they reached Bordeaux was a particularly well-equipped hospice dedicated to the apostle yet in the English version of his name, St. James, rather than the customary French Saint-Jacques. The Hôpital Saint-James was an English endowment, the main donor being the English King Henry II in the year 1181. It had been Henry's marriage to Eleanor, Duchess of Aquitaine, which had brought this region under the English crown.

The Pilgrim's Guide has surprisingly little to say about Bordeaux beyond the observation that the region is "excellent in wine and abundant in fish," with which any modern traveler would be happy to agree. By now it is beginning to give off a clear message to pilgrims, to the effect that the good times are by and large over, at least until they get deep into Spain. But then the likely author of this section of the *Guide*, Aimery Picaud, was a man fiercely parochial about his native Poitou and the larger region of Aquitaine of which it formed a part, to the extent of finding it grudgingly hard to write anything good about people unfortunate enough to live elsewhere. It is just as well that most medieval pilgrims were illiterate, or at this stage of their journey they would certainly not be looking forward to what was in store for them in the days ahead. Their immediate prospect was to tackle the Landes.

Today the Landes is a national forest, one of the largest in Europe, composed mostly of Maritime Pine. But in Picaud's day most of it was swamp, and about as inhospitable as a landscape could be. We know that he did this journey himself once if not twice; so his description of the Landes is unlikely to have been entirely colored by prejudice. "This is a desolate region deprived of all good," he begins. "There is here no bread, wine, meat, fish, water or springs; and villages are rare here." He then manages to offer a few words of praise in passing. "The sandy and flat land abounds none the less in honey, millet, grasses and wild boar. But if perchance you should cross it in summer, guard your face diligently from the enormous flies that greatly abound there and are called locally wasps or horse-flies; and if you do not watch your feet carefully you will rapidly sink up to the knees in the wet sand copiously found all over."

Nobody is spared the wasps and horseflies today, though the knee-deep swamps of the Landes have long been drained. But we know Picaud was not exaggerating. The modern traveler keen to savor what life was really like here in earlier times would do well to take time off in Bordeaux to visit the Museum of Fine Arts. Here there are romanticized paintings dating from the early nineteenth century, not long before the swamps were finally drained; and they show local shepherds coping deftly with the swampy terrain by going about their work on immensely tall stilts like well-practiced circus clowns.

The old pilgrim road slices through the center of the Landes from north to south, now flanked mostly by pine- and oak-forest, with the occasional clearing where geese bred for *foie gras* wander around in the company of pigs whose distant ancestors Picaud described. A few survivors of the medieval pilgrimage remain, dwarfed by the trees. At Cayac stand the remains of a priory and ancient hospice reminiscent of the one straddling the road at Pons. Here and there on the route beyond Cayac the occasional medieval chapel still lines the old road, long battered by sea-winds; then an ancient stone preaching cross rises up by the roadside, and another former priory church huddled among trees; then on to Belin where a burial mound is traditionally believed to contain the bodies of Charlemagne's warriors killed alongside Roland near Roncesvalles. Belin possesses one further claim to historical fame: it was the birthplace of Eleanor of Aquitaine, the heiress whose marriage to King Henry II of England in 1152, as we have seen earlier, was very soon to bring this entire region into the possession of the English crown, where it remained for the next three centuries.

Finally the vast forest comes to an end, and the landscape softens and opens out into fertile meadows. Across the River Adour and the traveler is now in Gascony—about which *The Pilgrim's Guide* has suddenly a great deal to say. At first Aimery Picaud seems to have had a good experience of Gascony: "rich in white bread and excellent red wine, and covered by woods and fields, streams and healthy springs." But then he encounters the local people, and the tone changes. "The Gascons are quick with words, loquacious, given to mockery, libidinous, drunk, prodigal in their eating habits, ill-dressed and rather careless in the ornaments they wear. However, they are well-trained in combat and generous in the hospitality they provide for the poor." In other words, they were protective and kind towards pilgrims, which almost redeems them in Picaud's eyes.

As the journey towards Santiago progresses *The Pilgrim's Guide* becomes as much a twelfth-century social document as it is a guidebook. And if the chapter on the people and places along the route, as well as several chapters following, are indeed the work of Aimery Picaud, then we find ourselves building up a character-portrait of this cantankerous, pleasure-loving monk from Parthenay-le-Vieux whose abbey church we so recently visited back in his native Poitou. The strength and appeal of his writing lies in the way we are drawn into envisaging him like any modern traveler venturing into unfamiliar lands, noting whatever strikes him and sometimes being taken aback by how different people's habits can be. Hence his extremely mixed feelings about the behavior of the Gascons: "Seated around the fire, their practice is to eat without a table and to drink all of them from a single cup. In fact they eat and drink a lot, wear rather poor clothes, and lie down shamelessly on a thin and rotten straw mattress, the servants along with the master and the mistress."

The eating and sleeping habits of the Gascons were probably the least of a pilgrim's concerns on the road towards the Pyrenees. This was a stretch renowned for being particularly dangerous, doubtless due to the greater concentration of travelers making their way to Spain across the few mountain passes. For all the papal threats of excommunication bands of robbers were everywhere, some of them canny enough to pose as pilgrims with the appropriate clothing. Travelers making the long journey to Santiago were especially vulnerable since they were likely to be carrying most of their worldly possessions with them, such as they were. There is even a record of a blind pilgrim traveling on horseback who was robbed of his money *and* his horse, to be found later wandering helplessly by the roadside.

The journey could be hazardous enough even without the threat of robbery, as Picaud's account of the Landes swamps makes all too clear. At the same time the idea of a hazardous journey was not entirely unwelcome. Whatever the personal motive might be, the pilgrimage was supposed to be a spiritual journey—an expression of penance and a quest for the remission of a person's sins. It was a path to salvation, and this was not a path that was ever intended to be rose-scented. On the contrary it was supposed to be stony. Mortification of the flesh was regarded as a passport to heaven. One of the attractions of the journey was that it was arduous and demanding: in the eyes of the medieval Church physical hardship was a necessary and healing aspect of that journey. In the early centuries of the pilgrimage at least, the long trek to Santiago was certainly not supposed to be a pleasure trip.

Parallel to the celebration of physical hardship lay the cult of poverty. Avarice was a cardinal sin in medieval Church teaching, and to hoard money was the act of a miser, or so ordinary people were told. Congregations were reminded in sermons of the teaching of Christ that it was "easier for a camel to pass through the eye of a needle than for a rich man to enter the kingdom of God." (The fact that churches and religious institutions frequently behaved otherwise was conveniently overlooked.) The personification of this poverty cult, early in the thirteenth century, was St. Francis of Assisi; and it would have surprised none of his followers that St. Francis chose to include Santiago among his numerous barefooted travels. There could have been no more influential a role model: the most celebrated pauper of the day became the most celebrated pilgrim of the day. By his own exalted example, and by his preaching of the virtue of poverty, St. Francis greatly broadened the appeal of the pilgrimage—ironically among the rich as much as among the poor. Furthermore the Franciscan Order, founded by the saint, proceeded to contribute to the pilgrimage movement, adding further to the support already given by other religious orders, in particular the Benedictines and Augustinians, and later the Cistercians.

The Pilgrim Guide's suspicions of the Gascons grow darker the further the author travels from his beloved Aquitaine, and ventures deeper into what he considers to be "barbarous" country in the approaches to the Pyrenees. How much his account of events derives from firsthand experience is impossible to tell; but there comes a point in his account when he feels it imperative to offer the severest warning to all pilgrims traveling this road. The place is the riverside village of Sorde, then a popular stopping point for pilgrims on account of its important Benedictine abbey, today in a semi-ruined state draped in ivy (perhaps held together by ivy), but dramatically situated by the water's edge, with a long curving weir stretching away towards the far bank.

There was no bridge, and the only means of crossing this river was by raft. And here, Picaud explains ruefully, the pilgrim was at the mercy of the local ferrymen—"and may they all be damned," he fulminates.

> They have the habit of demanding one coin from each man they ferry over; whether rich or poor, and for a horse they ignominiously extort by force four coins. Now, their boat is small, made of a single tree, hardly capable of conveying horses. Also, when boarding it one must be most careful not to fall accidentally into the water. You will do well to pull

your horses by the reins behind you in the river, outside the boat, and to embark with only a few passengers, for if it is overloaded it will soon become dangerous. Also, many times the ferryman, having by this time taken his money, now has so large a cargo of pilgrims in his boat that it capsizes and the pilgrims all drown; upon which the boatmen, having laid their hands on the spoils of the dead, wickedly rejoice.

Perhaps Picaud had observed such a scene himself, or more likely it was a favorite piece of juicy gossip being bandied about in the local inn over a jar of wine. We are reminded, reading his account, that it was intended only for the tiny minority of pilgrims who were literate, and in consequence likely to be better off than most, and probably making the pilgrimage on horseback—as Picaud himself would have done. They were, after all, supposedly a party of three pilgrims, the other two being a man and his (supposed) wife. Picaud's description of the ferrymen's murderous behavior at Sorde raises the question of whether he made not one pilgrimage, but two. If the account is based on a personal experience, or at least on stories told to him locally, then the journey to present the *Codex Calixtinus* manuscript to Santiago cathedral must have been on a subsequent occasion, since the *Guide's* account of evil-doings at Sorde are contained in the *Codex* itself.

From Sorde pilgrims had a choice of two roads leading south towards the Pyrenees, depending on whether they had risked the river crossing or wisely continued westwards until a bridge took them on to the old Roman road. Either route led them into the Basque Country, At this point Picaud offers some robust observations. "This land whose language is barbarous, is wooded, mountainous, devoid of bread, wine, and all sorts of food for the body, except that in compensation it abounds in apples, cider and milk."

Picaud regarded the Basque language as barbarous because it was incomprehensible to him. As a monk he would have understood Latin, and spoken any one or more of the local languages, or dialects, all of which derived from Latin—which Basque does not. It is one of the oldest languages in Europe, with its roots deeper than Ancient Greek and Latin and still largely mysterious. Picaud is quite likely to have thought of it as the language of the Devil. He may even have known the story of how the Devil, keen to win the Basques to his side, set himself to learn their language: and in seven years managed to learn three words. Today's travelers in the region may find themselves faring little better.

Village follows village very closely in the Basque Country, always with the snow-covered peaks of the Pyrenees drawn like a long curtain across the road ahead, drawing closer hour by hour. Here and there a house is still marked *Pelegrinia*; a reminder that we are now entering the region where pilgrim roads from far distant parts of France draw closer together and finally converge. At the height of the Santiago pilgrimage, in the twelfth and thirteen centuries, virtually every house in these Basque villages would have offered accommodation—like a modern bed-and-breakfast. It was a regular source of local income during the spring and summer months, and innkeepers would compete with each other for trade, often posting their male children in the street to invite passing pilgrims in. Picaud was as suspicious of such innkeepers as he was of ferrymen.

They were not the only locals he despised. There was a third category of predator of whom the *Guide* warns pilgrims to be especially wary. "In this land," writes Picaud,

> ...in the town called Ostabat ... there are evil toll-gatherers who will certainly be damned through and through. In truth they will actually advance towards pilgrims carrying two or three wooden rods, and proceed to extort by force a totally unjust tribute. And if some traveler should refuse to hand over the money on demand they then beat him with the rods and seize the toll money while cursing him and searching even in his breeches. These are ferocious people, and the land in which they dwell is savage, wooded and barbarous. The ferocity of their faces and likewise of their barbarous speech scares the wits out of anyone who sees them.

He goes on to point out that pilgrims were, in fact, exempt by law from such tolls. This exemption was widely known, though not always widely observed, one reason being that it was also widely abused. Merchants, as well as other kinds of traveler, were well known to pose and dress as pilgrims precisely in order to avoid payments; and it was this practice which may have offered a slender excuse for the behavior of the Ostabat toll-gatherers, at least in their own eyes.

But the main reason that they gathered in such predatory numbers in this small town was that Ostabat was the very first halt for travelers after the junction of three main pilgrim roads: the one from Paris and the Loire, the road from Vézelay and Burgundy, and that from Le Puy and the mountains of central France. Pilgrims would have collected here from every country in Europe, and for toll-gatherers they were easy prey. They were a source of easy profit for innkeepers as well: in the Middle Ages this small Basque town, hardly more than a village, contained at least twenty hospices. From spring through to autumn there would have been an exceptionally large number of pilgrims from all parts of Europe grateful for a roof and a bed of straw in Ostabat.

Today Ostabat gives out scarcely an echo of what the place must have been like in medieval times. For a more genuine sound of the past the modern traveler needs to leave the main pilgrim road for a short distance and head into the woodlands of the Bois d'Ostabat in search of a tiny community by the name of Harambels. Here survives a fragment of the medieval pilgrimage which the modern world has bypassed. The core of this tiny village consists of four houses and a rugged stone chapel dedicated to St. Nicolas, a fourth-century saint widely associated with the giving of gifts (and often believed to be the original Santa Claus). The chapel of Harambels is one of the earliest pilgrim churches in Europe, recorded in documents as early as 1039, at least a century before *The Pilgrim's Guide* was written.[18] The tiny community was built as an extended hospice for pilgrims,

18. See Plate 9.

created by four lay friars who brought their families here. They were all of them Basques, named Salla, Borda, Etcheto and Etcheverry: and such is the timelessness of life in these Basque woodlands that the four houses they built are still owned by the same four families. And the small graveyard beside the chapel is where generations of them have been buried for a millennium.

After the solitude and silence of Harambels the main pilgrim road can seem like a bustling highway—which, relatively speaking, it has been for more than one thousand years. It has been the pilgrims' highway to Spain. The road crosses and re-crosses a fast-flowing river deeply shaded by overhanging trees; and close to one of the bridges is another witness to the history of the pilgrimage—one that is even more secluded than the Harambels chapel. At first it is merely an ordinary track cutting through the undergrowth. Then, as we descend towards the river the track is seen to lead to a line of huge stepping-stones, each well over a yard in diameter, which have been laid in a long curve across the fast current from bank to bank. How old is a stepping-stone? Who can tell? Though there is one thing we *can* tell: these stones must have been put in place at least a thousand years ago, because this simple track, cutting through the forest on either side of the river, is one of the four main pilgrim routes—the one from Le Puy and the Massif Central. The road descends to the river on one side, then on the farther side rises out of the woods to disappear round the flank of the hill to the west.

Where it is heading—out of sight from here—is one of the pivotal points of the entire Santiago pilgrimage. It is a site so understated that you might pass by without even noticing. We are on the crest of a hill the name of which, Mont Saint-Sauveur (Mount St. Savior), at least provides a hint that this is no ordinary hill. The only other clue is a modern *stèle*, a marker stone, set on an ancient carved gravestone.[19] But then if we stand here and gaze southwards we can pick out three tracks threading their way towards us through the fields and the deep hedgerows from three different directions: and

19. See Plate 10.

from the point where they converge they are joined by a lane which runs down the slope to become a single broad track snaking its way through the rough terrain beyond and over the far hill towards the Pyrenees and the Spanish border now only a few miles away.

The three converging tracks are three of the ancient pilgrim roads—the first the road from Paris which we have been following, the second the road from Vézelay, and the third the road from Le Puy by way of the stepping-stones we saw on the far side of the hill. From this junction-point the three tracks become just one pilgrim road which leads over the Pyrenees—until on the far side of the mountains this will be joined by the fourth road, the one from Arles and Provence, from that point on to become the single Spanish road to Santiago, the *camino*.

But that must wait. Meanwhile today's travelers need to follow in the footsteps of their medieval forebears on the second of those "four roads to heaven," beginning many hundreds of miles further north among the noble vineyards and fertile valleys of Burgundy, always the heartland of the pilgrimage movement.

II. Via Lemovicencis:
The Road from Vézelay

7. The Magdalene and Sacred Theft

VÉZELAY IS THE GREAT CHURCH ON THE HILL—a beacon which has guided pilgrims and other travelers in northern Burgundy for almost a thousand years. Originally it was attached to one of the most renowned of Benedictine abbeys, founded by the local count in the ninth century. Today the abbey is no more, but the church remains. It is now the largest parish church in France, and a masterpiece of medieval art and architecture. Anyone keen to make a study of the Romanesque in France could do no better than to begin here. The former abbey church of Vézelay is quite simply one of the most beautiful buildings in the world.

Such was the prestige of Vézelay in its heyday that two crusades to the Holy Land were launched from here in emotional ceremonies attended by English and French kings. In addition, throughout the Middle Ages it was the starting point for the second of the four main pilgrim roads to Spain and the shrine of St. James in Santiago de Compostela.

The early history of Vézelay is as violent and stormy as that of all early abbeys in France. Founded in the mid-ninth century by Count Girard de Roussillon, it was first established in the valley below (now the village of Saint-Père). Within a very short time this first monastery was plundered by Viking invaders from what is now Normandy, whereupon Count Girard had it rebuilt in a location easier to defend, on the nearby hilltop, and here he installed a group of Benedictine monks. He also acquired the added security of making the new establishment a dependency of the papacy in Rome, so protecting it from grasping local landowners. Nonetheless barbarian invasions continued early into the tenth century, not only by the Vikings from the north, but by the Magyars from the east (the region of Central Europe which is now Hungary). It was the most anarchic period in European history, and the new Vézelay, along with much of Burgundy to the south of it, was repeatedly pillaged and wrecked.

The early fate of the abbey mirrors that of France as a whole during these years. This was the Dark Ages at its darkest. It is hardly surprising that few people ventured on pilgrimage in so unsafe a world. The Christian empire which Charlemagne established late in the eighth century out of the ashes of the former Roman Empire soon collapsed and disintegrated, resulting in a plethora of weak little European kingdoms which could do nothing to resist invaders hungry to pluck the flesh off the body of the old empire. France was especially vulnerable, being geographically placed to tempt Saracen invasions from Spain, Viking onslaughts from the north and the Magyars from the east. The economy was crippled: agriculture declined as villages became devastated and deserted, and arable lands returned to forest and wasteland. Life was at a standstill, at best a grim struggle, as people could do little more than rescue what they could from the flood-tide of anarchy. Nothing was safe, and nothing was sacred. Europe bled and burned.

What eventually turned the tide was a combination of various factors, all coming together at much the same time. Crucially, the Norman Vikings finally settled down: early in the tenth century a peace treaty was signed with the powers in France, and Normandy became Christianized, and a powerful independent duchy. One major menace was thus removed. Piety had taken over from piracy. Then, as outlined in Chapter 5, there was the impact of the new strict social order of feudalism. By structuring everyone's life from the highest to the lowest feudalism had the effect of stabilizing society as a whole. People knew precisely where they stood, and what their duties and obligations were, from monarch to count to humble serf. For the first time since the centuries of the Roman Empire society became ordered. One of the duties of a local lord was to be able to raise an army whenever necessary from the serfs who were at his disposal. As a result feudalism created a military structure within that society which at last proved capable of dealing with Saracen and Magyar onslaughts. Force could now be met with force. In turn, protection from repeated attack brought about a revival of agriculture, and hence of the economy in general: farmers could now grow crops and graze animals without the fear of them being burned or driven off.

By no means least, feudalism greatly strengthened the various institutions of the Church, supplying them with legal powers and with a moral authority which had the effect of making them pivotal to the day-to-day life and wellbeing of a community. Altogether life began to feel safer and more benign. A human condition scarcely enjoyed for centuries was suddenly available—a sense of peace. In the words of a distinguished medievalist, Professor Joan Evans, "it seems as if Europe settled on an even keel."

With relative peace and economic stability it became possible for people to travel without undue fear. Merchants could convey their wares. Craftsmen could ply their trades. And pilgrims felt able at last to visit the shrines of the saints.

From the earliest days following the arrival of peace Vézelay became an immensely popular venue for pilgrims. This was no doubt partly due to its location, dramatically placed on a hilltop in the heart of prosperous Burgundy. But there was always a special aura about the place. Its patron saint was the Virgin Mary, which endowed it with a special and feminine appeal. But then before long another female saint became associated with the abbey. She is perhaps the most intriguing saint in the entire Christian pantheon, as well as the most unlikely; one who has exercized the imagination of countless commentators, theologians, fantasists, artists and fiction writers. She is Mary Magdalene. And, improbable though it may sound, the abbey claimed for centuries to possess her body.

How this happened derives from a chain of events which veered from the naive to the shoddy. But it was a claim which in a remarkably short space of time brought a hitherto insignificant Burgundian abbey unimaginable wealth and fame, so much so that by the twelfth century only Jerusalem, Rome and Santiago de Compostela could match its appeal as a center of pilgrimage. The poor gave their pittance, and the titled and the wealthy their gold and their lands. The chronicler Hugh of Poitiers, who was a monk at Vézelay, wrote at this time that "all France seems to go to the solemnities of the Magdalene."

The magical appeal of the saint, and the universal fame of Vézelay, made it not only a shrine to be visited but also a gathering place for travelers heading elsewhere. Large numbers of pilgrims from northeastern France, Germany and the Low Countries now regularly assembled here as they prepared to set off for other shrines, in particular that of St. James in Santiago de Compostela. The road they would take was the *Via Lemovicensis*, the Way of Vézelay.

*

The association of Vézelay with Mary Magdalene was always tenuous and shrouded in mystery and concealment, perhaps deliberately so since there was no apparent evidence to support the claim. In the early days there seem to have been only rumors put about by the monks that the abbey possessed relics relating to the saint. Nothing was specific, just a half-veiled secret, though intriguing enough to arouse expectation and excitement. Before long these subtle whispers that relics of Mary Magdalene were held by the abbey had the effect of attracting pilgrims like bees to a honeycomb. Whether the abbot and his fellow monks actually believed their own story is impossible to establish. Nonetheless the result was a swift and dramatic improvement in the abbey's fame and fortunes.

Inevitably this vague association with the Magdalene soon required clarification, and this duly took place with the appearance, during the tenth and early eleventh centuries, of several lives of the saints which had been translated into Latin from the Greek, and which originated apparently from various monastic sources in the eastern Mediterranean or in Sicily and southern Italy.

The significance of these manuscripts was that they became vital to the creation of the Magdalene legend in France, and therefore to the fortunes of Vézelay abbey. Step by step they established an account of the life and travels of the Magdalene in the years following the Crucifixion and Resurrection of Christ, to which she had been a witness. The earliest of these manuscripts was the *vita eremitica*, a Life of the Hermits and early Desert Fathers, probably dating from as early as the ninth century. The list of hermits somewhat surprisingly included Mary Magdalene. It was claimed that she had fled Jerusalem after Christ's Resurrection to escape persecution and had spent the last thirty years of her life as a hermit in the desert.

This was the first chapter of her reinvented life. Next came the *vita evangelica,* a Life of the Teachers, which appeared about a century later. This manuscript consolidated the earlier account by offering a fanciful homily to the Magdalene, and was clearly designed to boost the prestige of the saint as the repentant sinner who was beloved by Christ, and who subsequently became in a sense an apostle to the other apostles. This was chapter two. Now Mary Magdalene was not only a dedicated hermit but a very special disciple.

Progressively a life—and a way of life—was being sketched out for the Magdalene, filling in the unknown years of which the Bible makes no mention at all. Only a number of apocryphal texts, including a Gospel of Mary Magdalene, make further reference to her. But the crucial group of manuscripts linking her story to France consisted of several Lives of the Apostle, *vitae apostolica,* which emerged during the course of the eleventh century. Now it was explained that Mary, together with several other followers of Christ including her brother Lazarus, escaped persecution in Jerusalem by fleeing on a vessel to the coast of France, arriving at what is now Marseille, whereupon they proceeded to evangelise pagan Gaul, at that time under Roman rule. Mary Magdalene was said to have preached near Aix, choosing the life of a hermit until at her death she was buried there by St. Maximin, another of the former followers of Christ who had been a fellow occupant of the vessel that brought them to these shores.

What these Lives of the Apostles successfully achieved was to transplant the scene of Mary's thirty years as a hermit from the deserts of the Bible Lands to Provence. And the story did not end there. In a further manuscript, almost certainly originating in Burgundy, it was claimed that in the year 749 the then Abbot of Vézelay had despatched one of his monks, a certain Badilus, to Provence for the express purpose of bringing the body of Mary Magdalene back to Burgundy, thereby safeguarding the saint's relics from Saracen invaders who at that time were regularly plundering much of Provence.

The legend was now complete. Vézelay had been the custodian of the saint's body for at least the past four hundred years, so people were led to believe. And for those who might question the likelihood of such a tale a commentator at the time had the perfect response. "Many people," he wrote, "ask how the body of St. Mary Magdalene, whose native land was Judaea, could have been translated to Gaul from a land so far away. To such doubters we may make a brief reply. All things are possible to God." To this observation a scholar of our own day, Professor Francis Haskell, has wryly added: "How Mary Magdalene's relics came to Burgundy, how she became a heroine of the church, and one of the most loved saints of the Middle Ages, is one of the great romances of the age of chivalry."

The rise of Vézelay to become one of the foremost places of pilgrimage in the Christian world, and her position as a major starting-point for pilgrims to Santiago, more or less coincided. And one common factor of huge significance united the two roles. In the year 1026 the pope placed Vézelay under the authority of another Burgundian abbey a short distance to the south. This was Cluny. And Cluny had a direct interest in both the Magdalene legend and—in particular—the Santiago pilgrimage.

It was stated at the time that the reason for the pope's decision was disciplinary; that the monks of Vézelay had been guilty of "wretched and lascivious behavior," though discreetly no details were offered. Lapses from the Rule of St. Benedict, the order's founder, were frequent at the time, and Cluny, as the champion of monastic reform, was the natural disciplinarian in such instances. In fact the pope's edict was relatively gentle, and Cluny's control of its junior house was never total, and did not last long. Even so, the partnership between the two abbeys was soon productive to an extent far beyond the expectations of either abbey. In 1037 Cluny appointed one of its own monks as Abbot of Vézelay. He was Abbot Geoffrey, and it was during the decades of his abbacy that the place flourished as never before. It was Geoffrey who specially promoted the cult of Mary Magdalene. As a result the flow of pilgrims soon became a torrent. Donations poured in from people in all walks of life. Among the powers attributed to Mary Magdalene was that of aiding prisoners who had been unjustly jailed: before long so numerous were the chains offered to the abbey in gratitude by prisoners released in response to their prayers that Abbot Geoffrey had them melted down and made into new altar rails placed round the shrine of the saint. Accounts of numerous miracles performed by her spread throughout France and further afield. The Magdalene cult was reaching a pitch of hysteria, and in the process the wealth and fame of the abbey continued to multiply.

The contribution of Cluny to this bonanza was incalculable. Geoffrey was in charge of Vézelay under the shadow of two of Cluny's most formidable abbots: the first was Odilo, described by Bishop Fulbert of Chartres as "the archangel of monks," and the second Hugh the Great. Both men were later canonized. Between them they held office spanning the entire eleventh century, overlapping it at either end—fifty-five and sixty years respectively. Both abbots shared a messianic ambition directed towards Spain; this took the form of employing the resources of the Church to push back Islam and so reclaim the land for Christianity. As we shall see in Chapter 9, and in the final section of this book, a vital arm of this ambition was to safeguard the pilgrim road to Santiago by establishing a powerful network of religious house along the route, both in France and in northern Spain. These would act as service stations for the benefit of pilgrims, offering a roof over their heads, a meal, a prayer, a psalm, the comfort of company, local information, even medical treatment. Santiago pilgrims were the Christian army without arms, but whose sheer numbers, and through the donations they generated, were a vital support for the beleaguered Spanish monarchs struggling to hold their territories against the Saracens.

No wonder Cluny, through the good offices of Abbot Geoffrey, did everything possible to promote Vézelay as one of the great Christian shrines which was also a major starting point for Santiago. Cluny even used its influence with the pope to obtain a papal bull in 1050 acknowledging that the saint's relics were indeed at Vézelay, and that the Magdalene, and not the Virgin Mary, was the abbey's principal patron. The Magdalene had become at once icon and fundraiser. As for the pilgrims collecting here from eastern France and beyond, as they attended their last mass in the abbey church before setting out on their long journey to Santiago, it was Mary Magdalene who pointed them the way. She had acquired a unique role in the Christian world. Of all the New Testament figures whose relics were believed to have survived, she was the woman closest to Christ. She had been embraced by Jesus both figuratively and literally. The medieval Church made much of the fact that she had been a whore. But now, embraced by Christ, her sexuality had been made respectable. Alone of all the female saints in the Christian calendar, she stood for a purity that is also sexual.

In the popular view of Mary Magdalene as the prostitute-turned-disciple, it was maybe the example of her own reformed life which accounted for her enormous popularity among pilgrims. She offered the promise that they too, as penitent sinners, could be accepted by God.

With her body now declared to be safely in Vézelay the triumph of the Magdalene cult was now complete. By the mid-twelfth century *The Pilgrim's Guide* could confidently claim that "the most worthy remains of the Blessed Mary Magdalene must first of all be rightly worshipped by pilgrims. She is that glorious Mary who ... watered with her tears the feet of the Savior, wiped them with her hair, and anointed them with a precious ointment while kissing them most fervently." The author continued by corroborating the story of the removal of the saint's body. "After a long time a distinguished and holy monk called Badilon transported her most precious earthly remains to Vézelay where they rest up to this day in a much-honored tomb."

By now her story had become official history, and Vézelay was the proud beneficiary. From the perspective of our own times the story seems bewilderingly improbable. Historians generally agree that the *vitae* on which the Magdalene legend is founded are works of pious fiction, either deliberately invented in the interests of generating support for the pilgrimage movement, or in the innocent belief that a story about biblical figures passed down from generation to generation must be of divine origin. The first *vita*, the Life of the Hermits, which established the Magdalene as a desert hermit, was likely to have been a confusion of identities between the Mary Magdalene of the gospels and a desert penitent known as Mary of Egypt. The subsequent *vitae* establishing her living presence in Provence relate to a popular legend dating possibly to Roman times of the Boat of Bethany which brought a crew of saints, including the Magdalene and her brother Lazarus, to these shores to found the first churches.

As for the story of Badilus, and how the Magdalene's body become transported to Burgundy, this seems likely to have been a tale put about by the parties benefitting most richly from the escapade, namely the abbey of Vézelay masterminded (as with the chronicle of Archbishop Turpin and the pseudo-history of Charlemagne's dream) by Cluny. The supposed removal of Mary Magdalene's remains from Provence to Burgundy was justified on the grounds that they had been saved from being destroyed by the Saracens. But in the eyes of the medieval Church there was a spiritual justification too, that the removal of her body must have been God's will. It was a practice piously known as *furta sacra*, Sacred Theft—to which there could be no answer or objection.

No doubt there were those in Vézelay who remained skeptical about the razzmatazz surrounding the cult of the Magdalene, and who viewed the accumulating wealth and glory of Vézelay as a triumph, not of Christian virtues, but of mammon. And maybe those skeptics found a certain justice in the fact that on the saint's day, July 21 in the year 1120, a disastrous fire broke out in the abbey church during mass, largely destroying the place and killing more than a thousand pilgrims, so it was claimed. Widely believed to have been an act of arson perpetrated by local people exasperated at being taxed to the hilt, the conflagration nonetheless brought about just one beneficial result. It necessitated the rebuilding of the entire abbey church—the result of which is the architectural and sculptural masterpiece we still have today.

*

Vézelay remains "one of the most beloved and beautiful of mediaeval sites," wrote the American medievalist and scholar of the pilgrim routes, Professor Kenneth Conant. The Basilique Sainte-Madeleine, as it is now known, proudly rides the spur of its hill overlooking the rich undulating landscape of northern Burgundy. To pilgrims making their way here from the north and east the great church may have looked like the stern of a mighty ship preparing to sail southwards towards Spain. Today's travelers are more likely to approach it from the west, making their way up the hill through the old town along a road studded underfoot with golden scallop-shells, symbols not just of the Santiago pilgrimage itself but of the golden wealth the abbey once attracted through the devotions of so many thousands arriving to worship here before setting off on the roads to Spain. The "large and beautiful basilica" mentioned in the twelfth-century *Pilgrim's Guide* must have been only recently completed after the disastrous fire of 1120. Reconstruction had begun immediately, and the new church was inaugurated by Pope Innocent II in 1132 after twelve years of intensive construction work.

Today the abbey itself has disappeared apart from fragments of the former refectory. As for the true glory of Vézelay, the traveler who finally reaches the open square after following the trail of golden scallop-shells through the old town is not immediately aware of it. The massive west front which dominates the square was constructed as recently as the mid-nineteenth century by the French state's official restorer of historic buildings, the architect Viollet-le-Duc. It is a worthy if somewhat soulless pastiche, the original facade having been wrecked during the wars of religion in the sixteenth century, and the vandalism completed in the French Revolution of the late eighteenth century. Then comes the revelation. We pass through the outer door into the spacious narthex, or assembly area, its very size being evidence of the numbers of pilgrims who would assemble here awaiting their final mass before setting off on the road. Crossing the narthex to the inner door we stand at the entrance to one of the longest and most handsome naves in France, its roof vaulted in alternating colors of local sandstone—from white through yellow to dark green—supported by ranks of columns whose capitals, no fewer than 135 of them, have been carved to illustrate scenes from the Old Testament.

But this is now the moment to step back, and gaze upwards. Overhead, spanning the entire double-entrance to the nave, is the huge tympanum, carved early in the twelfth century, which is the ultimate expression of that triumphant statement which will be repeated to pilgrims on countless churches all the way to Santiago: the iconic image of Christ in Majesty. The Vézelay tympanum, almost twenty feet in width, is one of the most astounding sculptural ensembles in the Christian world. In the center Christ stares straight at us, into us and beyond us, his imperious gaze suggesting an unchallengeable authority over all around him. From his outstretched hands and finger tips a blessing is transmitted to the four evangelists who surround him, the blessing being described as so many rays of light which are carved as though they were ribbons floating windblown across the golden surface—a magical touch making the stone itself feel as light as air. It is sculptural alchemy.[20]

Only the greatest artists have been capable of making intractable stone touch the human soul. We do not even know who the artist was who created the Vézelay tympanum, though there are intriguing clues pointing to where he is likely to have come from, and where he could have honed his remarkable skills. And these clues we shall follow up in a later chapter as we head south along the pilgrim road into the Burgundian heartland.

The Vézelay tympanum would have had a special meaning to

---

20. See Plate 11.

the scores of Santiago pilgrims gathering here for that final mass. Those ribbons of light spreading outwards from Christ's extended fingers represented not only a last blessing to his apostles after the Resurrection; it was also a final instruction that they should spread his message to all corners of the earth. This was a mission which St. James fulfilled by preaching the gospel in Spain, and by his body being returned there after his martyrdom to become the object of this pilgrimage. Christ's blessing on that carving above the heads of pilgrims as they left the abbey church was a blessing to them too—a blessing and a reassurance—as they set off for St. James' shrine.

*

The Abbot of Vézelay at the time of the devastating fire of 1120 was a Burgundian nobleman by the name of Renaud of Semur. Like several of his predecessors Renaud had been a monk at Cluny. Furthermore he was the nephew of Hugh of Semur, canonized as St. Hugh in this same year, who had been Cluny's all-powerful abbot for sixty years until his death eleven years earlier. Scarcely had the flames of the church fire been extinguished than Renaud launched a massive rebuilding operation. A church guaranteed to attract so many pilgrims needed to be built as swiftly as possible, with no expense spared. Renaud's plans for the new church were also on an ambitious scale, doubtless colored by the fact that at Cluny, his mother-house, the largest and most richly ornamented church ever built had recently been completed. It was one of the wonders of the world, and Renaud would not have wished to be outshone. The cult of Mary Magdalene was at its height, and so was the Santiago pilgrimage: accordingly the first requirement of the new church was that it should be capable of accommodating pilgrims in huge numbers. The nave itself was planned to be more than two hundred feet long, and the entire church almost four hundred feet in length.

For the next eight years until his death it was under the guidance of Abbot Renaud that the great pilgrim church we see today began to take shape. Throughout those years of intensive rebuilding, and over the decades following, the Cluny connection remained unbroken.

From the outset Abbot Renaud placed another Cluniac monk in charge of operations. He was Pierre de Montboissier. He held the position for only two years before being elected Cluny's abbot, known to us as Peter the Venerable, protector of the persecuted Peter Abelard, and alternately friend and combatant of the fiery St. Bernard of Clairvaux. Meanwhile at Vézelay the Cluny connection continued. As the new abbey church was near completion it was Peter the Venerable's own brother who became Vézelay's abbot. For a while the two greatest abbeys in Burgundy ran in tandem.

Few periods in French medieval Church history have been as decisive as the mid-twelfth century: it was as though everything a resurgent Church had been striving towards came about at more or less the same time. These were days of triumph. The abbey church of Cluny, already being described as "the greatest church in Christendom," was now completed. So for the greater part was Vézelay. Further north another ambitious building project was also on the verge of completion. This was the royal portal of Chartres cathedral, its west front, like the Vézelay tympanum, one of the greatest surviving treasures of medieval sculpture.

Then, in 1146, came another key event in the history of Vézelay. The French monarch, Louis VII, invited the most eloquent churchman of his day, Bernard of Clairvaux, to launch the Second Crusade to the Holy Land with a sermon to be delivered on Easter Sunday that year, in the presence of the king himself together with many of France's noblemen and leading churchmen. And the location selected was Vézelay. The choice could not have been more appropriate. The new abbey church, by now almost complete, was already the foremost center of pilgrimage in France, largely due to the fervent cult of Mary Magdalene, whose relics the abbey was now widely believed to possess. It was also one of the major starting points for the pilgrimage to Santiago, attracting penitents and adventurers alike from eastern France and far beyond. Both events, together with the charismatic presence of Bernard, assured a huge congregation and an auspicious launch for the new crusade, just half a century after the first one. Equally important was the fact that the two movements, the pilgrimage and the crusade, shared a vibrant common cause, the continuing struggle against Islam—the pilgrimage being directed towards Spain, and the crusade to Palestine. The launch of the Second Crusade in Vézelay brought the two movements close together, each giving strength and a sense of purpose to the other.

So great was the crowd gathered at Vézelay on that Easter morning that Bernard undertook to deliver his sermon not in the new church in spite of its size but in a natural amphitheatre on the northern slope of the hill below. The sermon was by all accounts a *tour de force,* and Bernard's passionate words brought overwhelming support for the new crusade. And at the climax of his oration Bernard tore a strip off his clothing to be made into a crusader's banner. It was Vézelay's finest hour. And afterwards scores of pilgrims would have set off on their journey south with the great man's words ringing in their ears. The appropriateness of Easter Sunday for Bernard's sermon would not have been lost on many of them. Here was a place dedicated to Mary Magdalene, who had been the first disciple to see Jesus after the Resurrection on that very first Easter morning. The event at Vézelay had a powerful ring of history about it

We can still savor something of that historic moment. A footpath leads down the north flank of the hill below the abbey church, finally emerging from the woods onto a natural platform projecting from the hillside. This was the scene of Bernard's famous address, today marked by a tall wooden cross mounted on a plinth of bare rock. Originally a stone cross stood here, destroyed like so many other religious symbols during the fury of the French Revolution. What does survive is a tiny chapel nearby, all but masked by trees. This is the Chapelle Sainte-Croix or the Chapelle de la Cordelle, touching in its perfect simplicity, erected here shortly after that Easter day to commemorate the occasion when the king and much of the nobility of France came to hear Bernard's passionate call to arms.

The fame of Vézelay as a place of pilgrimage lasted little more than a further hundred years. But it was to be a final glorious century. In 1190 the Third Crusade was also launched here, this time by the

French King Philippe-Auguste, accompanied by the King of England Richard Coeur-de-Lion. Half a century later the French monarch Louis IX (St. Louis) came to the shrine of the Magdalene several times to seek her blessing on two further crusades which he led, and during the second of which he died.

But then, abruptly, it all came to an end. In 1279 the Count of Provence, Charles of Anjou, led an excavation in the ancient crypt of the Church of Saint-Maximin where, according to legend, Mary Magdalene had been buried after her thirty years as a hermit. The count was determined to prove that the saint's relics still remained within his domain, and that the story of her body being transported to Burgundy was false. Charles led the excavation himself, removing soil and stones from around ancient tombs in the crypt with his bare hands. His efforts were duly rewarded. With the aid of unsubstantiated evidence and several forged documents, including a letter purporting to be from the Emperor Charlemagne, it was established to the count's satisfaction that the Magdalene's relics were indeed still there. It transpired that in anticipation of Saracen raids in the ninth century the monks of Saint-Maximin had taken the precaution of removing the saint's remains from her own tomb, swapping them for another body in a sarcophagus close by.

How Saracen raiders were supposed to be fooled by such a ruse, even if they had heard of Mary Magdalene or cared a jot about her bones, was not explained. The crucial significance of Count Charles' discovery was that the monk despatched by Vézelay to obtain the Magdalene's relics had clearly arrived too late, and so had unwittingly stolen the wrong body. The mistake was soon confirmed by the pope, and Saint-Maximin re-established by papal decree as the guardian of Mary Magdalene's body. And the role of Vézelay as one of the foremost pilgrimage shrines in Christendom for so many centuries was suddenly over.

But if one of the most celebrated acts of Sacred Theft in Christian history had finally proved to be a case of mistaken identity, it had nonetheless resulted in centuries of colossal wealth and prestige for Vézelay. Now, after the "departure" of Mary Magdalene, what remained was the abiding legacy of her cult, which by its very magnificence has continued to entrance and inspire pilgrims and visitors alike ever since. That legacy is Vézelay's incomparable abbey church—the great church on the hill—whose glorious carved tympanum depicts Christ extending his blessing to his apostles and to all corners of the earth, including of course that far outpost of Christianity in northwest Spain to which medieval pilgrims and their modern counterparts set off from here on the long road under the stars. The power of that imperious image, and that blessing extending like rays of light to all around, would have remained, as it does today, in the memory of pilgrims for every mile of the journey to come.

8. The Broader Picture

THE PILGRIM ROADS WERE RIVERS OF HUMANITY flowing inexorably west and south; and like all rivers they were fed by tributaries from far and wide, swelling their size and strength as they headed towards Spain and the far Atlantic. Besides the four principal pilgrim roads listed in *The Pilgrim's Guide* there were many others spread across the country like a giant spider's web, leading to innumerable local shrines. Pilgrims setting off for Santiago would make a point of visiting some of these local shrines along the way: they served as welcome diversions on a long journey: a pause for a prayer and a blessing to sustain their spirits on the road ahead—not altogether unlike the modern traveler making a diversion to take in Siena and Assisi on the way to Rome. As today, there was always an element of spiritual tourism about pilgrimages: celebrated and beautiful places had a natural allure.

By the twelfth century the number of such shrines in France attracting pilgrims was beyond count as the cult of holy relics reached fever pitch. We can even talk of a "relics industry" as churches and monasteries vied with each other to possess and display the most prestigious objects for pilgrims to venerate. Donations by grateful travelers whose prayers had been answered became a major source of income for religious houses fortunate enough to own such items. Relics were "realizable assets," as Professor Christopher Brooke has succinctly put it.

Not surprisingly, with such intense competition there were soon simply not enough saints to go round. And as demand outstripped supply, so theft, and faking of holy relics, became widespread.

Christian principles all too often gave way to greed and opportunism; and many a priest preserved his self-respect by conveniently turning a blind eye to unwelcome evidence of trickery.

In the midst of this relics "fever" the crusades made a welcome and important contribution. Crusading knights, many of them adventurers in disguise, began to bring back as trophies whatever fragments of the Bible Lands they could lay their hands on, together with colorful tales of what these objects were supposed to be and where they were supposed to have come from. An impressionable Christian world eagerly awaited them, and for the returning crusader it became a profitable trade. This new source of holy relics was particularly exploited as a result of the Fourth Crusade of 1204 in which the flower of European chivalry, having taken the cross and set out with heroic ambitions, proceeded to sack and loot Constantinople, the capital of the Eastern Church. One outcome was that relics began to reach France and elsewhere in Western Europe in a veritable flood; and the number of churches and priories claiming to possess sacred objects from the Holy Lands increased proportionately. Skeptics calculated how many sailing vessels could be constructed out of the many supposed fragments of the holy cross. And Chaucer's Pardoner, as we have seen, carried with him a fragment of the sail which St. Peter had "whan that he wente upon the see."

Of the numerous tributaries feeding the pilgrim "rivers" to Santiago one of the most important ran south from the spectacular island shrine and abbey of Mont Saint-Michel, off the coast of the English Channel on the western borders of Normandy. Here, unusually, there was no holy relic to be revered. The popularity of the site as a place of pilgrimage was due to a mystical combination of geography and legend. According to local tradition early in the eighth century—a time when most of what is now Normandy was in the hands of Viking pirates—the Archangel Michael appeared to a beleaguered local bishop, instructing him to build a church on a rocky islet a short distance offshore. Before long a cult of St. Michael grew up on the site, and after the eventual Christianization of Normandy in the tenth century successive Norman dukes undertook to finance the building of a magnificent abbey on the sacred island. The man responsible for designing the abbey and its church was that remarkable architect-monk from Lombardy, in northern Italy, by the name of William of Volpiano (discussed in Chapter 4). In the late tenth century William had been invited to Burgundy by the Abbot of Cluny for the express purpose of rebuilding abbeys and churches which had been desecrated during the centuries of barbarian invasions, principally by the Vikings. Together with his team of masons and craftsmen from Lombardy, William established in Burgundy what has become known as the "first Romanesque style" of church architecture. His reputation soon drew the attention of the rulers of Normandy, now an independent dukedom whose rulers, in the fervor of their new faith, were engaged in making Normandy the heart of Christian culture and monastic life.

Hence, early in the eleventh century William of Volpiano was given the challenging task of building an abbey on the bare rock of Mont Saint-Michel. William's courageous response was to place the transept crossing of the abbey church at the very peak of the rock, so necessitating the construction of underground crypts and chapels in the rock below in order to bear the weight of the new church.[21]

It was a daring and triumphant achievement, and it heralded the emergence of Mont Saint-Michel as among the most revered centers of pilgrimage in Europe. Its importance soon extended to England, and to English pilgrims setting out for Santiago, largely as a result of the Norman invasion and conquest of England by William the Conqueror in 1066. As a reward for its support for the invasion by Duke William—now also King of England—the abbey of Mont Saint-Michel was awarded lands on the English side of the Channel, among them a small islet off the coast of Cornwall which was a match for the abbey-island across the water. A Benedictine priory was duly built on the Cornish islet, which took the name of the mother-abbey, becoming Saint Michael's Mount. Soon English pilgrims, before embarking from Plymouth further along the coast, would begin

---

21. See Plate 12.

their journey by offering prayers to the Archangel Michael here in Cornwall, then repeat their prayers to the same saint on arriving at the far side of the Channel, at Mont Saint-Michel. Prayers in hope of a safe crossing were thus echoed by prayers offered in thanksgiving for having arrived safely across the sea.

From here in Normandy English pilgrims then headed south, some towards Chartres and Orléans, others keeping further west and finally joining travelers on the principal pilgrim road from Paris and Tours somewhere in the region of Poitiers, and before long linking up with the scores of other English pilgrims who had sailed from Bristol towards Bordeaux on those trading vessels soon to return home with a less pious cargo—wine.

Doubtless most pilgrims beginning their long journey with a visit to Mont Saint-Michel would have understood than an archangel had no earthly presence and hence no earthly remains. Even so, there are accounts of devout travelers arriving on the island fully expecting to be able to venerate the body of St. Michael. By way of compensation there grew up a popular custom among pilgrims to the abbey of gathering rocks and pebbles from the seashore at low tide as mementos, much as children on a seaside holiday today will collect shells and colored pebbles to take home.

An even more famous site attracting passing pilgrims was the cathedral of Chartres, some fifty miles southwest of Paris. Chartres is not situated on one of the main pilgrim routes, and it is not known when the cult of St. James first became attached to the cathedral; but by the early thirteenth century it was well enough established for the most beautiful assembly of stained glass in the world to include two windows that relate specifically to the apostle and to the discovery of his tomb in Spain. The first window is devoted to the Life of St. James, made up of thirty panels, one of which depicts Jesus giving James his mission to preach in Spain, the sequence ending with the apostle's martyrdom at the hands of King Herod. But it is the second window, in the choir of the cathedral, which is more directly related to the Santiago pilgrimage, and known as the Charlemagne Window. One of the twenty-four panels shows St. James appearing to the emperor in a dream, urging him to return to Spain to defeat the Saracens. Another panel depicts Charlemagne departing for Spain, and in a third he is gazing up at the Milky Way which will lead him to the tomb of St. James. Together the three scenes illustrate the popular variation of *La Chanson de Roland* in which Charlemagne's visitor in his dream is not the archangel Gabriel but St. James himself. In other words the Chartres window faithfully follows the version of the Song which was almost certainly conceived as a deliberate boost to the cause of the Santiago pilgrimage, probably on the initiative of the abbey of Cluny (as described more fully in Chapter 5).

That this version of the tale should have become enshrined in France's greatest cathedral nearly two centuries after it was composed is a testimony to how successfully the Charlemagne story had been adapted for public consumption. The medieval imagination lent itself to reinventing legend and presenting it as reality: it was a gift which supplies a key to the creative genius of the age—a gift manifest in its painting, its sculpture and, as here, in the magnificence of its stained glass.

Chartres had already enjoyed a long tradition of pilgrimage dating back at least to the early Christian era. The principal focus of veneration is made clear by the most celebrated of all the medieval windows at Chartres, one which would probably get the popular vote as the most beautiful example of stained glass in the world. This is the Virgin window with an incomparable intensity of lapis-lazuli blue. The cathedral itself is dedicated to the Virgin Mary (as is Notre-Dame in Paris), and it became a center of the Marian cult as a result of possessing the most revered of all the supposed relics of the Virgin, known as the *Sancta Camisa*, or Sacred Veil. This consists of about sixteen feet of cloth (today preserved in a reliquary in an apsidal chapel), believed to have been worn by Mary at the birth of Jesus. The origin of the garment is as obscure as that of most holy relics, though it was probably stolen from a Jewish community in Jerusalem and brought to Constantinople, where it was offered as a gift to the Emperor Charlemagne by the Byzantine emperor of the Eastern Church. Charlemagne's grandson, Charles II, then presented it to the Bishop of Chartres in about the year 876. In the succeeding centuries it became an object of fervent adulation by pilgrims, especially by women, in the widely held belief that prayers offered here would ease the pains of pregnancy.

The far-fetched superstitions attached to sacred relics like the *Sancta Camisa* may be deeply alien to rational minds today, as is the gullibility of the medieval pilgrim; yet the modern traveler might prefer to reflect that without the deep passions and huge popular support generated by such holy shrines a great many of the places we now throng to visit and admire would simply not exist.

Chartres itself is as case in point. The era of the great cathedrals had a quite different feel about it from that of the monasteries which preceded it. Cathedrals themselves performed a largely different role within a community. Monasteries were created to be isolated from the social world: the monastic ideal had emerged at a time when Christianity could only survive in sealed pockets within a largely hostile environment—or at least an environment which had no need of monasteries and to which they did not belong. They were the product of a siege mentality. But by the time the great cathedrals were being created, in the twelfth and thirteenth centuries, Christianity had won. The world was no longer alien. Hence cathedrals were built in the very heart of a community, and for the needs of that community. They were the most important buildings in a town, and accordingly played a substantially different role in people's social and spiritual life from that of an abbey. Townspeople used them for a variety of purposes other than religious services and private prayers. Because of its sheer size and dominant position a cathedral became the natural venue for such events as local markets, each of its huge carved entrances becoming the focus of a wide range of commercial activities. In Chartres there were four annual fairs which were held in the area around the cathedral; and these coincided with the four feast days in honor of the Virgin—the Presentation, the Annunciation, the Assumption and the Nativity. Pilgrimages were a key entity on such occasions: the commercial success of these fairs was assured by the presence of large numbers of pilgrims who had come to venerate one of the most revered holy relics in Europe. While the majority of pilgrims may have been poor, there was invariably a minority with money to spend.

As the new cathedrals became "people's palaces" it was natural that ordinary people should wish to play their part in maintaining them—particularly after the disastrous outbreaks of fire which were a perennial hazard in medieval churches where naked candles and torches provided the only artificial light, and roofs were often still of timber. Chartres suffered a sequence of devastating fires, necessitating constant rebuilding between the ninth and the thirteenth centuries.

It was precisely in response to such disasters that the widespread public affection for the local cathedral became demonstrated in spectacular fashion in what has become known as the "building crusades." In the year 1145 more than a thousand pilgrims arrived at Chartres dragging carts filled with building materials along with provisions for those who had lost everything when one of the fires spread through the town. It was recorded that for months, men and women hauled heavy wagons of stones up the slope to the cathedral site.

This degree of public involvement in the construction and life of a cathedral had the effect of altering to some extent the very nature of pilgrimages. Now that cathedrals themselves had become corporate institutions, so too were the pilgrimages which they attracted. The notion of a pilgrimage being one person's private journey in search of salvation was being replaced by an essentially collective exercise whose benefits might be described in today's jargon as a form of group therapy. Jonathan Sumption has described this change in the following words. "Building crusades reflected the view that pilgrimages performed *en masse* were more meritorious than those performed alone. We find the pilgrim-builders forming themselves into sects, or brotherhoods, performing their penitential rituals in common, and solemnly expelling those members who showed signs of returning to their old ways."

So a note of self-righteous intolerance creeps in at the same time. But it is the image of the wandering band of travelers, cheerful and high-spirited, which springs to mind most readily when we think of medieval pilgrims on the way to some distant shrine: a journey that has freed them from the toil and drudgery of daily life, and where penitence may count rather less than the enjoyment of good company and the pleasures of the open road. Chaucer may be partly responsible, with his motley crew setting out for Canterbury from the Tabard Inn, where many of them had no doubt spent a thoroughly jolly evening, then entertaining each other with robust stories along the way. A similarly colorful picture of the collective pilgrimage gazes out at us from illuminated manuscripts of the same era as Chaucer's great poem. Among the most delightful is contained in a fourteenth-century Burgundian Book of Hours now in the Musée Condé at Chantilly (illustrated on page XX). It shows a band of pilgrims returning from Santiago with all the time in the world to enjoy themselves, laughing and joking and playing games as they make their easygoing way through the pleasant country. In this bucolic scene we could hardly be further from the image of the solitary sinner plodding grimly to some holy shrine in the hope of being spared eternal damnation. It seems most unlikely that any of the pilgrims represented in the Burgundian Book of Hours had considered for one moment the possibility of hellfire.

Such records of pilgrimage as an enjoyable enterprise, whether from Chaucer of from illuminated Books of Hours, are both from the later Middle Ages. And inevitably they feel much closer to the spirit of the present-day cultural tourist than to that of the early centuries of pilgrimage when the journey to Santiago was still fraught with dangers and few physical comforts were to be found along the way.

It is tempting to make a further comparison with the modern traveler, and speculate on how much medieval pilgrims would have been struck by the beauty and sheer magnificence of the places they visited. Most pilgrims were from small communities where the only buildings of any size and grandeur would have been the church and perhaps the castle of the local lord. Few would ever have traveled far from home. Suddenly a great abbey, or a soaring cathedral, even a bustling town with tall mansions and cobbled streets, would have been quite unfamiliar. More bewildering still would have been the gold and glittering jewels decorating the shrines they had come to venerate. It is impossible not to believe that a sense of wonder and awe would have overwhelmed pilgrims, sustaining them on their journey, and remaining with them as a collection of golden memories for the rest of their lives.

*

After Chartres, the second of the magnificent Gothic cathedrals of France to honor St. James was that at Bourges, further to the south and situated on the main pilgrim road we have been taking from Vézelay. The two cathedrals, Bourges and Chartres, are more or less contemporaneous, the former's main structure as it exists today having been started in the final decade of the twelfth century.

With its five elaborately-carved portals, soaring nave and flamboyant flying buttresses, Bourges cathedral must have been another awesome sight for pilgrims. They would have made their way down from Normandy, perhaps taking in Chartres, then crossing the Loire at Orléans before heading south into the dukedom of Berry, of which Bourges was the regional capital. And at this point they became united with other travelers who had been heading south from Vézelay.

Unlike Chartres, Bourges offers no open propaganda for the Santiago pilgrimage itself: in its stained-glass windows there is no re-telling of the Song of Roland with St. James beckoning the Emperor Charlemagne in a dream to follow the Milky Way into northwest Spain. In Bourges the tributes to St. James are more low key: he is just one apostle among others. A thirteenth-century window in one of the chapels relates the customary legend of his life, concluding with his decapitation by Herod. In another chapel he appears as a bystander witnessing the Annunciation. And in a tall window on the south side of the cathedral, high up, his full-length figure is flanked by those of St. Philip and St. Thomas. Nowhere in the cathedral is there a call to pilgrims to set off for his shrine in Spain. But by this time, in the later Middle Ages, perhaps the urgency of the Santiago pilgrimage had already become a little blunted. The Saracens had long been reduced to a small pocket in southern Spain; and there was no longer any need for a rousing call to arms.

Pilgrims joining the road from Vézelay at Bourges would have missed out on the drama and hysteria surrounding the worship of Mary Magdalene described in the previous chapter. They would have missed, too, the impact of that peerless abbey church with its uplifting valediction for pilgrims who were setting off into the unknown in the form of the carved tympanum over the main entrance.

Historic roads often display fragments of their past, like mementos pinned here and there along the way. Today's travelers moving south from Vézelay are everywhere reminded of the history of the road they are taking. Modern highways tend to cut corners, whereas the original pilgrim road still threads its way more peacefully at walking pace from village to village. Following it here in Burgundy is to make a journey into the past—a journey signposted with names that speak of distant times, like the village of La Maison-Dieu, named after a long-vanished pilgrims' hospice. Or another village, Asquins, close to Vézelay, with its little Church of Saint-Jacques, and where there was once a ninth-century priory until it was sacked by Viking marauders along with other religious houses in the area, so causing the local count to found a new abbey more securely on a nearby hill—which became Vézelay.

Echo follows echo along the old pilgrim road as it meanders from village to village, through forests that have never changed and over streams that have always flowed. The traveler may notice a half-buried church on a hill, or stone slabs which are the remains of an ancient ford now replaced by a bridge. One village is called Metz-le-Comte, the *comte* being a descendant of the very count who built Vézelay back in the tenth century. So the echoes resound. Then, back on the modern road we encounter another ancient forest, and suddenly an isolated chapel comes into view dedicated to a saint widely venerated in these parts, St. Lazare—in other words Lazarus, the friend of Jesus whom he brought back from the dead, and who according to legend accompanied his sister Mary Magdalene on the open raft which brought them to the shores of Provence. This was a legend so closely interwoven with the Santiago pilgrimage that the worship of St. James throughout this area of France cannot be separated from it—as we have already seen at Vézelay, and as we shall see again in the next chapter.

The pilgrim road continues southwest through more wooded countryside towards the western border of Burgundy. The boundary here is defined by the Loire, the longest river in France, soon to flow northwards past Saint-Benoît and Orléans, from where it will accompany pilgrims on the Paris road towards Tours (as described in Chapter 2).

On this upper stretch of the river pilgrims made their first direct encounter with the religious house whose invisible presence had been felt all along the way so far. This was the great Burgundian abbey of Cluny. The monastery itself lay some distance away to the southwest; yet here on an island in the river was a priory which had come to be described as the "eldest daughter of Cluny." The priory was La Charité-sur-Loire. As pilgrims made their way south it was one of the major landmarks along this stretch of the road—a handsome cluster of buildings which included a magnificent church capable of holding a congregation of at least five thousand. The priory was surrounded by well-tended gardens, orchards and vineyards, as well as numerous houses, workshops and outbuildings. To pilgrims weary from the long road it would have seemed a paradise island.

La Charité had been the first monastery to be established as a dependency of Cluny—the first of what was to become an expanding empire of dependent abbeys and priories throughout Europe. And as Cluny's "eldest daughter," in the eyes of the mother-house, La Charité seems to have remained also her best-loved daughter. It was the greatest of Cluny's abbot, Hugh of Semur (St. Hugh), who in 1059 founded the priory here on what was then an island. Within a century it had grown massively, both in size and influence. Donors included King Henry I of England, son of William the Conqueror. While remaining under the authority of Cluny the priory also acquired at least seventy dependencies of her own, including religious houses in Chartres, Paris, and even in England.

As a key halt on the road from Vézelay, La Charité naturally attracted an abundance of pilgrims, and not entirely for spiritual reasons. The generous hand of Cluny ensured that hospitality matched the appeal of the place itself. It was a reputation for providing the best food and Burgundian wines that pilgrims were likely to enjoy on the entire journey which earned the priory its name—which has stuck ever since.

Today La Charité remains one of the gems of Burgundy—a gem that has been painstakingly repaired and polished. Historical events did not always treat it as charitably as it treated travelers. Fire caused extensive destruction of the whole complex in the sixteenth century, followed by further damage during the French wars of religion when it became—ironically—a Protestant stronghold, and suffered for it. Finally, in the nineteenth century, the local authorities delivered the *coup de grâce* by proposing to drive a road right through what remained of the priory church. In the nick of time help arrived in the shape of that angel of mercy, Prosper Mérimée, the French government's inspector of ancient monuments. Mérimée not only put a stop to the road scheme, but set in motion a restoration program for the church, his greatest triumph being to save the superb Romanesque tympanum belonging to the original west facade, which he discovered set into the wall of a local private house, and which subsequently found a new home in a surviving area of the church, where it remains today to the admiration of all.

The riverside town on its graceful sweep of the Loire has been lovingly restored along with much of the ancient priory. The whole town, which grew up around the priory, is still dominated by the proud Romanesque tower which has been a beacon guiding pilgrims on the road from Vézelay for more than eight centuries. The noble church towers of this region are among the glories of Burgundy. Rising elegantly from the center of so many towns and villages, they are a testimony to those incomparable stonemasons from Lombardy who brought their Christian vision and their building skills from northern Italy to this part of France in the early years of religious revival. They are towers which remain symbols of a resurgent faith and resurgent hope after the dark centuries of anarchy and insurrection. They stand tall across the Burgundian countryside as if God had planted his own standard across a newly-won land.

The greatest of these Burgundian towers were those that crowned the vast abbey church of Cluny which was Abbot Hugh's ultimate dream, only realized after his death, and of which sadly no more than one of the lesser towers remains. Political events and human rapacity struck Cluny even more cruelly than at La Charité. Yet if all too little of Cluny survives today, its contribution to the cause of St. James, and to the vast enterprise of the Santiago pilgrimage, remains inviolate, beyond the reach of vandals. Cluny is the great ghost that haunts the pilgrimage movement. And so, leaving the "eldest daughter," this is an appropriate moment to visit the mother-house itself.

9. Cluny and Autun

THE BENEDICTINE ABBEY OF CLUNY was never on one of the main pilgrim roads: yet being equidistant between two of them—the routes from Vézelay and from Le Puy—its long arms seem to embrace them both, extending even further to include the other principal roads as well. Cluny was the powerhouse of the Santiago pilgrimage—or, perhaps more accurately—its godfather. In the words of one of the eleventh century popes, Urban II, himself a former monk at the abbey, "Cluny shines like another sun over the earth."

Today Cluny is mostly a memory, a hollow shell filled with whispers of a glorious past. A single tower presides over the wreckage of a church which for five hundred years was the largest and most sumptuous house of God in Christendom.[22] Its accompanying abbey was for more than two centuries the spiritual heart of the Christian world, and its abbots among the most powerful figures in Europe, friends and advisers to Holy Roman Emperors, successive popes and the kings of England, France and Spain. To be Abbot of Cluny in the eleventh and twelfth centuries was to hold a position in the world as prestigious and influential as any president, statesman or business tycoon of our own day. As if temporal power was not enough, they had God on their side too.

It had all begun early in the tenth century when Duke William of Aquitaine decided to found a monastery on his own lands. The site chosen was in a secluded valley a short distance from the town of Mâcon, strategically placed between two major rivers, the Loire to the west, flowing north, and the Saône to the east, flowing south. Furthermore, in these lawless times its location made it relatively safe from raiders who posed a continual threat: Vikings from the north, Magyars from the east and Saracens from the south.

From the outset the young monastery benefited from the duke's decision to draw up a foundation charter which exempted it from all forms of local interference, secular as well as ecclesiastical. Neither the local bishop nor the local lord (including Duke William himself) would have any control over it. The duke placed it under the sole authority of the papacy in Rome. This shrewd political move sowed the seeds of a binding relationship between future popes and abbots of Cluny. It was a relationship which was to prove of inestimable value to both parties, especially in the years to come when two such leading figures of the Church would soon combine in a holy war against Islam—sponsoring crusades to the Holy Land, military action in Spain and facilitating the pilgrimages to Santiago. It was a triple campaign, in harness, with both institutions bringing their weight to bear on the most pressing issue of the day. It was a bond with Rome symbolized by the young abbey's coat-of-arms, composed of the twin keys of St. Peter and St. Paul.

From modest beginnings the abbey soon expanded in size and importance. Many monasteries had grown lax in discipline during the troubled centuries of lawlessness and insurrection, paying little respect to the Rule of St. Benedict laid down by the order's founder, and even less to rules of celibacy. Some priories swarmed with wives, mistresses and male lovers. Increasingly it became the role of Cluny to implement reforms. Priory after priory, abbey after abbey, became subject to the authority of the mother-house.

The burgeoning success of Cluny stemmed to a large extent from another farsighted clause in the duke's foundation charter, which stipulated that the abbey's monks held the right to elect their own abbot. Since the monks who chose to enter Cluny tended to be from Burgundian noble families, the abbey not surprisingly proceeded to be governed by a succession of aristocratic abbots who wielded extensive local power within this tightly-knit feudal society, both within the Church and in the secular world at large. And with such authority wielded on two fronts, ecclesiastical and secular, Cluny began to attract handsome donations of property and land from lords keen to invest in so powerful an institution whose monks were believed to have the ear of God, and would therefore be sure to include them in their prayers. Once again mutual benefit served the abbey well.

Less than a century after its foundation one of those aristocratic abbots forged the first link with Spain and the cult of St. James. He was Odilo (to become St. Odilo), and he was elected Abbot of Cluny in the year 994. He was to hold the office for the next fifty-five years, filling those years with extensive journeys to visit dependent priories, as well as with building projects for the new and much larger abbey church, and with pursuing a lifetime's ambition—doing whatever lay within his power to bring about the Reconquest of Spain from the Saracens.

Two events took place at the very outset of his abbacy which gave that ambition a special urgency. In the year Odilo took office the historic monastery of Monte Cassino in southern Italy, which had been founded by St. Benedict more than four centuries earlier, was sacked by Saracen raiders from Sicily. It would have seemed that the heart of the Benedictine movement had been plucked out. Then only a short while later came another devastating Saracen attack: this time the victim was Santiago, the church and community that had grown up round the apostle's tomb discovered almost two hundred years earlier. Santiago was devastated by the most feared of the Saracen warlords in Spain, Al-Mansur. It was the culmination of two centuries of Saracen assaults on the Christian kingdoms of northern Spain. And it was Abbot Odilo's' bitter inheritance.

Perhaps it was the shock and the pain caused by those two early disasters which spurred Odilo to devote so much of his boundless energy to supporting the Christian cause in Spain. It was under his abbacy that Cluny, with the invaluable backing of the papacy in Rome, began to use its weight to support the harassed Christian rulers south of the Pyrenees. And soon the tide began to turn. It was as if the passing of the feared Millennium without the world having come to an end had influenced events in favor of the Christian powers. In 1002 Al-Mansur died: and with his death the military power of the Saracens in Spain seemed to slacken, as though they were losing their self-assurance and their predatory will. Soon, with the backing of Cluny, the ruler of Navarre, the small kingdom bordering on the Pyrenees, felt able to extend his territory to include neighboring Aragon, Castile and León.

It is hard to establish Cluny's precise role in the re-establishment of Christian authority in northern Spain. What is certain is that under Abbot Odilo the abbey had a hand in promoting numerous military expeditions against Saracen strongholds, using its lordly connections to rally Burgundian knights to fight in Spain. It even seems likely that without the contribution of Cluny these campaigns might never have taken place. A measure of the debt of gratitude felt by the Spanish monarch, Sancho III, was the quantity of gifts showered on the abbey, mostly from loot seized by the conquering Christian armies. Burgundian knights also rewarded the abbey in a similar fashion. Hence the benefits of Odilo's relationship with the Spanish ruler were mutual: Sancho strengthened and enlarged his kingdom, while Cluny greatly enlarged its wealth and its power. There was now a greatly expanded abbey and a greatly expanded abbey church. A Roman priest who visited Cluny about this time in the company of the papal legate recorded his astonishment at "the great and vaulted church ... richly adorned with various precious things."

The close relationship between abbot and monarch continued until Sancho's death in 1035. By this time, and largely through the efforts of Odilo, the French Church had begun to extend its influence right across northern Spain, now cleared of the Saracen presence. Sancho himself had invited Cluniac monks to Spain, and in turn Cluny received and trained Spanish monks. One of these, by the name of Paternus, then returned to Spain to become abbot of Cluny's first monastery south of the Pyrenees, San Juan de la Peña.

Odilo's passion for the Reconquest of Spain needs to be

understood in the context of the Saracen threat to Christian Europe over the previous centuries, and its sheer magnitude. Like every well-informed churchman Odilo would have been all too familiar with such a relentless passage of events. The advance of Saracen armies had been dramatic and terrifying. The prophet Mohammed had died in the year 632. Within a century of his death a number of apparently disorganized Bedouin tribes in the desert had come together to form a disciplined military nation which proceeded to take over all the former Roman colonies in North Africa, cross over to Spain and put its Christian rulers to flight. They had occupied Sicily, harassed southern Italy and made pillaging forays deep into France as far north of Tours, within striking distance of Paris—all the time maintaining total control of the Mediterranean Sea to facilitate the passage of their armies. This in a mere hundred years.

Even after their historic defeat near Poitiers by Charles Martel in 732, Saracen marauding parties continued to raid Christian lands as strongly as before. In the mid-ninth century they repeatedly drove north over the Pyrenees, sweeping westwards across southern France as far as the Rhône Valley and Marseille, leaving burned towns and wrecked churches and abbeys in their wake. Finally, late in the following century, came the humiliating sack of St. James' city.

All this had been Abbot Odilo's dark inheritance: and it was a trend of history which he devoted much of his long life to reversing. By the time of his death in 1049 the Reconquest of Spain and the opening up of the pilgrim road to Santiago had become a twin cause indelibly identified with Cluny.

But if the banner was waved vigorously by Odilo his military contribution bore no comparison to the scale of the campaigns mounted by his successor as abbot. Hugh the Great (St. Hugh) was only twenty-four on his accession, and was to remain Abbot of Cluny for a further sixty years, during which time he became one of the most powerful political figures in Europe as well as the most influential churchman of his day—head of a vast monastic empire made up of almost 1,500 dependent abbeys and priories across the continent.

A number of the most important of these, both in France and

northern Spain, were acquired by Hugh specifically in order to further the cause of the Santiago pilgrimage. They were situated along the various pilgrim roads in France, and acted as invaluable staging-posts for travelers: Saintes, Saint-Jean-d'Angély, La Charité and (most notably) Vézelay, we have seen already; others, including Limoges, Moissac, Toulouse and Saint-Gilles lay on the various roads to come.

They were all of them signposts to Spain. Hugh's personal connections with Spain, and with Spanish monarchs, were even closer than those of his predecessor. They were strengthened—in the familiar fashion among the European ruling classes—by family ties, in particular with the Spanish King Alfonso VI (grandson of Odilo's patron Sancho III). Alfonso's Queen Constance was a niece of Abbot Hugh. Their daughter and heiress, Urraca, at the age of eight was married to another member of Burgundian nobility, the immensely wealthy Count of Burgundy, whose brother became none other than Pope Calixtus II, the supposed author of part of *The Pilgrim's Guide*. In this way political power, money, family ties and unchallengeable religious authority all came together to create an alliance which proved to be of immeasurable advantage to all parties concerned.

Hugh's abbacy of sixty years began in what was the golden dawn of the pilgrimage movement. By the time it finally ended, in the second decade of the twelfth century, it was already high noon. The years immediately following Hugh's abbacy were those of the pilgrimage's greatest popularity: the era of *La Chanson de Roland*, of the Charlemagne cult, of the pilgrim hymns recorded in the *Codex Calixtinus*, and of course of *The Pilgrim's Guide* and the racy travelog of the Cluniac monk Aimery Picaud on his colorful journey to Spain. And there was also the glorious finale of the cathedral at Santiago completed later in that same century. But that was all to come.

Meanwhile the dependent abbeys and priories which Hugh established in France were a prelude. They were part of a larger plan. Soon he set about extending a similar chain of religious houses right across northern Spain. These new outposts, strung along the pilgrim road from the Pyrenees all the way to Santiago, acted as service stations for the ever increasing flow of travelers heading for St. James' city—now recovered from the assault by Al-Mansur. Well-equipped and well-endowed, these new establishments offered pilgrims a degree of security and comfort, even medical attention, which had hitherto been lacking. They were now able to set off on their ambitious journey in the reasonable hope of actually getting to their destination without being attacked, robbed or starved. The pilgrimage was beginning to becoming almost genteel.

The financial rewards to Cluny and to the Church in Spain were proportionately rich. As facilities improved and the road itself became less hazardous the wealthy and the highborn began to undertake the pilgrimage, accompanied by their retinues of servants and attendants; and as a result donations and bequests began to pour in, swelling the coffers of the Church. For Cluny the Santiago pilgrimages was becoming not only a cause but a lucrative investment.

More directly beneficial to the Burgundian abbey was the prodigious gratitude of the Spanish monarch Alfonso VI, soon to become Hugh's nephew by marriage. Like his father and grandfather before him Alfonso committed himself to making an annual tribute to Cluny of one thousand pieces of gold. We have no means of knowing how much this meant, but it was clearly an enormous sum; and no doubt it contributed to Hugh's decision to cross the Pyrenees expressly to meet his royal benefactor. The meeting took place in 1077, and it was Hugh's first visit to Spain. No account survives of what actually took place between abbot and monarch: nonetheless a whole range of events was clearly set in motion. On Hugh's part the meeting was a confirmation of Cluny's unequivocal support for the strengthening of the Church in Spain in the wake of Saracen incursions. In parallel ran an equal commitment by Cluny to rally military support for continued campaigns against the Saracens. In return Alfonso offered lavish quantities of gold in addition to the annual tribute, and—especially welcome to Hugh—beneficent gifts of abbeys and priories in Spain, mostly set on the pilgrim route.

These included the ancient abbey of Sahagún, already a key link in the chain of religious houses on the road to Santiago, and soon to become known as "the Cluny of Spain." Appropriately it was a Cluniac monk whom Hugh appointed to be its abbot.

In an era when archbishops and even popes donned armor and rode into battle it is hard to assess precisely what contribution Cluny may have made to Alfonso's military campaigns. There is no question of Hugh himself, or indeed any of his monks, actually taking part; yet in the feudal world of which Cluny was a part Hugh's aristocratic connections would certainly have played a vital role in persuading local lords to assemble armies from among their serfs. Alfonso's most dramatic military victory came just eight years after Hugh's first visit to Spain, when his army captured the key city of Toledo after almost three centuries of Saracen rule. The man credited with having led the successful assault was a French knight from Burgundy, Eudes de Bourgogne, who as it happens was Abbot Hugh's cousin.

Alfonso's reaction to this military triumph was another overwhelming gesture of gratitude towards Cluny. Toledo was once again the seat of the Primate of Spain; and now, presumably at Hugh's request, the monarch agreed that the Cluniac monk by the name of Robert, who was then Abbot of Sahagún, should become Archbishop of Toledo and head of the Church in Spain.

At a stroke Cluny had effectively become the spiritual ruler of Christian Spain.

Alfonso's gratitude did not end there. He decided to double the annual tribute paid to the abbey, from one thousand gold pieces to two thousand. The immediate effect of this vast increase in Cluny's fortune was the realization of a dream which Hugh had cherished ever since his election as abbot thirty-six years earlier. His dream was of a church that was larger and more splendid than any yet built anywhere. Now Alfonso's tribute made it possible. Hugh wasted no time. The first stones of Cluny's new church were laid in 1088, just three years after the capture of Toledo. In spite of the nave roof collapsing at one stage, work on the enormous building progressed steadily over the following decades. In 1130, after forty-two years, it was finally completed—almost 600 feet in overall length. It was the greatest church in Christendom, and would remain so for many centuries. One of Hugh's biographers described it as "so spacious that thousands of monks could assemble there, and so magnificent that an emperor could have built nothing finer."

Hugh never lived to see his dream fulfilled, the abbot having died in the year 1109 after sixty years at the helm.

There are surviving drawings and engravings of the abbey church, made before almost all of it was destroyed in the angry wake of the French Revolution. They manage to convey something of the sense of awe and grandeur it must have inspired. It was not only the largest and most magnificent church ever seen at the time; it remained a true wonder of the world right up to its destruction in the 1790s.

It is easy to see the mighty abbey church as Abbot Hugh's outstanding legacy. Yet the sheer scale of the achievement too easily overshadows other developments relating to Cluny which arose directly out of the fruitful relationship between Abbot Hugh and King Alfonso, several of which had a powerful bearing on the Santiago pilgrimage. Like the roads in France, the Spanish pilgrim road was a venue for many interests and many kinds of traveler. Once free of Saracen attack, and with bridges, hospices and other services now in place, the road attracted a growing number of merchants and tradesmen of widely varying skills, all of whom found lucrative work along the way. In a hitherto impoverished land they brought skills which were sorely needed, and soon the Spanish road became a flourishing trade route as well as a vehicle for pilgrims. Suddenly there was new wealth, much of it generated by these craftsmen and merchants from north of the Pyrenees who had traveled with their goods and their skills along the pilgrim road. That it became known as the *camino francès* was not only due to the pilgrims being predominantly from France: the craftsmen who found work opening up the road were mostly from north of the Pyrenees too. King Alfonso recognized their value and saw to it that these skilled foreigners were encouraged to stay and settle down: as a result a number of towns along the Spanish pilgrim road became to a large extent French settlements—among them Estella, Logroño, Burgos and (as its name suggests) Villafranca.

In all this rapid social change taking place in northern Spain Cluny had a firm hand. The abbey was a presiding presence. Already in Hugh's day Santiago had a Cluniac monk as bishop. Now Cluniac abbeys and priories were dotted along the pilgrim road as far as Santiago itself; and the hospices where pilgrims and other travelers found food and rest were for the most part attached to those same religious houses. With Alfonso's benevolent encouragement the abbey had come to control the Church in the whole of northern Spain. And since the Church was the only organized body in the land Cluny effectively administered the entire area. The power of the French Church shored up Alfonso's fragile kingdom, while the Spanish monarch's patronage of Cluny made the abbey rich.

In the perspective of our own time it is hard to grasp how a monastery could exert quite so much political power and influence on an international scale. The monastic tradition was vastly different: the early monasteries had been enclosed cells established as far away as possible from the materialistic world which was regarded as "Babylon," godless and corrupt. Monks were walled in, isolated with their prayers, their rituals and their holy texts. The contrast with eleventh- and twelfth-century Cluny could hardly be greater: from a remote cell in the desert to a sumptuous Burgundian abbey at the heart of a new feudal world, with its aristocratic connections and its abbots esteemed as power-brokers to kings and popes.

The transition from desert hermits to rulers had its roots many centuries earlier. It was the monasteries which produced some of the first thinkers in the Christian world, applying their intellect to explain and define the nature of their faith and its practice. They were men who began to make an impact on the wider world beyond monastery walls by setting out to establish basic rules on how a Christian life should be led: men such as St. Martin of Tours in the fourth century, St. Augustine of Hippo and St. John Cassian in the next century, and above all St. Benedict of Nursia, founder of the Benedictine Order and author of the *Rule of St. Benedict*, in the sixth century. But it was not until after the dark centuries that followed, and the establishment of a new political order in Europe, that the intellectual role of the monasteries could find its key position within a refreshed Christian world.

By the eleventh century feudalism, with its clear definition of everyone's position and responsibilities in life, had supplied a social structure to Christian Europe in which the new monasteries played an invaluable part. They harbored an intellectual elite within those protective walls, as they had always done; except that now their role as men of learning had a vital political dimension. Because the leading monks were literate, and spoke Latin, they could communicate with each other right across the Christian world. Unlike their rulers, who were mostly illiterate, they enjoyed the gift of tongues. This made them invaluable to those rulers as ambassadors, political advisers and (as we have seen) as the sponsors of armies.

Add to all this their spiritual role, and it becomes clearer why the monasteries, Cluny above all, should have enjoyed such spectacular success. Its abbots had not only the ear of God, but that of the pope and of just about every political leader in Europe. And within their own world they presided over an empire so regulated that it imposed a structure—a sense of stability and order—in which religious practice could flourish. The pattern of how a Christian could lead his life was now clear. And without that structure it is impossible to imagine how a social phenomenon as sophisticated and widespread as the Santiago pilgrimage could possibly have taken place. Cluny's empire laid down a vast network of tracks across the continent of Europe along which pilgrims could set forth, in confidence, and in good faith.

They had become the roads to heaven, drawing travelers magnetically from shrine to shrine, each with its own story to tell— its own legend, its own special magic. Medieval society made little distinction between truth and folklore. The concept of "evidence," or "proof," was largely unknown or considered unnecessary. Doubts could easily be quelled if some dream or miracle revealed that God had willed it so. Accordingly legends were widely accepted as historical facts; and in medieval Christian Europe the most popular legends were those which brought people closer to the source of their Christian faith—already more than a thousand years ago and far away geographically. Legends relating to events in the life of Christ and his followers bridged those gulfs, extending a hand of comfort and reassurance to those living in a bleak world, as well as bringing a certain moral structure to people's lives. The pilgrimage movement of the Middle Ages was to a large extent the expression of a widespread yearning for such reassurances.

Legends relating to the birth of Christianity confronted pilgrims wherever they traveled. Those beginning their journey at Vézelay found themselves immersed in one of the most poignant of them. The convoluted tale of Mary Magdalene and her supposed life and death in France was told in Chapter 7. The intense popularity of her shrine as a place of pilgrimage was undoubtedly why *The Pilgrim's Guide* could list Vézelay as one of the four natural starting points for those setting off to Santiago. Once again two Christian legends had intertwined—that of St. James and that of Mary Magdalene—each enhancing the popular appeal of the other.

But the Magdalene legend had a further appeal to pilgrims traveling through this region. The miraculous raft (or boat) which had borne Christ's persecuted relatives and followers safely to the shores of Provence had also carried Mary Magdalene's sister Martha and her brother Lazarus, Christ's friend whom he had brought back from the dead in Bethany. The various legends relating to Lazarus are even more confused than those surrounding the Magdalene; nonetheless the one most widely circulated was that on arriving at these shores Lazarus became the first Bishop of Marseille and subsequently the city's first Christian martyr. Meanwhile his sister Martha evangelized elsewhere in Provence, performing astonishing deeds including the taming of a man-eating monster, the Tarasque, on the banks of the River Rhône—an event still celebrated in the town named after the beast, Tarascon.

Thereafter the story of Martha and Lazarus, like that of Mary Magdalene, appears to have lain dormant for many centuries, until the new blossoming of Christian faith in the early Middle Ages led to widespread church building and a fervent quest for shrines and holy relics which the faithful could venerate: in other words, the arrival of the age of pilgrimage. The event which seems to have re-awakened the legend of Lazarus took place early in the eleventh century when the Duke of Burgundy made a public display of presenting what he claimed to be the head of Lazarus to a collegiate church in Avallon, in the northern region of his duchy.

There followed an immediate and indignant response from the cathedral authorities of another Burgundian city further to the south, Autun. The saint's entire body, they insisted, had been in their possession for many centuries. The age of ecclesiastical rivalry had begun. Pilgrimage was becoming big business, and Church authorities all over France were now vying with each other to attract the lion's share of it.

For Autun the principal rivalry was never with Avallon, but with a far more important establishment, the great abbey of Vézelay. Here, by the end of the eleventh century the cult of Mary Magdalene had firmly taken hold, and the news that the abbey was in possession of the saint's body was attracting pilgrims from far and wide. The cathedral authorities at Autun decided to follow suit, having no wish to be outdone by Vézelay. They too had holy relics, equally prestigious. The decision was made, early in the following century, to build a new cathedral dedicated to St. Lazarus, whose body would in the course of time be placed in a magnificent tomb to be set behind the high altar.

The story of Autun in many respects runs parallel to that of Vézelay. Both religious houses rose to become among the most popular pilgrimage centers in France. The romantic legend of the saints from the sea landing on these shores made a powerful impact on the medieval mind: it was as though the miraculous raft guided here from the Holy Land by the hand of God symbolized the arrival in France of the true faith, and the whole land was moved to celebrate the event. There was a distinctive difference between the popular appeal exerted by the two saints. At Vézelay the Magdalene was the repentant sinner beloved by Christ, whose feet she had washed with her tears and dried with her long hair. At Autun the appeal of Lazarus was altogether more stern. Not only was he Lazarus of Bethany, friend of Jesus, but he was also an early martyr at the hands of the Romans. As if this was not credit enough the identity of Lazarus had by now become confused with the figure portrayed in Christ's parable of Lazarus (in St. Luke's gospel) of the beggar living outside the rich man's gate whose sores are licked clean by dogs. Hence Lazarus became cast as the patron saint of lepers, who soon began to flock, or hobble, to Autun to receive the saint's blessing.

It must have been inevitable that pilgrims assembling at Vézelay before setting off for Spain and Santiago would have become drawn into the highly emotional cult of Mary Magdalene. The saint herself may even have been seen as a key figure guiding them on their long journey ahead. How many pilgrims, having left Vézelay on the road, then felt compelled to seek out the shrine of the Magdalene's brother Lazarus, who had been a fellow survivor of that miraculous raft, is impossible to know. But Autun was no more than a two-day trek to the east of the main pilgrim road as they traveled south: besides, any traveler who had made the longer diversion to Cluny is almost certain to have paused at Autun on the way.[23]

Today's travelers are more likely to embrace all three sites— Vézelay, Autun and Cluny—as a single group. Together they represent a trio of incomparable artistic masterpieces. Even though so little remains of Cluny itself, without the imaginative vision of its abbots and its school of carvers and stonemasons there would have been no Vézelay and no Autun. Cluny was the birthplace of some of the finest medieval art and architecture ever created. And in the case of Autun we have the rare experience of knowing the name of its creator.

It was an uncommon mixture of luck and prudishness which brought about this discovery. Early in the nineteenth century the local authorities responsible for the maintenance of Autun's medieval cathedral decided to remove an innocuous slab of plaster which in the previous century had been laid over the original tympanum above the cathedral's west door. The eighteenth century had been an era which

23 See Plate 14.

was happy to regard medieval art as crude, even barbaric, entirely inconsistent with the prevailing taste for elegance and good manners. Ironically, by covering over the entire area above the west door these eighteenth-century fathers of the church managed to preserve what would very likely have been wrecked by the anti-clerical mob during the French Revolution a few decades later. When the nineteenth-century restorers set to work they revealed a vast sculptural panel on the theme of the Last Judgment, with the figure of Christ in Majesty seated imperiously on a throne, his arms outstretched towards the damned on one side and the saved on the other. And below the feet of Christ emerged a Latin inscription, boldly engraved in the stone, which read *Gislebertus hoc fecit*. "Gislebertus made this."

Only an artist held in the highest esteem would have been permitted to trumpet his achievement quite so proudly, and in such a place where it would catch the eye of every parishioner and pilgrim entering the cathedral—right above their heads as they gazed up at the majestic figure of Christ in Glory. Here was self-congratulation writ large, endorsed what was more by the highest possible authority. And yet, but for that inscription the identity of a sculptor of genius would never have been revealed. Curiously, the effect of that name— even of a man about whom we know absolutely nothing else— does more than simply banish anonymity: it manages to impose a distinctive personality not only on this remarkable tympanum but on to the vast body of work which we know the same sculptor undertook throughout the cathedral.

Few artists in any era have been solely responsible for decorating an entire house of God. Gislebertus was one of that rare breed. His spirit inhabits the cathedral of Autun much as Michelangelo's inhabits the Sistine Chapel in Rome. The world of Gislebertus is one of extremes—of light and dark, joy and pain, gentleness and cruelty, pathos and tragedy. It is a restless world, populated by people who seem constantly on the move, expressed by the sculptor's simple device of making their clothing ripple and flow as if caught in a wind. Stone is made to appear light and fluid, as if molded and caressed by the human hand rather than carved. Pressed together into one

expanse of stone are so many little stories, each one a cameo, and
with so many perceptive touches of humanity it is easy to imagine
pilgrims gazing up at such a rich panorama of life and recognizing it
as a world they knew.[24]

Autun's great cathedral, and the way Gislebertus has humanized
it, explains as movingly as any place in Christendom the nature and
spirit of pilgrimage and people's urge to venture forth and follow
their star. At Autun we come to realize that not all religious art of
the Middle Ages was a punitive sermon in stone, and that human
love and compassion, even ribaldry and laughter, had their place
alongside prayer and punishment.

As it happens pilgrims entering Autun's cathedral were generally
spared Gislebertus' vision of the Last Judgment. The west door was
rarely used as the main entrance for the understandable reason that
it opened directly on to the city's main graveyard. On the other hand
it may be that the bishop realized that the north door would always
be the preferred entrance, which is why he commissioned Gislebertus
to create here a second tympanum on the theme of the saint to whom
the new cathedral was dedicated, St. Lazarus—whose relics were the
main reason why pilgrims came here in such large numbers. Sadly
only a few fragments of it remain, the tympanum having fallen victim
to the same mindless iconoclasm by the eighteenth-century cathedral
authorities as the west portal; except that in this case they were not
content merely to plaster it over, they destroyed it—or most of it. From
a brief description in the fifteen century we know that on the central
supporting pillar the sculptor carved a tall standing figure of the saint,
while the theme of the Raising of Lazarus from the Dead occupied the
main area of the tympanum itself. Alas, neither has survived.

It may seem surprizing that the heads of the cathedral chapter
should take such violent exception to images of their own saint,
and one who attracted so many pilgrims. But the true reason may
lie elsewhere. At the base of the tympanum, right above the central
pillar, Gislebertus carved the lintel as a horizontal frieze depicting

24. See Plate 15.

the Temptation of Adam in the Garden of Eden, one of the favorite themes of the medieval Church. It spanned the entire width of the north door, making it necessary for the two figures of Adam and Eve to be horizontal—approaching one another head to head at the center by the Tree of Knowledge, Adam from the left, Eve from the right accompanied by the serpent and grasping the apple. It is an extraordinarily interpretation of the Temptation—vivid and distinctly erotic. Furthermore, because of the width of the lintel both figures were virtually life-size, and except for a few stray wisps of grapevine entirely naked. In other words pilgrims making their way into the cathedral to celebrate mass would be required to pass beneath a startlingly realistic enactment of mankind's primal sin, which looked far too enjoyable to be regarded as sinful. And while such a scene was clearly acceptable to Bishop Etienne and his fellow-churchmen in the twelfth century it was evidently too much for their eighteenth-century successors in an age of periwigs and buckled shoes. Accordingly, in the interest of propriety and good manners they had it ripped it out.

The figure of Adam disappeared, along with that of Lazarus and most of the entire north tympanum. Eve—by the skin of her teeth—has survived. She was discovered by chance in the middle of the nineteenth century during the demolition of a house which had been built in 1769, a mere three years after the cathedral authorities had torn down the north door. They had not even bothered to clear away Gislebertus' wrecked tympanum, but had left the various sculpted panels lying around to be taken away for some nominal fee. Eve had become useful building material—a fate matched a few decades later and on a larger scale by Cluny's abbey church.

Eve is unlike any other medieval carving. She is enigmatic and entirely unforgettable. Her sexuality is presented as dangerous (as religious teaching of the day would have insisted), but also as irresistibly appealing. In this respect Gislebertus made her transcend the prevailing misogyny of the medieval Church; and she becomes a forerunner of a great European tradition of the female nude in art. Today she is the prime exhibit in Autun's Musée Rolin only a short distance from the cathedral which she graced for more than six centuries.

10. On the Miracle Trail

IT WOULD BE HARD TO EXAGGERATE the impact of sacred places such as Vézelay, Cluny and Autun on the ordinary pilgrim in twelfth-century France who in all probability had never experienced any religious ceremony more splendid than Sunday mass in the local village church. The opulence of the abbey buildings themselves, the grandeur of the church services with their elaborate liturgy, and of course the shrines of the saints with their jewel-studded reliquaries and so many vivid accounts of miracles performed there: all of these, put together, would have amounted to an overwhelming experience imbued with mystery and magic—an introduction to a totally unfamiliar world, even a glimpse of heaven on earth.

Now, after so many highly charged events, those pilgrims who had gathered at Vézelay only a week or so before might have anticipated a less demanding stretch of road as they left Burgundy and headed southwest along the fringes of the central mountains in the direction of Limoges, Périgueux and the distant plains of Gascony. But it was unlikely to be so. Protected though they might be by their scallop-shell and their status as spiritual travelers, pilgrims were still exposed to the political climate of the day; and towards the end of the twelfth century those hoping to make their peaceful way south from shrine to shrine would have found themselves drawn into two of the major social upheavals of the day—the crusades to the Holy Land on the one hand, and on the other the burgeoning war between England and France over the English-held territory of Aquitaine.

The pilgrim road had become a battle zone as well as a sanctuary for returning crusaders who had been captured or wounded in battle; and for travelers at that time there were reminders of both events everywhere they went. At Issoudun, just a few miles along the road from Bourges, the main street today known as the Rue Saint-Jacques leads to the Tour Saint-Jacques which originally formed part of a massive fortress built by the King of England, Richard the Lionheart, on his return from the Third Crusade in the last decade of the twelfth century. The tower was a bulwark in his heavy-handed campaign to secure his territories against the French King Philippe-Auguste, who had laid his hands on them in Richard's absence. The pilgrim road passes through the very town where the English monarch was fatally wounded by a bolt from a crossbow. Nearby, in the small town of Neuvy, pilgrims would pause to offer their prayers in a church dedicated to St. James which had been built by an earlier returning crusader as a copy of the Church of the Holy Sepulchre in Jerusalem on the site of Christ's tomb. A few miles further still another small settlement, Cluis, lay on the very frontier between territory ruled by the Kings of France and the Duchy of Aquitaine which was ruled by England.

Altogether this was hardly the tranquil progress through the gentle countryside which pilgrims may have expected. In addition to the customary hazards of the road they would have found themselves continually assaulted by echoes of crusading fervor mixed with the roar of battle. Today's traveler in this historic region of France can still hear those same echoes rebounding from the medieval churches and castles that survive along the road, overlaid by further echoes of the great pilgrimage itself as it took place throughout these troubled centuries.

Relations between the English and French kings had for a time seemed amicable. Richard and Philippe-Auguste were united in their commitment to the Third Crusade, and had spent months together in 1189 at the abbey of Vézelay planning the great adventure in which they were to lead their respective armies. It was during the course of their campaign in the Holy Land against the Muslim leader Saladin that the two monarchs fell out. The French king decided to return early, to the contempt of Richard, leaving the English king to continue the campaign. After conducting a peace pact with Saladin Richard set sail for home, only to be shipwrecked on his return journey and held captive by the Duke of Austria until a vast ransom was raised to obtain his release and return to England in 1194.

This precis of one of history's most famously romantic melodramas would have little relevance to the Santiago pilgrimage had Richard quietly remained in England. But he was English only in name. Neither of his parents, Duchess Eleanor of Aquitaine and King Henry II, was English; and Richard himself did not even speak the native language: Where he was born, and where he mostly lived, was France: and now, as soon as possible on his release from captivity his ambitions turned to France, and in particular to the security of Aquitaine which his mother's marriage to Henry had placed under English rule. And as he led his army to secure its borders, his opponent was naturally his former fellow-crusader and companion at Vézelay, King Philippe-Auguste of France. The ensuing conflict turned the pilgrim road into a battlefield. And one longs for a traveler's eye-witness account of how pilgrims fared in such threatening circumstances.

Today Richard the Lionheart's presence is never far from this stretch of the pilgrim road. One of the most striking landmarks in the region is the slender spire and bell tower of the Church of Saint-Léonard-de-Noblat, rising high above the river valley to the east of Limoges. The English monarch is known to have contributed to its building costs in thanksgiving to St. Leonard for having facilitated his release from prison in Austria on his ill-starred return from the Third Crusade, so the king believed. In fact the connection between Saint-Léonard and the crusading movement dated back to its foundation a century earlier in the aftermath of the First Crusade, when one of its chapels was sponsored by a returning crusader and named after the Holy Sepulchre in Jerusalem.

Christian saints in the Middle Ages tended to rise from obscurity in response to particular needs and circumstances of the day; and St. Leonard seems to have been a case in point. There is no record of his very existence until the eleventh century: then his emergence as a focus of popular veneration coincides with the beginning of the crusading movement as well as with the military campaigns of Reconquest in Spain that was heavily supported by Cluny. In both spheres of operation the saint was reputed to have performed numerous miracles in which prisoners captured in battle were released: From nowhere Leonard became widely recognized as a patron saint of prisoners; hence Richard the Lionheart's gesture of thanks in Aquitaine following his release from captivity.

Characteristically the fame of St. Leonard soon attached his name to the list of saints whose shrine were required places of veneration for pilgrims generally, in particular those bound for Spain, where many cases of prisoners miraculously released from Muslim hands had been reported. *The Pilgrim's Guide* includes a fulsome entry on St. Leonard, who is reputed to have been a sixth-century nobleman who abandoned a life of riches for a hermit's existence in the forests around Limoges, "enduring cold, nakedness and unspeakable labors [before] passing away in a saintly fashion."

The overheated style of writing suggests that the author may well have been the same Benedictine monk from Parthenay-le-Vieux, Aimery Picaud, whose ribald earlier comments on the behavior of the Gascons we encountered in Chapter 6. His account of St. Leonard's legacy continues in the same rich vein: "His extraordinarily powerful virtues have delivered from prison countless thousands of captives. Their iron chains, more barbarous than one can possibly recount, joined together by the thousand, have been placed in testimony of such great miracles all around his basilica."

One of the saint's miracles mentioned in the *Guide* concerned a crusader by the name of Bohemond, Prince of Antioch and son of the Duke of Apulia, who had been taken prisoner by the Turks in the aftermath of the First Crusade. The author attributes his release from captivity entirely to a miracle performed by St. Leonard, In fact the "miracle" consisted of the emir being moved to accept a huge ransom for Bohemond's release—a fact omitted from the *Guide*. Such was the universal belief in the power of miracles that Bohemond himself attributed his release to the saint's intervention, even traveling halfway across Europe to the newly-founded Church of Saint-Léonard-de-Noblat in order to offer his effusive prayers of thanks.

The pilgrimage movement was founded on a cult of the saints; and in turn the cult of the saints rested on a widely held belief in miracles. The canonization of a new saint by the pope would only take place if the proposed candidate could be shown to have performed an impressive number of miracles. And since a church or abbey in possession of relics of a newly canonized saint stood to gain massively as a result by donations and bequests of land, not surprisingly there grew up, as in the case of relics, something of a "miracles industry" in eleventh- and twelfth-century Europe. Candidates for sainthood were virtually queuing up. Centuries later, at the time of the Counter Reformation, Protestant jibes induced the Vatican to appoint a canon lawyer specially to mount an argument against the canonization of a candidate—a role popularly known as the "Devil's Advocate." But in the Middle Ages no such corrective system existed, and papal judgment was a more haphazard affair depending largely on the strength of each case as presented to His Holiness, and on the pope's personal inclinations.

In modern terminology it all sounds like an exercise in skilful public relations. Priests and monks who were guardians of an important shrine are known to have taken pains to maximise its profitability by drawing up lists of miracles which their saint was claimed to have performed. Sometimes the eagerness of an abbey to claim a miracle led to an outcome at odds with public interest as well as the rule of law, as when the monks of Vézelay keenly supported the release of a prisoner due to the miraculous intervention of Mary Magdalene, in spite of the fact that the man was a convicted murderer.

The process of claiming and authenticating a miracle was not always as naive or grasping as it may appear to our eyes. Any number of events which today would be attributed to some identifiable natural cause did not appear so to the medieval onlooker. If there is an eclipse of the sun we know without hesitation what shadow has caused it, and that it will duly pass. If there is a severe flood in the plains we know that excessive rainfall in the mountains has caused rivers to burst their banks. If there is an outbreak of malaria we know that mosquitoes are the cause, just as that the Black Death was brought by fleas carried by the black rat. But in those times there was no such understanding of the workings of the natural world. Our life on earth was full of wondrous and fearful mysteries many of which lay outside the natural order of things, and could not be explained except in terms of divine will: that God—and sometimes a punitive God—had willed it so.

But God was also perceived to be merciful: hence it seemed natural to appeal to Him for forgiveness and help. What He had willed He could be persuaded to ameliorate; and here a channel of communication existed through which such an appeal could be made. That channel was through the good offices of the Christian saints, who had God's ear. They were the essential intermediaries: accordingly their shrines were of supreme importance in the medieval order. They were God's listening-posts. An appeal to divine authority through the intercession of His saints could lead to a radical transformation of human fortunes, whether in matters of illness, some natural disaster, an unjust imprisonment or personal misfortune, a crisis in battle, or whatever. And that divine intervention *was* the miracle.

Belief in miracles lay at the heart of Church teaching and of the Christian faith in medieval Europe. It was a belief universally held, and tended to invite scenes of public celebration that were sometimes far from godly. Yet, for all the trappings of funfair and fairground, without an unquestioned core of belief in the miraculous there would have been no such occasions: there would have been no cult of the saints, and no shrines to attract worshippers. And of course without those shrines there would have been no pilgrimages; and those of us today who love majestic church architecture, and the masterpieces of painting and sculpture which accompany them, would be sorely deprived. For many centuries the pilgrim roads in France and Spain were the inspiration for some of the most glories achievements of western civilization; and it is places like Vézelay, Autun, Aulnay, Chartres and Santiago de Compostela itself which to our eyes may be seen as the real "miracles" of the Middle Ages, whose magnificence transcends all cults and matters of faith. They lift the human spirit, bearing witness to the soaring power of the creative imagination.

*

This stretch of the pilgrim route along the edge of the central mountains was the frontline in territorial battles between England and France for more than three hundred years. As a result it is hardly surprising that so many of the saints revered in this region should have been honored for miracles in which victims of war were healed of their wounds or released from captivity. In the medieval church at Saint-Leonard-de-Noblat, with its massive bell tower and spire pointing so confidently to heaven, kings, dukes and countless more humble soldiers came to offer prayers of thanks to the saint whom they believed to be responsible for their deliverance from enemy hands.

Medieval pilgrims who paused here to pay their respects to those miraculous events then set off westwards across the valley to the city which gave its name to this road. The city was Limoges, and in *The Pilgrim's Guide* the road is named the *Via Lemovicensis*, and two thirteenth-century bridges still lead the traveler along the ancient trail. Limoges was the city of another miracle-worker, St. Martial, who surprisingly gets no mention whatever. The little we know about Martial is that he was a preacher in pre-Christian Roman Gaul who defied persecution and achieved popular acclaim for his miraculous powers. It is reasonable to suppose that he might have been forgotten like St. Léonard but for the fact that seven centuries later the newly-founded monastery of Cluny began to extend its influence southwest from Burgundy, sponsoring abbeys and priories which serviced the pilgrim road to Spain. And the Abbey of Saint-Martial was one of them.

With the growing popularity of the pilgrimage movement, spurred by the legend of St. James and the emotive cause of Reconquest in Spain, the Abbey of Saint-Martial soon became richly endowed, and by the eleventh century the abbey church could offer pilgrims the dramatic sight of a reliquary of the saint as a seated figure entirely encased in gold, with hands outstretched in a gesture of blessing. The impact on pilgrims entering the candlelit church after days and weeks on the road in this war-torn countryside must have been hypnotic. Here was the golden saint himself, the master of miracles, arms extended in welcome to the weary traveler. It would have been close to being blessed by God. Alas, the reliquary-statue no longer exists, but travelers on the third of these pilgrim roads, the one from Le Puy, will soon come face to face with a similar reliquary which *has* survived, the extraordinary Majesté de Sainte-Foy at Conques, about which a great deal will be said later.

The Church of Saint-Martin in Limoges was one of the five archetypal pilgrim churches which conformed to the same plan, designed to accommodate large congregations capable of circulating freely along the broad aisles and behind the high altar in order to venerate the relics placed in full view on the altar itself (as described in Chapter 3). Such was the power and appeal of the great Church of Saint-Martin, with its iconic reliquary-statue, that pilgrims threw themselves to the floor before the miracle-working reliquary in the most extreme states of self-abasement. There is even a record in the fourteenth century of pilgrims arriving at Saint-Martial and stripping off their clothing in order to appear before the saint entirely naked and therefore bereft of all worldly trappings. Yet late in the same century devotional practice shifted from the dramatic to a more prosaic form of veneration: the golden statue with the outstretched hands was removed (and presumably the gold melted down) and replaced by a mere box with a little door which pilgrims could open in order to observe a hand of the saint. It is hard not to think that a certain magic had been sacrificed in the interest of being able to set eyes on the relic itself. Miracle-working had come down to earth.

Today almost everything here has gone—golden statue, reliquary box, saintly relic, church and abbey too—for the most part victims of the furious iconoclasm of the French revolutionary mob with its loathing of all monastic institutions in the late eighteenth century. But then in 1960, during building work in the city center, a small but important discovery was made. Excavations revealed the crypt of the original church, part of which dates from the fourth century barely a hundred years after the death of St. Martial, in the very first phase of the conversion of the Roman Empire to Christianity.

Like so many places along this pilgrim road Limoges bore the scars of battle from the protracted conflict between England and France known as the Hundred Years' War—which actually went on for three hundred years. In 1370 Edward the Black Prince, son and heir of the English King Edward III, occupied the city after a siege and proceeded to massacre three thousand inhabitants in an act of reprisal, or so it was claimed by the French chronicler Froissart. The fact that the figure may have been a considerable exaggeration scarcely alters the picture we have before us of a brutal time and place when the ordinary traveler or harmless citizen constantly feared for his life. Not surprisingly, miracle-workers were greatly in demand, and through much of the Hundred Years' War the Abbey of Saint-Martial, like that of Saint-Léonard-de-Noblat a few miles to the east, was the scene of numerous appeals and prayers of thanksgiving on behalf of prisoners miraculously released from captivity on both sides of the conflict. And if little remains in Limoges itself as witness to those violent times, today's travelers continuing south along the pilgrim road soon find themselves in the small town of Châlus where a gaunt stone tower stands as a monument to another grim event in that conflict between England and France. It was from this tower in the year 1199 that the English King Richard the Lionheart was fatally wounded by a bolt from a crossbow. The monarch who had survived a crusade and years of imprisonment eventually fell to the unlikely shot of a distant marksman.

Echoes of the crusades continued to follow pilgrims on their journey southwards from Limoges and Châlus. The next major halt on the road was the city of Périgueux. And here the modern traveler may be confronted by the same sense of disorientation as the medieval pilgrim would have experienced; because the huge sprawling cathedral of Périgueux, built in the twelfth century, is likely to make any visitor feel transported far away to the eastern Mediterranean. The building is crowned by a huge white dome, quite unlike any other cathedral in France, its design having clearly been drawn up at the behest of French crusaders or merchants familiar with the churches of Byzantium. Opinions vary as to its precise origin, but the consensus favors either the Church of the Holy Apostles in Constantinople, or possibly that it may have been conceived as a copy of the Basilica San Marco in Venice.

To the medieval pilgrim this was the shrine of another miracle-working saint. The appropriate entry in *The Pilgrim's Guide* makes this quite clear: "After the Blessed Léonard one must visit in the city of Périgueux the remains of the Blessed Fronto." The saint whose name now translates as St. Front is reputed to have been the first Bishop of Périgueux, and the *Guide* gives a colorful account of his life. It seems he was ordained in Rome by no less a figure than St. Peter himself, who then sent him to preach in newly conquered Roman Gaul together with a colleague by the name of George. When the latter died Front returned to Rome and was given Peter's staff which had the magical power to restore a man to life. "And so it was done," we are assured. Front proceeded to convert the city of Périgueux to Christianity, "rendering it illustrious through many miracles, and died there in all dignity." It was a story widely enough circulated for the scene of St. Peter presenting the saint with his staff to have been carved in stone for the main portal of the original cathedral, the carving now displayed in the local museum.

This account of St. Front's life is an eloquent illustration of how the medieval Church sought to link the shrines along the pilgrim roads to the era of Christ and his apostles: it was a continual process of bridge-building, designed to bring the pilgrimage movement close to the fountain-head of the Christian faith across a gulf of more than one thousand years. Whatever the origin of the *Guide*'s version of the St. Front story, we know that there were other versions in circulation at the same time. From as early as the eleventh century St. Front's companion, George, is named as one of the early Bishops of Le Puy in the late fifth century, almost five hundred years after he was supposed to have been brought back to life by St. Peter's magical staff. The vast discrepancy in dates between the two versions is probably best explained by the relative requirements of two different Church bodies. The authorities in Le Puy, which was an important bishopric, would have been keen to establish a respectable list of early founders who had enjoyed close links with the Christian saints. The Cluniac promoters of the Santiago pilgrimage, on the other hand, would have been principally concerned to make a direct connection with the era of their patron, St. James, whose shrine was the pilgrims' ultimate goal.

Resolving the discrepancy of five hundred years between the two versions required a small, but vital, twist in the story. In the Le Puy version St. Front was merely given the apostle's staff on his return to Rome in order to bring his companion back to life; whereas in the *Guide*'s story St. Front is given the staff by St. Peter in person. Given the thread of propaganda running through the *Codex Calixtinus,* of which *The Pilgrim's Guide* forms a key part, it is hardly surprising that this should have been the version chosen by its compilers. Pilgrims arriving at the shrine of St. Front in Périgueux could feel confident that they were walking in the footsteps of those who had been close to Christ and his apostles.

The story of St. Front, dramatic as it is seen to be, is a variation on the tales told of so many of these revered early saints who were rediscovered by a revitalized Christian world after so many centuries in obscurity, and whose relics were now venerated all along the pilgrim roads in France and elsewhere in Europe. The saints themselves, as described in medieval texts, were generally isolated figures living in Roman Gaul, often hermits, who bravely preached the gospel in the face of persecution and who frequently suffered martyrdom. Crucially in the eyes of the medieval Church (particularly so in the case of St. Front) for a religious house to possess the relics of these early saints provided a direct link with the time of Christ and his disciples, and thereby created an invaluable bridge between the "modern" Christian era in Western Europe and the source of the Christian faith in the Bible Lands so very far away in time and place.

The pilgrim route from Vézelay is especially rich in shrines, churches and other monuments which offer that bridge between the early centuries of embattled Christianity and the medieval world, as well as a further bridge to our own time, a total span of more than two thousand years. Traveling along this rural stretch of road is a constant re-acquaintance with the past as we journey through different time zones—a trail of miracles and memories which threads its colorful way from Burgundy to the Pyrenees and Spain.

The shrines and weathered churches along this route are like signposts directing the pilgrim onwards. Soon after Périgueux the squat fortified tower of Sainte-Marie de Chancelade is a further reminder that here was an Anglo-French battleground. This was once another of the numerous abbey churches established by religious orders for the wellbeing of passing pilgrims. The former abbey was (for once) not part of the chain of religious houses set up by Cluny, but was Augustinian, following the Rule of St. Augustine of Hippo, whose greatest contribution to the pilgrimage movement was to establish the great monastery of Roncesvalles high in the Pyrenees at the gateway of Spain (see Chapter 15).

A short distance further, and the village of Sainte-Foy-la-Grande can feel as though it fell asleep in medieval times and has scarcely woken since. And then, at La Réole, pilgrims reached the Garonne, the broadest river they would encounter on this route. River crossings, we have seen, were a constant hazard for pilgrims, since they were entirely at the mercy of local ferrymen, who were mostly not there for charitable reasons but to make money out of travelers who had no choice but to employ their services. Aimery Picaud's graphic account in *The Pilgrim's Guide* of the murderous ferrymen at Sorde, on the route from Tours and Paris, was quoted in Chapter 6. Here at La Réole, at least, travelers were relatively safe: between spring and autumn, when pilgrims were generally on the road, the River Garonne was shallow enough for them to wade across it. Picaud's lugubrious warnings would have been unnecessary. On the other hand, having crossed the river, pilgrims were now entering that region of swamp and vicious insects known as the Landes, about which his pronouncements were almost as dire. "This is a region deprived of all good" was his blunt warning. Today this eastern area of the Landes is no longer the hostile swamp of earlier centuries, being largely forested or agricultural. But before setting out to cross it today's traveler is reminded that the miracle trail is still very much with us. On the edge of the Landes the handsome town of Bazas, small though it is, nonetheless possesses a cathedral which was modelled on the great Gothic cathedrals of northern France. Its surprising grandeur is due to its dedication to John the Baptist, a phial of whose blood was said to have been preserved here, unlikely though it may seem, and was the object of much veneration by pilgrims and—not surprisingly—the source of a great many miracles.

The Pyrenees are now in sight, and the road from Vézelay is drawing closer every mile to the Paris road, which at this point runs almost parallel only a short distance to the west; while closing in from the east is the third pilgrim road, from Le Puy, which we shall shortly be taking. Soon all three roads will converge in the Basque Country just this side of the mountains, while the fourth and most southerly road, from Arles and Saint-Gilles, will keep its distance and not join the other three until the far side of the Pyrenees.

Meanwhile the Vézelay road continues southwards across the eastern expanse of the Landes. One of the few human settlements of any size along this lonely stretch of highway is the small fortified town of Roquefort. And here, in a modest way, is the response of feudal lords to the scathing description in *The Pilgrim's Guide* of the Landes as a region "deprived of all good" because Roquefort is an example of the "new towns," known as *bastides*, which were established later in the Middle Ages specifically to bring together into viable communities those scattered inhabitants of the swamps with their pigs and waterlogged hovels. Aimery Picaud would have appreciated Roquefort with its massive ramparts and towers, and a hospice where pilgrims could enjoy "bread, wine, meat, fish" and other delights of the flesh which the area, as he knew it, so sorely lacked.

After reaching the far side of the Landes yet another in the twelfth-century chain of abbeys founded by Cluny awaited pilgrims. This was Saint-Sever. All that survives today is the handsome abbey church, distinguished by a group of capitals startlingly painted in bright colors which seem incongruous at first until we realize that most stone carving in medieval churches was once similarly painted—even the great tympana of Vézelay and Autun. Then, a short distance beyond Saint-Sever an even earlier historical sight greeted the traveler. At Hagetmau in the year 778 the Emperor Charlemagne founded an abbey to house the relics of another early preacher, St. Girons, who had suffered persecution in Roman Gaul. Though the abbey, greatly enlarged at the hands of Cluny in the twelfth century, was destroyed in the sixteenth-century wars of religion, the crypt with its fourteen magnificently carved capitals survives. And perhaps because we come across it unexpectedly in this unsung rural region it gives out an especially powerful echo of those times when medieval pilgrims would arrive at this same place after a long day on the open road, to admire these same carvings in this enclosed vaulted room and to offer their prayers.

Finally there was one more river to cross, the Gave de Pau. And today at Orthez a medieval bridge supports an imposing central watchtower which for eight hundred years has presided over the passage of pilgrims as they made their crossing. And from here it is less than a day's walk into the Basque Country, about which Aimery Picaud reserved some of his most vituperative epithets, as recounted in Chapter 6. And soon, on the crest of the gentle hill called Mont Saint-Sauveur, is a place where we have been before; because on this hillside with the broad flank of the Pyrenees not far ahead is where our road joins the road we took from Paris. It is also where we shall find ourselves yet again in a later chapter—because on this lonely spot, commemorated by a modern stone cross—is where the third of the four pilgrim roads also joins the first two.

And that third road is our next journey.

III. Via Podiencis:
The Road from Le Puy

11. A Gift of Many Gods

O F THE FOUR PRINCIPAL PILGRIM ROADS IN FRANCE the one from
Le Puy-en-Velay, the *Via Podiensis* is the least well documented
geographically. A few key churches and abbeys along the route are
recommended in *The Pilgrim's Guide*, but they are far apart from
each other, and there is no Aimery Picaud to fill in the gaps and
indicate which precise route travelers should take in order to reach
them: neither are there any trenchant comments on the delights
and horrors which pilgrims could expect on their journey. Wayside
chapels and former hospices now provide invaluable markers of the
likely route. As a result, the road from Le Puy has now become easily
the most popular of the four roads for the traveler on foot, especially
now that backpacking has become almost an industry. In the words
of Robert Louis Stevenson, who famously traveled not far from
here, in the Cévennes, with his donkey, "I travel for travel's sake.
The great affair is to move." Today the track clearly marked GR65
has become generally identified as the original pilgrim road. And its
popularity is understandable: the countryside of the Velay region, on
the southern slopes of the Auvergne mountains, is wild and glorious;
and as the traveler continues southwest several of the artistic gems
of the pilgrimage are welcome staging posts. Altogether, the pilgrim
route from Le Puy is the walker's road to heaven.

Much of the landscape of the Velay was formed by volcanoes,
and geologically this has given Le Puy an eccentric appearance, a city
of sudden hills and surprising pinnacles of rock rising out of the plain,
like gnarled fingers pointing to the sky.[25] It is a natural environment

25. See Plate 16.

which has always lent itself to myth-making, and Le Puy is known to have been a sacred place long before its association with the Santiago pilgrimage, and indeed long before the conversion of Roman Gaul in the fourth century. The Emperor Charlemagne celebrated mass here on at least two occasions, in 772 and 800. And a century and a half later, in 950, the local bishop by the name of Godescalc made a pilgrimage to the shrine of St. James at Santiago. This was a little over a hundred years after the first reports of the discovery of the apostle's tomb; and we know from a visiting traveler's report that a sizeable pilgrimage from far and wide had grown up in the intervening years. Even so, Godescalc's journey seems to have been the first "official" visit to the shrine by a senior churchman, and it is widely credited as a historic landmark in the story of the Santiago pilgrimage.

In 951 Bishop Godescalc commemorated his safe return to Le Puy by building a chapel on a tall pinnacle of rock within the city, and dedicated it to the Archangel Michael. It is unclear why his pilgrimage to Santiago should have prompted him to undertake such a thing, but in consequence the Chapelle Saint-Michel d'Aiguilhe became one of the earliest of the "pinnacle" churches in Europe. In the centuries to follow a number of others would be raised on similar isolated hilltops. Not being an earthly being, and therefore having no relics to be enshrined, St. Michael became traditionally venerated in chapels loftily perched and pointing to the heavens, Mont Saint-Michel in Normandy and Saint Michael's Mount in Cornwall being the most celebrated examples. Disciples of mystical geometry have found deep significance in the fact that shrines to the archangel are in a straight line from Cornwall and Normandy in the north, Le Puy in central France, and finally to his earliest known shrine on the slopes of Monte Gargano in southern Italy.

In Le Puy pilgrims to the Chapel of Saint-Michel d'Aiguilhe faced a formidable ascent of 267 steps. Worshippers in large numbers were clearly undaunted by this challenge, keen to express their devotion the hard way, and the chapel rapidly became a popular pilgrimage site. For several centuries it matched the appeal of the cathedral on a neighboring hill only a short distance away.

The Cathédrale Notre-Dame dates from the late eleventh and twelfth centuries. Today's travelers, familiar with the glow and lightness of Chartres and Bourges cathedrals, may find it oppressively gloomy, being roofed by an almost windowless dome which seems to smother the daylight. The whole building has a distinctly oriental flavor, owing a great deal to the early churches of Byzantium, which are invariably dark. It was evidently inspired by crusaders returning from the Holy Land, and in particular Constantinople. What they wanted back in their own homeland was a place of worship which took them back spiritually to the Bible Lands which they had recently helped to "liberate."

At the same time there are equally powerful influences of that other crusading arena, Spain and the campaign of the Reconquest. The interior of the cathedral, and particularly the ribbed vaulting, carry strong references to that masterpiece of Muslim architecture in southern Spain, the Great Mosque in Córdoba. And here there may seem to be a startling contradiction—an important place of Christian worship inspired on the one hand by churches in the Holy Land that had been built in defiance of Islam, and yet inspired equally by one of the finest of all places of Islamic worship.

It is an enigma which perfectly illustrates the ambivalent relationship that existed between Christian Europe and Muslim Spain during these early centuries before the might of Christian arms eventually expelled Islam from the Iberian Peninsula. Until that moment Islamic Spain was the enemy which Christian leaders—at least the more enlightened among them—could not fail to admire, even envy. On the public level the appeal of a Holy War was irresistible to European leaders: the early military campaigns of the Reconquest and the burgeoning pilgrimage movement to Santiago carried huge political and emotional weight throughout the continent, boosting the morale of Europe and lifting Christian spirits so long battered by continual Saracen raids right across France.

At the same time there were other less demonstrative currents that flowed between the Islamic and Christian powers. From early in the tenth century the Muslim caliphate of al-Andalus (modern

Andalusia) in southern Spain, with its capital the city of Córdoba, became the wealthiest and most sophisticated political state in Europe, far outstripping the attainments of Christian civilizations in all the practical sciences, in mathematics, in philosophy and other intellectual pursuits. The library in Córdoba, with its Arabic treatises on medicine and the sciences, and its wealth of translations of Greek, Latin and Hebrew texts, was richer and far broader in scope than any library in the Christian world. As a result, even while armies in northern Spain were battling for territorial gain there existed a great deal of diplomatic and intellectual contact between the two regions of Spain, and even more so between the Muslim south and Christian scholars and Church leaders throughout Western Europe. Córdoba was something of an international city, open to foreigners, and many a Christian visitor returned from his travels with glowing accounts of what he had encountered. Islam might be the enemy in matters of faith and on the battlefield, but it possessed a priceless jewel—learning.

Then, at just the time when the present cathedral of Le Puy was being built, access to that Islamic world of learning was dramatically increased by one of the key events in the Christian Reconquest of Spain, the capture of the city of Toledo in 1085. Here was a city second in importance in Muslim Spain only to Córdoba, and it possessed a library equally rich. Now, with Toledo in Christian hands, that library was open to scholars from all over Europe, and free access became available to that rich store of learning and scholarship which travelers had hitherto only glimpsed in Córdoba. There began a veritable fever to explore those new fields of knowledge which had been opened up. Muslim scholars with an understanding of Latin were now in great demand as translators of Arabic texts on a variety of subjects from mathematics to medicine. One of the most enlightened churchman and theologians of the day became an enthusiast for Islamic writings. He was Peter the Venerable, the last of the great abbots of Cluny and a brave defender of Peter Abelard. He made a special visit to Spain in the year 1142 in order to seek out translators. The abbot had a particular ambition, remarkable for that time, which was to commission a translation of the Koran, though his primary motive for doing so was hardly ecumenical in spirit. It was in order that he could refute it. Unlike so many of his contemporaries in the Church he believed in the power of reasoned argument. Alas, it was an ambition he never had the chance to fulfil. The cathedral of Le Puy is an embodiment of these contrasting influences—Byzantine churches on the one hand, Muslim mosques on the other. And it is far from being a lone example: other churches along the pilgrim roads, and this road in particular, send out similar signs of diverse cultures. In every case the result is magically harmonized; and it seems appropriate to speculate that a shared desire to create an environment designed for worship may transcend even the most profound religious differences. Indeed, maybe all places of prayer possess a key element in common, regardless of faith. If true, that is true ecumenicalism.

There are other gods, too, whose presence can be felt in Le Puy. Those preparing to climb the broad flight of sixty steps to the striking facade of the cathedral with its alternate bands of white limestone and black volcanic rock, might appreciate a charming story before they set off. It concerns a priest by the name of Voisy who was to become the first bishop of this cathedral late in the fourth century, and who made this same climb, at that time up the slope of what was a bare hillside. It was a midsummer day, yet to the priest's astonishment he found the crest of the hill to be blanketed in snow. Furthermore, on the snow-covered ground rose a solitary standing stone, a menhir, which had been raised to some pagan god long ago. Voisy then noticed that encircling the menhir at a certain distance were the tracks of a deer. It was instantly clear to the priest that the deer had been a divine messenger, and that its footprints were meant to trace the outline of the church to be built on the site. And so, the legend goes, the first cathedral of Le Puy was erected here, and dedicated to the Virgin Mary. The menhir became Christianized in the process, and was given a place of honor in the sanctuary of the new cathedral, becoming known as the Throne of Mary. Hence Le Puy (together with Chartres) became the earliest shrine to the Virgin Mary in France.

By the ninth century the cathedral authorities had grown more purist, uncomfortable with this association of Christ's mother with pagan gods; and they broke up the menhir, incorporating the fragments into an area of the cathedral floor. Christian Europe was increasingly under threat by this time—from Norman invaders and Magyars sweeping across France from the north and east, and from Saracens in the new Muslim territories in Spain. Other gods were no longer to be tolerated by the Church, and the menhir had to go. But the older deities were not to be disposed of that easily, and the mystique of the menhir remained. The section of the church containing fragments of the ancient stone came to be known unofficially as "the Angel's Room," and miracles continued to be reported. And when the present cathedral was constructed during the eleventh and twelfth centuries—a great deal larger in order to accommodate increasing numbers of pilgrims—it became part of the new narthex.

Thus Le Puy's long history of honoring many gods persisted across the centuries.

*

The four principal pilgrim roads listed in *The Pilgrim's Guide* all had points of departure which were traditionally centers of pilgrimage in their own right before the cult of St. James took hold in France. It is easy to see why they were selected. They were all of them places which already catered for pilgrims: hence they acted as ideal platforms from which those preparing to undertake their long journey to Santiago could be sent on their way.

From the earliest days in the twilight of the Roman Empire the saint widely venerated at Le Puy was the Virgin Mary. It was to her that the first bishop, Voisy, had dedicated his church after the revelation of the deer's footprints in the snow. The subject of the Marian cult in the medieval Church is a contentious one which lies safely beyond the parameters of this book. It is relevant here largely because it came to contribute greatly to the popularity of Le Puy as one of the main starting points for the Santiago pilgrimage. In the year 1254 the French King Louis IX, later canonized as St. Louis, returned from the Seventh Crusade, of which he had been one of the leaders, bringing with him a number of sacred objects which he had obtained in Palestine. One of these was a statue of the Virgin Mary, believed to have been of Egyptian origin and perhaps carved in dark cedar-wood (probably from Lebanon), or, more likely, in ebony (from Africa). The king duly presented the statue to Le Puy, appropriately so since the cathedral was dedicated to the Virgin.

The carving was soon known as the Black Virgin of Le Puy. Dressed in gold brocade and set on the high altar, it immediately became the focus of intense devotion by pilgrims from far and wide. As so often with medieval shrines the cult of one saint became merged with another: pilgrims gathering at Le Puy before their long trek to Santiago would naturally attend their final mass in the cathedral, and find themselves offering their prayers not in the presence of a statue of St. James but before this hypnotic black figure in gold on the high altar. Not surprisingly, so powerful an image remained with them on their journey, and the aura of the Black Virgin was carried by Santiago pilgrims westwards from Le Puy along the mountain route towards the Pyrenees, until the road became studded with wayside churches and chapels dedicated to Notre-Dame du Puy. In the minds of pilgrims the image of the saint was carried as a blessing on their journey, in much the same way as the blessing of Mary Magdalene was carried by those on the road from Vézelay.

The original statue of the Black Virgin brought back from the Holy Land by the French king was destroyed—like so many sacred objects throughout the country—during the French Revolution, and was replaced in the cathedral by a copy. The original statue seems to have been naturally black, or very dark, due to of the nature of the wood: yet it came to be associated with other effigies of the Virgin which were deliberately painted black.

It is these "blackened" effigies which represent a puzzling variation of the Marian cult, one that attracted a wide following throughout the Middle Ages. It has been calculated that at least five hundred such statues of a Black Madonna existed in medieval

Europe, nearly two hundred of them in France. The most celebrated of these was the Black Virgin of Rocamadour, to the west of Le Puy in central France. This is a statue that still exists, and is still widely venerated, and in the Middle Ages it was the focus of one of the most popular pilgrimages in Christendom, King Henry II of England and St. Bernard of Clairvaux being among those who journeyed specially to kneel before it. The statue was first mentioned in chronicles as a place of popular pilgrimage in the year 1235, just nineteen years before the French king's gift to Le Puy, though the Rocamadour statue is many centuries earlier, and the cult surrounding it was already long established. *The Pilgrim's Guide* makes no mention of it, presumably because Rocamadour was not on one of the four recommended roads. Yet the fame of the place, and its proximity to the route from Le Puy, must have beckoned many a traveler to make the necessary detour in the days ahead—and in a later chapter we shall follow them.

How the cult of the Black Madonna came about in the first place has exercised the imagination of theologians and historians for many centuries, giving birth to a variety of explanations of varying credibility. One favored view is that she was a pre-Christian goddess, probably Isis the Egyptian goddess of rebirth, with whom the early Christians, especially the desert fathers, would have been familiar. And since the Le Puy statue is supposed to have originated in Egypt, the link is even closer. Another line of pursuit has led to the Old Testament, which has been widely quoted in order to identify her with the enigmatic figure in the Song of Solomon: "I am black and beautiful, O daughter of Jerusalem." More prosaic explanations that have been offered include the view that the figure was black, or at least dark, since this would have been Mary's natural skin color in that part of the world. More humdrum still is the conclusion that these Black Madonnas had simply become darkened over the years by exposure to candle smoke in dimly lit churches.

Whatever the truth of her origin and identity, the Black Virgin of Le Puy hugely enhanced the popularity of the city as a place of pilgrimage, as well as the wealth obtained by scores of innkeepers, merchants, craftsmen and tradesmen who profited from the annual flood of pilgrims here to venerate the celebrated Madonna, as well as to equip themselves for the long trek across the hills of the southern Auvergne and the plateau of the Velay. And as travelers do today, they left the city of pinnacles and many gods through the medieval town along the cobbles of the old pilgrim road now appropriately named the Rue Saint-Jacques. Some of the most uplifting landscape in France lay ahead of them.

12. Golden Majesty

L IKE TODAY'S TRAVELERS, medieval pilgrims gazing down from the plateau of the Velay could take a parting look at the impressive city they had just left, Le Puy. Those armed with a simple faith may have found themselves carrying confusing memories of the place where they had assembled a few days before. They would have been memories of a house of prayer which seemed to speak as much of Islam as of Christianity; of a chapel on a pinnacle reaching for the heavens and dedicated to an archangel; of a massive stone altar set within the cathedral yet erected originally in honor of some pagan god; and above all the memory of a painted wooden image of the Virgin Mary who was as black as ebony.

Who knows what they made of it all as they trudged westwards across the mountains? Maybe, like many of Chaucer's pilgrims on the road to Canterbury they were driven to go on pilgrimage by a spirit of adventure rather than by any pious urge, so all those strange and unexpected sights came as welcome offerings bringing color to a monochrome life. Pilgrimage, after all, was perhaps the only permitted adventure of the day in which everyone could participate, regardless of rank or circumstance. Whatever the spiritual motive, pilgrimage was also the great escape and the great leveller.

Departing from Le Puy, pilgrims could choose one of several tracks across the high plateau. One road after barely an hour from the city led to Bains, where a priory (whose church still exists) carried the name of Sainte-Foy. Here was a foretaste of one of the great medieval abbeys of France, Conques, likewise dedicated to the martyred virgin, and which many pilgrims taking this road would have been anticipating as one of the highlights of their journey, now no more than three days ahead. The priory at Bains was one of the dependencies of Conques, and its monks had come from there, no doubt keen to fill travelers' ears with the wonders of the famous mother-house.

An alternative track led to Montbonnet, where there is still a medieval chapel originally dedicated to St. James. The two parallel roads then join into a single track, becoming what is now the celebrated Grande Randonnée 65 (GR65). From now on modern travelers are constantly reminded of whose footsteps they are following. Every few miles another boxlike chapel meets the eye, where pilgrims would have paused to offer a prayer and refresh themselves from a stone water basin. It was always rough going on this road, in terrain with few creature comforts, a fickle mountain climate and persistent dangers which included wolves, bears and—in particular—bandits. Some of these made a habit of posing as *bona fide* travelers, even as pilgrims. And the fact that pilgrims tended to carry their few valuable possessions with them made them especially vulnerable. They might be protected by law, but bandits by definition lived outside the law.

So pilgrims needed all the protection they could get; and one body of men in particular took it upon themselves to provide it as an act of Christian duty. They were the feudal knights: not the knights who rode out to fight the infidel in Spain or led the crusading armies in the Holy Land. They belonged to one of the religious orders of knighthood which were emerging at this time, and whose members chose to observe a monastic way of life dedicated themselves to protecting vulnerable travelers as well as the sacred places where they worshipped.

In no other stretch of the French pilgrim roads were their services more urgently needed than in this wild mountainous region. The aptly named Domaine du Sauvage, now a farm, lay in a particularly rugged and inhospitable area a little further to the west along the road from Le Puy. Here in the twelfth century the Knights Templar established a hospice as a resting-place for pilgrims, together with a chapel dedicated to St. James. The Templars, the "poor Knights of Christ and of the Temple of Solomon" had been established a century earlier in the wake of the First Crusade to protect holy places in Palestine. Now their role had been extended to providing safe passage to pilgrims, a sizeable proportion of whom on this road would have been making the journey to Spain and the tomb of St. James. And in this capacity the Templars were joined even more prominently by their brother-knights, the Hospitallers, the Knights of St. John of Jerusalem, whose fortified settlements, or *commanderies*, remain a prominent feature of the wilder upland regions of France today. The *commanderies* were half-monastery, half-barracks, but they were also "safe houses" for pilgrims, usually established at strategic points along the road, such as a spring in the midst of a dry region, or—as at Les Estrets to the west of the Domaine du Sauvage—by a river crossing.

Here was the last fresh water pilgrims would come across for a while, because now the pilgrim road—the modern GR65—climbs up on to the Plateau of Aubrac, an expansive stony region nowadays populated largely by imperious long-horned cattle as white as the snow which carpets the Aubrac plateau for most of the winter months.

In medieval times, however, there were less amiable residents of the plateau. This, even more than the upland closer to Le Puy, was bandit country. Ironically, this persistent menace to travelers was to a considerable extent the product of a new political stability and social order in France which had made it possible for people to travel as never before. Politically and socially, Europe was no longer the anarchic battle-zone it had been for the past two centuries and a degree of law and order now existed. The mass movement of pilgrims from shrine to shrine throughout the twelfth century symbolized this new "settled" state of Christian Europe. Once pilgrims had obtained permission from the local lord, and a blessing from the parish priest, they were free to travel as never before in their lives, secure in the knowledge not only that they were protected by law, but that their property back home was likewise legally protected. Furthermore, being *bona fide* pilgrims they were temporarily released from a whole raft of feudal obligations of service to their lord. No wonder pilgrimage was becoming an attractive prospect, particularly to those who had spent a hard working life toiling in their master's vineyards. To be able to travel free as air from one hospice to another, punctuated by the occasional prayer, may well have seemed like the holiday of a lifetime.

On the other hand, to be free as a bird was also to be vulnerable to birds of prey. And in this new climate of social change and ease of travel there were plenty of predators waiting to swoop.

In the heart of the plateau of Aubrac the diminutive Romanesque church at Nasbinals is the surviving record of one of the first communities set up to safeguard pilgrims from such predators and offer them shelter. This was as early as the eleventh century, and the priory and accompanying church were established by monks from the ancient Abbey of Saint-Victor in Marseille, which had been founded by one of the key figures in the early Church, St. John Cassian, early in the fifth century during the final decades of the Roman Empire.

The missionary tradition of the great Marseille abbey was legendary. Due to the teaching and zealous drive of its founder Saint-Victor became the fountain-head from which Christianity had spread throughout the southern region of Roman Gaul in the form of numerous far-flung priories. Now that missionary zeal was extended to the role of protector and watchdog. Along with the monks sent out by successive abbots of Cluny to found priories along the pilgrim roads, these monks of Marseille were among the first to establish a religious house and hospice specifically for the benefit of pilgrims, preceding the contribution of the Templars and Hospitallers by almost half a century.

Some fifty years after the arrival of the monks from Marseille a dramatic event took place a short distance to the west of Nasbinals. An eminent feudal lord, the Viscount of Flanders, was crossing this desolate region on his return from the pilgrimage he had made to Santiago. At the highest point on the Aubrac plateau he was attacked by robbers. It seems unlikely that so wealthy and important a man would have been traveling alone, but perhaps in the spirit of humble pilgrimage he was accompanied by no more than one or two attendants. Whatever the precise circumstances the viscount survived the assault, and as a gesture of thanksgiving set about building a priory and hospice for pilgrims at the place where the attack had taken place. It was designed "to receive, welcome and comfort the sick, the blind, the weak, the lame, the deaf, the dumb and the starving." The result of his vow was an elaborate cluster of fortified buildings—a typical knights' *commanderie* maintained by monks, nuns and knights, the latter becoming the nucleus of a new chivalric order of knighthood, the Order of Aubrac. This became closely associated with the Hospitallers, yet remaining proudly independent of their powerful brothers.

The Dômerie, as the principal establishment was called (after the title given to the head of the order), was attached to a church in which monks would ring the "Bell of the Lost" to guide travelers when snow had fallen heavily across the surrounding plateau. Both monks and knights acted as a police force by patrolling the roads approaching the settlement, while the wellbeing of the *commanderie* and the travelers it cared for was maintained by large parties of lay brothers who supplied an extra labor force working in the fields and tending the hardy Aubrac cattle. It became a flourishing mixed community typifying the spirit of Christian charity represented by the new chivalric orders of knighthood. This was benevolent feudalism at its most humane. Here was a community with many diverse requirements, demanding a high degree of flexibility among its inmates, who might be required to tackle robbers one moment and milk the cows the next.

Much of this flexibility flowed from their observance of the Rule of St. Augustine of Hippo. This code of conduct had been laid down by Augustine in a brief document written about 400 AD, more than a century before St. Benedict set down his more precise and detailed Rule for how a monk should lead his daily life. St. Augustine's more generalized vision of a religious community was what may have allowed Aubrac to accommodate the multiple demands of a mountain outpost more satisfactorily than would a Benedictine community more strictly bound by its unerring routine of liturgy and prayer. Even the great abbey of Cluny, ever among the most liberal of Benedictine houses, would have found it hard to interrupt the flow of Gregorian chant in order to deal with bandits.

Today, amid summer pastures alive with wild flowers and grazed by the ubiquitous long-horned cattle, Aubrac is a small mountain town and holiday resort which has grown up around the viscount's original settlement. Alongside sections of old cobbled street the medieval church has survived, and so has an impressive section of the former Dômerie which managed to withstand the ravages of the anti-clerical mob during the French Revolution. It seems that the banditry which led the Viscount of Flanders to found Aubrac as a safe haven early in the twelfth century did not die out with the Middle Ages.

*

The pilgrim road descends from the heights of Aubrac to cross the River Lot at Espalion. The handsome eleventh-century bridge in the heart of the medieval town creates a setting so unchanged that travelers may well wonder what century they have wandered into. The illusion of being a time-traveler is reinforced a short distance beyond the town when a church of the same period, resplendent in pink limestone, confronts the eye with an image so fundamental to early medieval Christian teaching that we are dragged back to that dark era when the compulsion to seek redemption resounded like a fearful gong in people's ears all their lives. These were the benighted years when fear ruled, and the Church maintained its power over people by offering the only slender ray of hope. The portal of this former priory church is carved with undisguised relish on the theme of the Last Judgment. And just in case those entering the church for morning mass were still unfamiliar with the precise meaning of Hell, here it was graphically spelled out, torture by torture.

Pilgrims who had taken other roads, particularly the Burgundian road, would have grown accustomed to more uplifting visions of human destiny. Among the outstanding achievements of the Burgundian school of church carving, at Vézelay above all, and certainly at Cluny before its destruction, was to give the Christian story a human touch, creating images which appealed directly to the onlookers' sense of hope, rather than despair; with the central figure of Christ in Glory replacing that of Christ in Judgment. It was a radical change in the prevailing religious spirit of the time—from darkness into light—inspired to a large extent by the dynamism of successive abbots of Cluny and subsequently that of St. Bernard of Clairvaux as they set about revitalizing a battered Christian world.

It was a change of heart and mind which altered the very nature of the pilgrimage movement: from being a journey undertaken in a spirit of penitence to becoming one of celebration and discovery.

Here in the southern Auvergne we are in a world still untouched by that Burgundian optimism. The eleventh-century church close to Espalion belonged to an earlier and darker era, and was attached to a priory founded as a dependant of the most celebrated monastery in this mountainous region of France—the historic abbey of Conques. This was the next major goal, just as it is today for those thousands who each spring and summer follow the GR65 across the highlands of the Rouergue.

*

No place on all the roads to Santiago offers quite so powerful a sense of a medieval pilgrimage as Conques. The village itself feels trapped in the Middle Ages as if a spell has been cast over it. Today's travelers arriving from the direction of Le Puy find themselves gazing down on a dense tangle of gray slate roofs clinging to the wooded hillside, a prospect that has scarcely changed since the earliest pilgrims trod this same path across the hills and caught their first glimpse of this hallowed place. In the middle of the jumble of roofs rises the massive abbey church on a scale quite out of proportion to the buildings that press round it. The houses and the embracing hill seem to close round it, holding it in their grip. *The Pilgrim's Guide* explains that "in front of the portals of the basilica there is an excellent stream more marvelous than one is able to tell." And there it still is, bubbling into its stone basin—a "magic" spring like so many others around which holy shrines were established.

The particular character of pilgrimage towns is mostly lost: rebuilding, road widening and the destruction of town gates being largely responsible. A few survive, like Parthenay (see Chapter 3) and—pre-eminently—Conques. That special character is best appreciated today by an approach from the west, as travelers would have done on the journey home after completing their pilgrimage. From the ancient bridge over the River Dourdou the old road climbs steeply towards the village, appropriately named the Rue Charlemagne after the supposed founder of the first Benedictine monastery here. The winding cobbled road squeezes between half-timbered houses until passing through the fortified town walls at the Porte du Barry. From here it curves up into the center of the village between low fan-tiled roofs on either side. Finally, around a sudden corner, it emerges into the open square in front of the vast rugged facade of the abbey church.

The impact is the greater for being out of scale with everything thus far. We step into a building that is labyrinthine and of colossal height. Conques is one of a group of five huge churches which all conform to roughly the same architectural design since they were built specifically to hold large numbers of pilgrims collecting at any one time, generally on a saint's day. An important requirement of these churches was that they possessed aisles of double the normal width so that pilgrims could process on either side of the nave towards the east end of the church where a broad ambulatory allowed them to view a saint's relics placed prominently on the high altar. As outlined in Chapter 3, these pilgrim churches were established on each of the four roads described in *The Pilgrim's Guide*, the fifth church being the pilgrims' final destination, the cathedral of Santiago de Compostela itself.

But here at Conques there is one noticeable difference from the other four churches: it is much shorter. And it is this relative horizontal concision which has the effect of making its vast height feel even more vertiginous. Arches are stacked in tiers one above the other in an intricate balancing act, until high up the palest of sunlight is filtered from invisible windows, falling in slender shafts to create patterns on the grey stone in an intriguing ballet of light. Down below we wander through a forest of columns, our eye always drawn upwards by the soaring height of those columns.

Outside, the massive west door of the church supports a carved tympanum which no medieval pilgrim would have been likely to forget. If there had been any skepticism about the nature of divine punishment then the Last Judgment of Conques would have dispelled it. The future of the human race is graphically spelled out. On one side is paradise, and on the other damnation. In the center sits the figure of Christ in Majesty, an impassive figure of authority who calmly blesses with one outstretched hand and pronounces judgment with the other. The virtuous, who include Charlemagne, look a trifle self-satisfied, while the damned helplessly suffer—some are beaten, others cleft with an axe or hauled upwards by their legs. It is all somewhat matter-of-fact: an illustrated manual of official Church teaching, too impersonal to be horrifying, too stereotypical to be moving.

One significant figure is among the company of the blessed. It is a woman praying, and about to be touched by the hand of God. She is St. Foy, to whom this abbey and its great church are dedicated. And thereby hangs an extraordinary tale. *The Pilgrim's Guide*, in its chapter on important saints to be revered, gives a brief and rather anodyne account of her life and death, omitting a number of essential facts. Those who "proceed to Santiago by the route of Le Puy," it explains, "must visit the most saintly remains of the Blessed Foy, virgin and martyr, whose most saintly soul, after the body had been beheaded by her executioners on the mountain of the city of Agen, was taken to heaven in the form of a dove ... At last [she] was honorably buried by the Christians in a valley commonly called Conques. Above the tomb a magnificent basilica was erected by the faithful."

The author carefully leaves out how St. Foy's body came to be in Conques at all. It is conceivable that he may not have known, although two written accounts dating from a century earlier than the *Guide* tell the story in full. Foy (meaning Faith, in Spanish Santa Fe) was a young Christian girl living near the town of Agen during the last century of the Roman Empire, and who openly refused to worship the official Roman gods. Like many a Christian martyr of that period she was accordingly tortured and finally beheaded. With the Roman conversion to Christianity shortly afterwards a cult grew up round her, and her relics held in the abbey of Agen became widely venerated and the source of numerous miracles—all of which helped to establish the fame and prosperity of the abbey, much to the envy of neighboring abbeys, notably Conques, at that time a religious house of small importance buried away in a mountain village. Eventually, in the mid-ninth century the monks of Conques could stand it no longer, and hatched a plot. They selected a plausible member of their community to offer his services to Agen posing as a secular priest. The man's evident piety so impressed the abbot that he was appointed guardian of the abbey's treasury, which of course included the tomb of St. Foy The story goes that he then waited ten years before seizing a unique chance of being alone in the church at night while the monks were all feasting in celebration of Epiphany. He duly smashed an opening in the marble tomb and made off with the relics of the saint—to be received with the utmost joy and jubilation at Conques.

In consequence Conques thrived, while Agen declined. It was another example of *furta sacra*—Sacred Theft, justified on the grounds that it must have been God's will. It was not as spectacular a robbery as that of the supposed relics of Mary Magdalene in Provence by the abbey of Vézelay at much the same time; nonetheless it was equally remunerative. Relics of the saints were by far the most rewarding investment a religious house could make, and the body of this martyred girl from Agen somehow possessed an aura which touched the hearts and the purse strings of Christians throughout Western Europe. In the words of Jonathan Sumption, "A pilgrim was expected to be as generous as his or her means would allow, and there were some who asserted that without an offering a pilgrimage was of no value." Stories of miracles at Conques abounded, and these were repeated enthusiastically as an inducement to further contributions. They included tales of women failing to give their jewelry to the abbey and falling ill until the gift had been made.

One special attribute of St. Foy was her power to release soldiers taken prisoner by the Saracens in Spain. The carved tympanum over the west door of the abbey church includes the image of iron fetters hanging from a beam right behind the figure of the saint. In a gesture of thanksgiving former prisoners presented their chains to the abbey in such quantities that in the twelfth century they were made into an iron grille which still separates the ambulatory from the choir. This is an exact repetition of the miracles attributed to Mary Magdalene at Vézelay, where prisoners' chains were likewise made into an iron grille for the abbey church (see Chapter 7). Here at Conques its original function was to protect the abbey's treasures which on special days would be displayed round the high altar for pilgrims to marvel at, and were visible to everyone in the nave.

The treasures of Conques were regarded in the Middle Ages as among the wonders of the Christian world. They were also glittering proof of the enormous wealth which St. Foy had brought to the abbey over the past two centuries. Equally extraordinary is that the treasures have survived to this day, having been carefully hidden by the inhabitants of the village from the fury unleashed in the French Revolution which accounted for the destruction of so many sacred objects in other churches and monasteries throughout France.

Today the Conques treasure is housed in a special museum in an area of the former abbey cloisters. It represents centuries of gifts by feudal lords in gratitude for whatever miracles St. Foy was believed to have performed. Among the earliest and finest pieces of craftsmanship is the ninth-century reliquary of Pepin, the first King of the Franks and father of the Emperor Charlemagne (the supposed founder of Conques abbey). The reliquary is a jewel-encrusted box bearing a crucifix in gold relief surrounded by intricate floral motifs in gold filigree. Another dazzling object is the so-called A of Charlemagne. This free-standing letter in gold and silver bears a significance which comes down to us in legend: Charlemagne is said to have ascribed each of his religious foundation with a letter, awarding A to Conques to indicate that it was the most important of them all.

But the object which pilgrims came principally to see—as travelers in their thousands do today—is the golden reliquary statue of the saint herself, known as the Majesté de Sainte-Foy. Here is her sculpted effigy. She wears a golden crown. She is robed entirely in beaten gold which is studded with precious stones. She is seated on a golden throne with her arms outstretched as if in blessing. And she gazes with a hypnotic stare into some other world. Jewels are all over her, emeralds, sapphires and opals among them, interspersed with polished chunks of rock crystals. Everything about her shines and glitters. She is both the source of the abbey's wealth and the boldest possible demonstration of it.[26]

So, what are we to make of her? And what did medieval pilgrims make of her? By any account the Majesté is one of the most striking and memorable of all religious artefacts of the Middle Ages. Here is an image that would turn heads anywhere, as it has done for the past thousand years. At the very least she is an astonishing example of opulent medieval craftsmanship. But she is also so much more. To the early pilgrim setting eyes on her on the high altar of a vast candlelit church the impact must have been stunning— bewildering and perhaps disturbing. Who, and what, was she? She was a queen on her throne. She was one of God's saints, adorned in the most precious materials this world could provide. She was a golden apparition gifted from heaven. She had the power to make miracles—curing the blind, healing the sick, releasing prisoners from their chains. Perhaps most important of all she may have embodied the sense of divine magic which was at the core of religious belief in the Middle Ages, a magical power that was capable of lifting people above the pain and dross of their daily lives.

And in this capacity the Majesté de Sainte-Foy symbolizes what pilgrimages were all about: the search for a better life. Certainly there would be nothing quite as spellbinding as this golden image until the end of the road and the ultimate goal of Santiago de Compostela itself.

26. See Plate 17.

13. Tall Tales and Wild Beasts

Leaving Conques is like breaking a magic spell—though only slowly: the illusion of having wandered into the past lingers on. The old pilgrim road out of the village, the Rue Charlemagne, still belongs to the Middle Ages. So too does the fortified gate leading out into the countryside and down the wooded hillside into the valley. And even here the spell still holds as we cross the medieval bridge over the River Dourdou and prepare for the long climb up the mountain beyond, with a last glance back at Conques' abbey church rising from its cluster of gray roofs on the far hill across the valley.

There is one further link with events of almost a thousand years ago. The gold and bejeweled treasure of Sainte-Foy, which astonishes the modern traveler to Conques, is the very same assembly of glittering artifacts as medieval pilgrims marvelled at as they processed round the high altar of the abbey church on a saint's day mass all those centuries ago. The Conques treasure is unique in having survived wars and revolutions intact, while its centerpiece, the hypnotic Majesté de Sainte-Foy, makes much the same unforgettable impact today as it did when it was the most famous sacred reliquary in medieval Europe.

After a descent into the further valley the next halt for pilgrims was the riverside town of Figeac. Here there were once six hospices to accommodate them, a measure of the remarkable number of travelers regularly using these roads in the eleventh and twelfth centuries—many of them pilgrims, but also merchants, craftsmen and traders of one kind or another, and all contributing to what was an era of economic recovery in rural France after the long centuries of anarchy and insurrection. Provided you could escape the robbers, the toll-gatherers and the murderous ferrymen whom Aimery Picaud mentions in *The Pilgrim's Guide* these were good times to be out on the open road.

Besides the six hospices Figeac possessed an abbey dating back to the eighth century, and which in the eleventh century came under the capacious umbrella of Cluny, along with so many others across Europe. Like the six long-vanished hospices, nothing survives of the abbey, only a much battered and much restored abbey church, Saint-Sauveur.

From Figeac the pilgrim road followed the valley southwest until it joined the River Lot. Here the fortified town of Cajarc possessed a hospice much frequented since it was close to an important bridge over the river. There were few such crossings: the Lot flows through deep gorges, and the terrain to the north and south consisted of high plateaux even more inhospitable than the landscape between Le Puy and Conques. Pilgrims were wise to hug the river valley.

But there was one diversion which many would have taken, and it took them over the wildest of these bare uplands north of the River Lot. This was a limestone plateau at an altitude of more than three thousand feet, and known today as the Causse de Gramat. It was certainly a great deal more physically demanding than the diversion to Cluny which many travelers on the Vézelay road would have made, and only the most devout pilgrims would have undertaken it. Nonetheless theirs was a devotion that was widely shared, because their objective was one of the most popular shrines in the whole of Europe, attracting almost as many pilgrims annually as it does tourists today. The shrine was Rocamadour, set dramatically within a gorge of the Causse de Gramat.

A number of disparate threads have been woven together to create the mystique of Rocamadour. The principal focus of pilgrimage has remained the chapel of Our Lady of Rocamadour, the local variation of the cult of the Virgin Mary which swept through Christian Europe in the early Middle Ages. But there were other strands to the story which have contributed greatly to the emotional appeal of the shrine as a place of veneration. Its dramatic location has always helped, high on the wall of a river gorge, and which made the approach to it a physical challenge. Pilgrims often chose to respond to this challenge by stripping off their clothing and in a display of self-abasement climbing the 216 steps to the chapel on their bare knees. Pain and piety were good companions, as was frequently the case in medieval Christianity.

But the mystique of Rocamadour had a deeper resonance than this, due to an alluring legend which vastly enhanced the special appeal of the shrine. In the mid-twelfth century, when the local cult of the Virgin was at its fervent height, building works close to the chapel of Notre-Dame revealed an ancient tomb containing a body reported to have been miraculously preserved from decay. The events that followed read like an exercise in wishful fantasy. With the aid of numerous miracles the identity of the body was satisfactorily established as that of a biblical figure by the name of Zacchaeus, recorded in St. Luke's Gospel as being a tax collector from Jericho, a disciple of Jesus and believed to have been the husband of Veronica who wiped Jesus' blooded face with her kerchief on the road to Calvary.

How the chance discovery of a tomb on a French mountainside should have led to such an improbable identification can be explained by following the thread of a story put about at this time. Whether by coincidence, or possibly by a shared origin long lost, the story bears an intriguing resemblance to that of the supposed arrival of Mary Magdalene on these shores (see Chapter 7). And it is this. Following Christ's crucifixion Zacchaeus and his wife Veronica are said to have fled persecution in Jerusalem and left Palestine in a flimsy vessel which was miraculously guided to the shores of Roman Gaul, where the two vagrants sought the seclusion of the mountains. Then, after the death of Veronica some years later, Zacchaeus became a hermit known locally as Amadour, living high above a steep gorge where eventually he died and was buried beneath a rock. A religious community grew up around the hermit's tomb, incorporating his name into that of their priory: Rock of Amadour—which became Rocamadour.

So far the thread of the tale manages to make the gigantic narrative leap between the shrine in rural France and Palestine at the time of Christ. A further chapter in the story supplies a link of equal importance in establishing the aura of the place. At some date early in the Middle Ages, probably in the ninth or tenth century, the Benedictine community which now occupied a monastery on the site had come into possession of a wooden carving of the Virgin Mary seated with the infant Jesus on her knee. The statue may well have been the origin of the Virgin cult here in Rocamadour, though the cult only began to acquire fame with the discovery of the supposed tomb of Zacchaeus in the mid-twelfth century. From this moment onwards miracles galore were reported, and within a decade of the discovery *The Miracles of Our Lady of Rocamadour* was composed, a list of miracles that vastly increased the popularity of the shrine. Miracles were universally held to be revelations of God's will, and the medieval mind was certainly not given to questioning their reliability as evidence. Out of this deluge of miraculous events emerged the conviction that the carving of the Virgin and Child had actually been brought personally from the Holy Land by Zacchaeus himself.

Shrines could hardly get more sacred than this. Not only was Rocamadour seen to have been specially favored by the mother of Christ, it now possessed a link directly to the place and time when Mary was still alive. It was a link achieved through the offices of this saint—Zaccheus who became Amadour—and who had actually known Jesus and whose wife had wiped his brow. As a result, the fame of Rocamadour became second to no other shrine in France. Soon monarchs and leading churchmen right cross Christendom were bending a knee before the altar of the "Miraculous Chapel" above which the sacred statue was displayed.

But the story of Rocamadour grows even more complex. At some stage—it is unclear when—the legendary statue was described as the "Black Madonna," and so became incorporated into that bizarre subculture centered on the veneration of "black" images of the Virgin (described in Chapter 12). If the Rocamadour statue was ever black this is likely to have been largely due to centuries of exposure to altar candles: even so, its association with the Black Virgin which the French king, St. Louis, had presented to the cathedral of Le Puy, would have been a compelling reason for Santiago pilgrims to have chosen to make the detour across the *causse* to pay homage to the saint.

Other pilgrims, perhaps in even larger numbers, would have made a detour from the opposite direction, traveling south across the mountains from Burgundy and the Vézelay road. And here, on the northern outskirts of Rocamadour, stands a settlement by the name of L'Hospitalet, so-called after the large hospice (the ruins of which survive) where pilgrims once sought accommodation before undertaking the climb—whether on their knees or not—up the south slope of the gorge to the famous Chapel of Miracles.

*

The story of Rocamadour is a tangled web of fantasy, fable and the most naïve credulity, all brought together by an overriding yearning to establish a direct link between the beleaguered world of medieval Christendom and that of Christ and his disciples—an urgent longing to drink from the spring where Christianity was born. Rocamadour itself was never on one of the primary pilgrim routes to Santiago, and is not even mentioned in *The Pilgrim's Guide*. For most Santiago pilgrims the place may have been hardly more than a sideshow: after all, they were pursuing their own link with the time of Christ far away in northwest Spain. Yet, from the broader perspective of our own times, Rocamadour succeeded in encapsulating much of what made the pilgrimage movement as a whole so powerful a social phenomenon in the Middle Ages, and so vibrant a creative force.

From the viewpoint of today's travelers it is this creative force which may engage the eye and the mind most strongly as we follow the pilgrim roads across France and into Spain. It is the experience of witnessing the astonishing richness of that legacy: there is always something new to surprise us—churches and abbeys, feats of engineering, masterpieces of sculpture and painting, entire towns and villages shaped by the needs of pilgrims over the centuries, all of them created in the service of this unending tidal wave of humanity on the move. It is this that makes a study of these pilgrim routes so rewarding to the historian and art lover alike.

One such feat of engineering soon presented itself to pilgrims, both those who had made the detour to Rocamadour and those who had followed the principal route along the river valley and gorges from Figeac. It is a landmark which still exists: the remarkable Pont Valentré, a medieval bridge at Cahors with three lofty watchtowers which spans the Lot at the head of the long bend of the river that all but encloses the town in a narrow peninsula.

Cahors also offers a taste of what is to come further along the pilgrim road. The twelfth-century Cathedral of Saint-Etienne possesses a magnificent carved tympanum on the theme of the Ascension. This huge sculptural ensemble sits handsomely over the north door as though it has always been there. In fact it was originally set above the main entrance to the cathedral on the west front, and was then moved, stone by stone, when the west facade was rebuilt little more than a century later. In the fourteenth century Cahors was besieged by the English during the Hundred Years' War—but unsuccessfully: being protected by the River Lot on three sides kept the English army out and probably saved the cathedral from the customary pillage.

Further along the road to the southwest, one of the greatest abbeys in France was less fortunate: Moissac. And in view of its blighted history it seems little short of miraculous that the finest area of the abbey Church of Saint-Pierre should have survived virtually untouched: this is the incomparable south portal, which is a medieval masterpiece and one of the glories of the pilgrim roads.

Disasters haunted Moissac from the beginning. Founded as one of the earliest Benedictine abbeys in the seventh century it had scarcely begun to function when it was sacked successively by Saracen invaders from Spain, Vikings from the north and Magyars from the east. In 1030 the church roof fell in, and in 1042 a fire in the town spread to parts of the abbey. Such treasures as the abbey possessed continually went missing; a later chronicler described the place at this period as a "robber's cave."

Five years later a rescuer appeared. The Abbot of Cluny, Odilo (St. Odilo), on one of his frequent missions to bring lapsed and failing monasteries into the Cluniac fold, was traveling from Burgundy and passing through this region on his way south. The authority of an Abbot of Cluny being close to that of a monarch, no opposition was raised to Odilo's proposal that Moissac become a dependency of Cluny, adopting the reforms in monastic life which the Burgundian abbey had instituted and was putting in place throughout its expanding "empire."

There were several remarkable features of the Cluniac takeover of Moissac. Odilo had by this time been Abbot of Cluny for fifty-three years, and must have been at least in his late seventies by this time: to be riding across France at such an advanced age, even with a retinue of attendants, was an astonishing feat even for a man of Odilo's redoubtable energy and sense of purpose. He was by now as powerful a figure in the Christian Church as the pope and the Holy Roman Emperor. He was also a dedicated champion of the cause of Reconquest in Spain, and accordingly of promoting and safeguarding the pilgrimage movement to Santiago. As we saw in Chapter 9, in the early days of his abbacy Santiago itself had been overrun and sacked by the celebrated Saracen warlord Al-Mansur, and for Odilo the wound had never healed. The fact that Moissac lay on one of the principal pilgrim routes to Spain would undoubtedly have contributed to his determination to annex the abbey and make it an important landmark and stopping-place for pilgrims on their way to the Pyrenees and Spain.

Odilo immediately appointed one of his most trusted monks at Cluny, Durand de Bredon, to become Abbot of Moissac. And so began a century of extraordinary fame and prosperity for the abbey, during which Moissac received lavish donations of land and property, acquiring in the process more than seventy daughter-houses of its own, and enjoying such prestige that in the Cluniac hierarchy the Abbot of Moissac ranked second in importance only to the Abbot of Cluny himself.

Odilo lived only two more years after the acquisition of Moissac, and it was left to his young successor, Hugh, to oversee its expansion during the second half of the eleventh century and into the twelfth. In fact Hugh managed to exceed his predecessor's period of office by five years, remaining Abbot of Cluny for sixty years until 1109. They were sixty of the most creative and influential years for Cluny, for Moissac and for the western Church in general.

In 1063, just sixteen years after the Cluniac takeover of Moissac, Abbot Durand had secured the finances and administration of the abbey sufficiently to begin the building of a new church and cloister. Over the next three-quarters of a century Durand and his successors were responsible for creating some of the finest works of sculpture and architecture of the entire Middle Ages.

But before looking more closely at the treasure which is Moissac it is worth recounting the bizarre sequence of events which followed this intense period of creativity. Combined with the centuries of insurrection which had preceded the Cluniac takeover they made these few intervening years an island of peace and prosperity in the midst of hurricanes and tidal waves. Troubles broke out little more than a hundred years after Durand became abbot. In the late twelfth century Moissac became a battlefield on which the disputes over boundaries between France and English-owned Aquitaine were fought. In the following century the town was seized and the abbey damaged by the heretical Cathars opposed to orthodox Catholicism. In the fourteenth and fifteenth centuries it was once again a battleground in the protracted Hundred Years' War between England and France. A century later still it was caught up in the French wars of religion, and again severely damaged. Then, in the early seventeenth century the abbey was officially secularized, and much of the abbey building and church allowed to fall into decay. What remained managed to escape the worst of the mob's fury during the French Revolution; but immediately afterwards the new National Assembly put both the Church of Saint-Pierre and cloister up for sale. The final insult came in the nineteenth century when the cloister was on the point of being demolished to make room for a new rail link between Bordeaux and the Mediterranean. Only last-minute intervention by the Beaux-Arts authorities in Paris forced the cancellation of the project, and the railway was diverted a short distance away.

Moissac—or at least the best of it—had just survived.

The cloister which was so nearly a railway track had been begun at the behest of Durand de Bredon in the last decades of the eleventh century. It was mostly built after his death in 1072, but he still presides over the place in the form of a full-length portrait carved in bas-relief on one of the marble columns standing amid the rows of arches on all four sides and at the four corners. He is in exalted company, the other full-size figures being Christ's apostles—a measure of the esteem in which Durand was held. Together they make up a remarkable group of carvings, executed with a simple gracefulness and delicacy, and a human touch, unlike the relatively crude sculpture we have seen on the Le Puy road so far. But then Abbot Durand came from Cluny, and would have been aware of the sculptures being produced in Burgundian workshops, far ahead of their time, and which were then beginning to enrich the churches of that region as well as the new abbey church of Cluny itself.

Whether Abbot Durand actually brought Burgundian stone-carvers to work here in Moissac is a matter of speculation and dispute. Cluny's empire by this time was expansive and far-flung, and the Cluniac love of rich decoration and fine workmanship was resulting in a high demand for men with the appropriate skills, particularly stonemasons and sculptors. Hence craftsmen and their workshops tended to travel to wherever those skills were most needed, creating temporary villages—families and livestock included—often set up along the pilgrim roads where so many new priories and abbeys, including Moissac, were being established.

Moissac's link with Burgundian workshops were matched by connections that were equally strong in quite the opposite direction, with northern Spain. This was once again through the agency of Cluny, which was in the process of establishing a chain of dependent abbeys along the pilgrim road to Santiago in lands newly secured from attack by Saracen armies. These links with Spain are particularly apparent in the Moissac cloister, where the capitals on all four sides are carved with figures combined with intricate geometrical patterns, the latter clearly the work of craftsmen operating in the Islamic tradition in which the representation of living forms was forbidden. What seems likely is that as the Reconquest of northern Spain progressed many Saracen craftsmen chose to pay lip-service to Christianity and found ready employment for their stone-carving skills in the decoration of new churches and abbeys, both in northern Spain and north of the Pyrenees, often working side by side with stonemasons from French workshops. Altogether the Moissac cloister is a supreme example of this cross-fertilization of Christian and Islamic cultures.

Once the cloister was completed, about the year 1100, the monks turned their energies to creating a new church, including an elaborately sculpted main entrance, soon to be as massive and awe-inspiring as any church portal yet seen in France. Today, as we stand gazing at this extraordinary edifice we can see how the task unlocked the floodgates of the imagination as the sculptors drew on ideas and imagery that reflect the breadth and variety of the Cluniac empire itself. At the very center—appropriately—is the debt to Burgundy and probably to the portal of the new abbey church of Cluny itself (which alas no longer exists). This is the monumental tympanum on the theme of the Vision of the Apocalypse, made up of twenty-eight blocks of stone, and with the grave figure of Christ in Majesty in the center. Attending him on either side are the symbols of the four apostles, Matthew, Mark, Luke and John; while below him and on either side the twenty-four old men of the Apocalypse, familiar to many a pilgrim from so many portals throughout France, except that here the drama of this vision is intensified by the way they have involuntarily lowered their musical instruments to gaze up in wonder at the revelation of Christ. The entire semicircle of figures within the tympanum gives out an air of stunned silence.[27]

To the modern traveler the Moissac tympanum may be enjoyed as a monumental example of early Romanesque sculpture on a traditional theme more usually associated with the workshops of

---

27. See Plate 18.

Burgundy, where stonemasons were at that very time creating the majestic portals of Cluny, Vézelay, Autun, Souillac and so many others in the region. The medieval pilgrim, on the other hand, is likely to have viewed the tympanum with its dominant figure of Christ in a very different light, as a sign of reassurance from God, delivered by the hands of His representatives here on earth, that all might after all be well for troubled mankind at the end of the long road.

But if the great tympanum speaks of Burgundy, then the lower area of the portal echoes a place of origin altogether wilder and more violent. The twin doors of the entrance are flanked, left and right, by massive columns each carved with jagged shapes that give the impression of an open jaw set sideways, with teeth bared, waiting to close. The effect is dramatic and distinctly menacing, and one wonders what early pilgrims must have felt entering the abbey church through that portal. Then, in the center of the double entrance is another image which must have surprised the medieval pilgrim: this is a huge marble pillar more than twelve feet in height, carved with three pairs of rampant lions, each pair crisscrossed one above the other like some acrobatic circus act.

Where did all this exotic imagery come from? Opinions abound. Scholars have been keen to point out the similarity between the pillar of lions and the sculpture on early Assyrian temples in the Middle East, long pre-Christian and even longer pre-Islam; though it is hard to understand how a twelfth-century stonemason in France could possibly have known about them. The probable answer is a good deal closer to home. The overriding theme of the Moissac portal is taken from the Book of Revelations, the apocalyptic vision of St. John the Divine. The visionary account of events in the book made it a favorite source of material for medieval artists, and one work in particular tapped the further reaches of their imagination. This was a commentary on the Book of Revelations written by an eighth-century Spanish monk, St. Beatus of Liébana, at exactly the time of the Saracen invasions of Spain (as described in Chapter 1). The immediacy of the Muslim threat gave Beatus' interpretation of the Apocalypse a vividness and an urgency which inspired artists to illustrate it with a wealth of exotic imagery in illuminated manuscripts produced in monasteries mostly in Spain during the early years of the Reconquest as if in a spirit of exorcism.

These richly imaginative illustrations, writhing with real and apocalyptic beasts of all descriptions, were widely circulated between monastery and monastery, where they became invaluable source material for sculptors and stonemasons throughout Europe whose job was to decorate the new churches with images designed to amaze and often terrify an innocent congregation; because images *were* the word—at least for most people in a largely illiterate world. Images possessed a hold on the mind more powerful and more lasting than any spoken words from the pulpit.

It was these exotic illustrations to the celebrated *Commentaries on the Apocalypse* which are the most likely source of the imagery on the Moissac south portal. Hence, in two quite different moods, the tympanum above and the carvings around the doors below both relate directly to the Book of Revelations, so becoming the theme of the entire portal—a vision of justice and order up above, and of wildness and disorder below. The two contrasting visions encapsulate the struggle which dominated the minds of medieval churchmen, and was a spur to the crusading movement as well as to the Santiago pilgrimage. It was no accident that the twin themes should have been chosen to decorate the entrance of a major church on one of the most popular pilgrim roads.

Then comes a surprise, and another revelation. On either side of the central column of lions are two full-size figures carved in low relief which become visible only as we pass under the lintel supporting the tympanum to enter the abbey church. On the left of the column is the figure of St. Paul, while on the right is the prophet Jeremiah. Paul looks appropriately ascetic and stern, Jeremiah gentle and reflective. Both carvings are thought to be a little later in date than the figures on the tympanum above, and certainly they have a delicacy and sophistication which the former lack. Both figures are elongated, flattened into the stone as if with the lightest of touches. Jeremiah's body is twisted, his head turned slightly so that he looks away from us, lost in thought. He is the prophet who foresees only disaster, and can do nothing about it.

On these two figures at Moissac I wrote in my own book on the history of Cluny: "This is sculpture that searches deep, laying bare something of the human soul. Only the greatest artists in any era have made inarticulate stone expresses the inner man, and the sculptor of these two Moissac figure is one of them. Few examples of European art express so eloquently the passionate and mystical nature of the mediaeval religious experience."

No wonder a later abbot, gazing at the portal of his church, declared that the carvings before him were so beautiful that they must have been created "miraculously rather than by a man."

14. Crosses on a Mountain Top

A T MOISSAC PILGRIMS CROSSED THE BROAD RIVER GARONNE as it
flows westwards towards Bordeaux and the Atlantic: then they
entered the region of Gascony for which Aimery Picaud had already
reserved some of his most bilious comments about the domestic
habits of the locals. A short distance further, and the old pilgrim
road crosses a tributary of the Garonne by means of a handsome
five-arched bridge, the Pont d'Artigue, which still exists. This was
essentially a Roman bridge, and the pilgrim road is now the original
Roman road towards the Pyrenees and Spain. It is also specially
marked out as the route taken by pilgrims bound for Santiago since
it was now referred to in contemporary documents not simply as
the *Via Podense,* the road from Le Puy, but as the *Via Publica Sancti
Jacobi,* the public road of St. James.

There may have been a special reason why at this particular
point the road should have been identified in this way. Close to the
bridge in medieval times was a hospice for the benefit of travelers,
like so many on these roads, except that this one had been established
and administered by a body of feudal barons in Spain known as
the Knights of St. James, or Knights of Santiago. Nor was this the
only establishment set up by them along the *Via Sancti Jacobi*: there
would be others along the route between here and the Pyrenees.

The Spanish knights came together as a fighting force in support
of the Reconquest; but their presence here in quite a different role
is further evidence of the close connection between the military
operations in Spain and the pilgrimage movement. Like the more
celebrated religious orders of knighthood in the Middle Ages, the
Templars and the Hospitallers, the Knights of Santiago were a

product of the crusading spirit of the time. Half-monk, half-soldier, knights swore an oath of dedication to safeguard Christian shrines and to protect those who came to worship at them: hence the dual role.

It was in 1164 that a group of local barons in southwest Spain, a region still largely occupied by the Saracens, agreed to collaborate in assembling an armed force capable of recapturing towns and villages which were still in Muslim hands. In northern and central Spain the Reconquest was already at its height, the Spanish capital of Toledo having been recaptured more than seventy years earlier. Now the Christian armies were pushing further south. The barons formed themselves into an order of knighthood operating within monastic disciplines, which received papal recognition eleven years later, and they began their military operations in the two adjoining kingdoms of León and Castile.

The fervent religious spirit of the new order had a strongly symbolic flavor: it was said that the original group of knights numbered thirteen, this being a biblical echo of Christ and his twelve apostles. Membership of the order expanded greatly, as did the activities they undertook. Their pastoral role soon ran parallel to their military activity, and the order began to establish hospices, often attached to their own *commanderies*, and run by canons and canonesses whose task was to see to the needs of pilgrims and other travelers.

As the Reconquest pushed the Moors further and further south the order began to acquire ever-increasing quantities of territory and wealth, largely through endowments by landowners overflowing with gratitude towards the knights for having helped to drive out the Saracens and return them their property. After little more than a century since its foundation the order had established no fewer than eighty-three *commanderies* in Spain, and were in possession of nearly two hundred villages and small townships. And by the fifteenth century this number had more than doubled. In the meanwhile their activities and influence had spread far beyond Spain, particularly into France just across the Pyrenees where, as we have seen, they built hospices on the pilgrim roads. And so, like the Templars and the Hospitallers, and like the Cluniac and Augustinian monasteries, the Knights of St. James played their part in aiding the flow of that powerful current of humanity which was the Santiago pilgrimage.

There was one significant difference between the Knights of Santiago and the other religious institutions championing the pilgrimage. Their full title was "The Order of St. James of the Sword." The word "sword" was pertinent. Their patron, the apostle James the Greater, was perceived not only as a saint revered as the cousin and disciple of Christ, and who had supposedly evangelized Spain: James was also held to have been the country's military savior, personally responsible for leading Christian armies on the battlefield to expel the Saracens from Spain. Hence the sword which the knights took as the emblem of their order. It was designed as a cross, symbolizing the order's dual religious and military role; and in case there should be any misunderstanding about the military aspect, the sword was colored blood-red. The knights wore this insignia of a sword/cross as a badge on their white capes along with that of the pilgrims' scallop-shell.

The Knights of Santiago saw themselves as the heirs and avengers of their patron saint. Their chosen emblem of the bloodied sword held an extra significance: St. James had himself had been beheaded by a Roman sword, but had now returned to the land where he had evangelized to lead the Reconquest. He had become *Santiago Matamoros*, St. James the Moor-Slayer; and in doing so he became a hero in the eyes of every Spanish Christian, represented on churches and public buildings along the Spanish pilgrim road as a warrior on horseback, sword raised, the Infidel crushed beneath the stallion's hooves.

The dual role of St. James as spiritual and military savior of Spain was epitomized by the activities of the knights of Santiago, who were soldier-monks. They wore the mantle of their patron saint in both capacities, as evangelist and warrior. In addition, by establishing hospices along the pilgrim roads such as those here in Gascony they performed a third role, protecting pilgrims on their travels to the apostle's tomb.

The cult of St. James came to acquire an irresistible mystique, which the knights helped to perpetuate. It was a mystique to which an extra touch of glamor was added when the legend of the saint became associated with the military deeds of the Emperor Charlemagne. Historically, there could have been no possible connection between Charlemagne and St. James since the former's campaigns in Spain took place in the century before the discovery of the apostle's tomb. Folklore then came to the rescue. The Charlemagne story, centering on the heroic death of his nephew Roland in the Pyrenees near Roncesvalles, became a popular theme from the twelfth century onwards through the wide dissemination of the first French epic poem, *La Chanson de Roland*. Then, in the same century St. James' own role in liberating Spain from the Infidel becomes intimately linked to the Charlemagne legend through an equally popular story which was included in the very manuscript containing *The Pilgrim's Guide*, the *Codex Calixtinus*. The *Guide* itself has little to say about Charlemagne, being more concerned with martyred saints to be venerated along the way. But elsewhere in the *Codex* (as described in Chapter Five) is the supposed chronicle of Archbishop Turpin, who had accompanied Charlemagne on his campaigns. And it is this chronicle, entirely fabricated though it is, which makes a direct connection between the Emperor Charlemagne and St. James.

From now on pilgrims setting out for Santiago could follow the beckoning arm of the Milky Way confident that they were traveling under the divine protection of two warrior-heroes. And on this road, the *Via Publica Sancti Jacobi*, the medieval pilgrim would soon be in sight of the Pyrenees and anticipating the famous mountain pass where the dying Roland blew his ivory horn three times, and too late, to alert Charlemagne a few miles away to the north of an ambush. And on that same mountain pass, so the chronicle of Turpin maintains, St. James appeared in a dream to the emperor, urging him to lead his army westwards towards the apostle's tomb under the guiding path of the Milky Way.

That place of legend, now only a few days away, was Roncesvalles.

*

As the landscape opens out and horizons expand journeys on foot become more easily flagged by landmarks. Places are more visible from afar. Today, as yesterday, it is often the village church which rises in the distance beyond the flat fields and the winding river. The stone tower and the tall spire are the welcome beacons that pilgrims and other travelers on the road have been following for centuries. They measure the progress of a journey at a pace that has never changed. Nothing unites the modern backpacker and the medieval pilgrim more closely than the experience of putting one foot after another, hour after hour, day after day, in all weathers. Walking is a great leveler.

In this part of the world they are united too by the places to be visited along the way—the churches and shrines that were newly built at the time of Aimery Picaud and *The Pilgrim's Guide*, and are now historic monuments carefully preserved and accompanied by postcard stalls and information leaflets, yet still giving out powerful echoes of their former role. Their names tell stories, like the twelfth-century collegiate church with a fine octagonal tower a short distance from the Pont d'Artigue, called Le Romieu. The word means literally "of the Romans," and referred originally to pilgrims who had been to Rome, or were on their way there. Then, with the subsequent discovery of St. James' tomb in Spain and the popularity of the Santiago pilgrimage, "Romieu" became a generic term for all pilgrims no matter where their goal.

The old Roman road continues across Gascony, its progress punctuated every so often by landmark churches often visible half a day's journey away, just as they were in the Middle Ages: like the eleventh/twelfth-century church at Nogaro with its carved tympanum of Christ in Majesty that is worthy of Moissac; or Sensacq with its eleventh-century church of touching simplicity originally dedicated to St. James; then Pimbo with its medieval church on the site of an abbey founded by Charlemagne; La Sauvelade and another monastery church dedicated to St. James. And so on—village after village, church after church, with the far wall of the Pyrenees growing ever closer and higher.

By now we have left Gascony and the architecture of the region has changed. Whitewashed village houses are held together by solid oak timbers picked out in rich red, while local churches are simple and painted white with deep roofs and squat conical spires, like the appealing little country church at Olhaïby. And with this change in architecture comes a change in language. This is the Basque Country which we have met on the previous two pilgrim roads, and whose people speak what may be the oldest language in Europe, one that sounds and reads like no other. In *The Pilgrim's Guide* Aimery Picaud dismisses the language as "barbaric," which, since it is long pre-Christian, is literally true.

A short distance further on and the pilgrim road descends through woods to a fast-flowing river and a long curve of stepping-stones that feel as ancient and well-worn as the Basque language.[28] And in this hidden place the sound of water rushing between rocks evokes the reality of everyday travel over countless centuries. These are stepping-stone we have visited in an earlier chapter, because it is at this point on the journey that the pilgrim road emerges from the woods and climbs towards the crest of a hill where it is seen to join up with two other roads, the one from Paris and the north, and the other from Vézelay and the northeast. Today they are no more than rough paths; yet this would have been a dramatic meeting point in the Middle Ages. Here on the southern slope of Mont Saint-Sauveur our three of the "four roads leading to Santiago" described in *The Pilgrim's Guide* finally join up to form "a single road."

And there, as we gaze down across the valley, is still that single road that Picaud wrote about nine hundred years ago, and which he would have taken on his own pilgrimage to Santiago just as millions before and after him have done. It is hardly more than a broad track cutting a swathe through the scrub and shale of the far hillside, until it disappears out of sight in the direction of the Pyrenees and Spain.

The tone of *The Pilgrim's Guide* grows more dramatic as we are led towards the high Pyrenees. We are entering a region of magical

28. See Plate 19.

appeal for pilgrims, because this is where the legends of Charlemagne, of Roland and Oliver, and of the Christian Reconquest of Spain, all joined together to attach themselves to the cult of St. James. It was propaganda so powerful that much of the crusading energy driving the pilgrimage movement derives from it. Picaud's own account of this stretch of the road becomes almost euphoric as he describes what must have been his personal experience of the place from his own journey. "In the Basque Country there is on the road of St. James a very high mountain which is called Port-de-Cize ... Its height is such that he who climbs it may feel he is about to touch the sky with his hand."

Picaud goes on to recount the story of the Emperor Charlemagne's arrival here on his expedition to Spain in 778. "On the summit of this mountain there is a place called the Cross of Charles, because it was here that Charles, setting out with his armies for Spain, opened up a passage by means of axes, hatches, pickaxes and other implements, and where he erected a sign of the cross of the Lord, and falling on his knees and turning towards Galicia he addressed a prayer to God and St. James." It is significant that Picaud's account of Charlemagne's prayer is not taken from the Song of Roland, which makes no mention of St. James in this context, but from the version of the story in the so-called chronicle of Archbishop Turpin, a spurious account of events created—as we have seen—specifically to promote the Santiago pilgrimage, and which is contained in the *Codex Calixtinus*.

And so, by masterly sleight of hand the *Guide* has established the defining moment in the advocacy of Charlemagne as the first patron of the Santiago pilgrimage, awarding the emperor God-given prior knowledge of the whereabouts of the apostle's tomb, not as yet discovered.

Thereafter all pilgrims arriving here at the gateway to Spain would be following the emperor's example—as Picaud describes: "Falling on their knees they would turn towards the land of St. James and offer a prayer, each planting his own cross of the Lord like a standard. Indeed one can find there up to a thousand crosses;

and that is why this place is the first station of prayer to St. James." In all, it was a promotional triumph. The message lit a fire in the heart of a million Christians.

This mountain route to Spain, so high that a traveler might feel he could "touch the sky with his hand," was one of two ways across this stretch of the Pyrenees described by Picaud. In the earlier centuries it was the preferred route, perhaps because it was more arduous and more testing of a pilgrim's zeal; also because it was the old, well-trodden Roman road leading from Bordeaux to northern Spain. (Seven centuries later it also became the route Napoleon chose to lead his troops across the Pyrenees.) From the small town of Saint-Jean-le-Vieux the road led, as it does today, to the village of Saint-Michel and then into the foothills of the Pyrenees along a valley where the occasional outcrop of well-worn stone slabs bear witness to long use by travelers over the centuries. Finally the metaled road gives out and the traveler takes a stony track to the mountain pass at an altitude of almost 4,000 feet: and from this perch in the sky the meadows beyond and below are where Picaud's "thousand crosses" once dotted the landscape. Further beyond still, and when the clouds lift, the eye can make out the broad roofs of another eagerly awaited landmark on a pilgrim's journey, the Augustinian monastery of Roncevaux—or, since we are now on Spanish soil—Roncesvalles.

The second road across the Pyrenees became the more popular route for pilgrims later in the Middle Ages, perhaps because it was easier, and the pilgrimage movement had by now lost its ascetic drive and become more of a passage from comfortable hospice to hospice. Already in Picaud's time it had become the alternative route: "Many pilgrims who do not wish to climb the mountain go that way," he wrote a little dismissively. Today it is an attractive route, beginning at the frontier town of Saint-Jean-Pied-de-Port (St. John at the foot of the pass), with its cobbled street and Porte Saint-Jacques where pilgrims entered the town, and its elegant Roman bridge over the River Nive leading to the Porte d'Espagne where they departed.

This route, like the first, is signposted by historical landmarks that had already become legendary by Picaud's day. "Near the mountain, towards the north," he writes, "there is a valley called Valcarlos where Charles himself encamped together with his armies after his warriors had been slain at Roncesvalles." This was the celebrated incident on Charlemagne's return from his Spanish campaign when his rearguard was attacked and destroyed, and the emperor's nephew Roland blew his ivory horn with his dying breath. This was stirring stuff for pilgrims, who would mostly have known the story well from the Song of Roland and from other pilgrim songs recited and sung by minstrels accompanying them along the way. They would have been well primed for this journey.

Valcarlos, named after Charlemagne, is today the border village. The famous battle and slaughter on August 15, 778 took place a short distance further south along the road at the highest point of the pass. Here at Ibañeta the ground-plan of a church, San Salvador, is all that remains of the monastery the emperor is reputed to have established on the site of Roland's death. In Picaud's time there was a celebrated hospice on this historic site, where a bell would be rung to guide pilgrims when the pass lay under heavy cloud; because here was also where the two pilgrim roads met—the valley road up from Saint-Jean-Pied-de-Port, and the mountain path with its one thousand wayside crosses.[29]

We are over the pass now and heading south towards the great Augustinian monastery of Roncesvalles, built early in the twelfth century at about the time Aimery Picaud was preparing his text. He may even have witnessed it under construction. The Augustinians had spread their net widely: the roads leading to the Pyrenees were punctuated by small convents and priories owing allegiance to Roncesvalles, each with a hospice at the service of pilgrims. Today the old pilgrim track from Ibañeta keeps its distance from the modern road and picks its way through the dense beech woods that cloak the place.

Roncesvalles was the first major religious house since the junction of the three pilgrim roads in the French Basque Country; and accordingly the monastery might welcome up to three times the

---

29. See Plate 20.

number of pilgrims any of the French abbeys would expect to receive. Not surprisingly it is a huge complex of buildings. A monk would regularly stand outside the main gate offering bread to exhausted arrivals. There were separate houses for accommodating men and women, and inmates enjoyed the rare experience of sleeping on a real bed instead of the customary straw mattress on the floor. Two hospitals cared for the sick, and in the large refectory the food served was equally superior. Not surprisingly Roncesvalles was one of the highlights of a pilgrim's journey.

There was a more noble reason for the abbey's reputation and one which would have preceded it along the entire stretch of the pilgrim roads. More even than the site of the battle itself Roncesvalles was steeped in the Charlemagne legend; and any pilgrim arriving here would have identified the place with the emperor's campaign against the Saracens in Spain culminating in the heroic death of Roland. Roncesvalles had become the shrine to that legend. To this day the abbey's treasury displays what are claimed to be Roland's war clubs, while for many centuries the nearby Chapel of Saint-Esprit possessed tombs reputed to be of Roland's fellow-soldiers killed in the famous ambush on Charlemagne's rearguard. As for the "enemy" that launched the attack from the hills, the *Chanson de Roland* describes them as Saracens, whereas historians are generally agreed that the attack was by local people of Navarre seeking revenge on Charlemagne for having pulled down the protective ramparts of their capital city of Pamplona.

Pilgrims arriving here had their own special chapel *de los peregriños* a little beyond the gate of the abbey. In contrast to the heavy grandeur of the monastery this is a modest building of elegant simplicity dating from a century after Picaud's time, though he would have recognized the sound of its bell because it was originally the celebrated bell that once guided pilgrims from the mountainside when it was rung from the long-vanished hospice of San Salvador at Ibañeta.

It was also the first chapel dedicated to St. James that pilgrims would have encountered in Spain. They were now in what was then the Spanish kingdom of Navarre; and any pilgrims who had access to

Picaud's *Guide* might now have severe misgivings about the hospitality they were about to receive from the Navarrese. "They dress most poorly and eat and drink disgustingly," he begins. Then it gets worse. "If you saw them eating you would take them for dogs or pigs … This is a barbarous nation, distinct from all other nations in habits and ways of being, full of all kinds of malice, and black of color. Their faces are ugly, and they are debauched, perverse, perfidious, disloyal and corrupt, libidinous drunkard, given to all kinds of violence, ferocious and savage…" and so on, in a floodtide of abuse. It was just as well that most twelfth-century pilgrims were illiterate.

From the high Pyrenees the pilgrim road makes a gradual descent to the Navarrese capital of Pamplona, a city whose reputation has been redeemed by a writer more widely read than Aimery Picaud, namely Ernest Hemingway, for whom Pamplona was the city of bullfighting and bull-running. The old pilgrim bridge over the River Arga still exists, over which medieval travelers approached the city, passing through the massive ramparts (rebuilt since Charlemagne's day) by means of the gate named in their honor: the Puerta de Francia, the Gate of France. Soon, next to the cathedral, they would find a hospice where the welcome and the accommodation were the very opposite of the *Guide*'s dire predictions. Here they enjoyed proper beds, and a meal a great deal more wholesome than the customary bread with a little wine, including fresh salad and—if they were lucky—meat.

So this was Spain. And the next day, after a modest hike of fifteen miles, there would be another reunion. To quote the opening words of *The Pilgrim's Guide*, "There are four roads which, leading to Santiago, converge to form a single road at Puente La Reina, in Spanish territory." And there, in that compact medieval town with its handsome bridge built specially for pilgrims as early as the eleventh century, travelers would find themselves mixing with others who had journeyed from Italy and from Provence, taking the most southerly of the four roads across France, the one leading from the ancient Roman city of Arles—which is where our next journey begins.

IV. Via Tolosana:
The Road from Arles

15. Roman Footprints

ARLES WAS JULIUS CAESAR'S CITY, capital of his newly conquered "province of Rome," hence the name Provence. Four centuries later it became the favorite city of the Emperor Constantine, who had converted the Roman Empire to Christianity.

Of the four roads listed in *The Pilgrim's Guide* the road from Arles seems to have been the most frequented by Santiago pilgrims. Many of them came from northern Italy and from Eastern Europe, either choosing the coast route or crossing the lower reaches of the Alps to Avignon or Aix-en-Provence. Climate, too, may have played its part in making the Arles route the most popular. Spring was always the favored time of year for setting out on pilgrimage—returning in the autumn—and both seasons were a great deal more congenial for foot-travelers in Provence than in central and northern France. It is a reasonable assumption that pilgrims by and large found gratification of the flesh a lot more appealing than mortification, whatever the Church might preach.

Provence was especially rich in shrines to be visited along the way, mostly associated with the legends of early Christians arriving on these shores to proselytise the local "heathens" who were living under Roman rule and required to worship a sequence of strutting emperors as deities. Also, Provence as the final home and resting-place of Mary Magdalene and her brother Lazarus, or so it was believed, had a particularly powerful appeal, endowing the whole region with a certain magical sanctity as though Provence itself was a natural extension of the Holy Land.

The entrance to Arles from the east was by way of the magnificent gate in the Roman ramparts built by the Emperor Augustus, Julius

Caesar's successor, in the first century AD. But this was not all: travelers first needed to pass between a long double-row of Roman tombs and monuments interspersed with Christian chapels, which lined the avenue on either side. To any visitor to the city it was like being welcomed by a guard of honor. It would certainly have been an impressive experience for any pilgrim arriving here tired and weary after a hot trek across the Provencal plains.

This long ceremonial approach to the city was the former Roman cemetery known as Les Alyscamps, meaning the Elysium Fields. In the Roman world Elysium was held to be the final resting-place of heroes; and here in the third city of the Empire was one of the most splendid of all Roman burial grounds, famous throughout Europe. Families would vie for the privilege of being buried here, and one imagines that the waiting list was long. With the conversion to Christianity Les Alyscamps underwent a smooth transition to become a Christian cemetery—on an equally grand scale. Aimery Picaud had no reservations about the majesty of the place: "In no cemetery anywhere except this one can you find so many and so large marble tombs set up on the ground." Christian and Roman heroes lay side by side as if there had never been such a thing as religious persecution.

Les Alyscamps has a gracefully romantic air about it nowadays—a venue for picnics rather than for mourning. It is no longer any kind of entrance to the city, but a gated park inviting a gentle stroll among the monuments. The great avenue is presided over by the handsome bell tower of the Church of Saint-Honorat, which in medieval times was especially appealing to pilgrims on account of the numerous relics of early Christian martyrs on display. Today Saint-Honorat, semi-ruined, is a solitary survivor: in Picaud's day there were no fewer than seven churches and chapels greeting pilgrims as they made their way towards the gate of Augustus and into the heart of the city.

There was one particular story relating to Arles which was a cause of great celebration among pilgrims arriving here; and it was one that created a direct link between the ancient Roman cemetery of Les Alyscamps and the medieval city itself. *The Pilgrim's Guide* makes it all clear. "First of all pilgrims must visit in Arles the remains of the Blessed Trophime, the confessor. It was he whom the Blessed Paul, writing to Timothy, remembers, and it is he who was ordained bishop by the same Apostle and who was the first one to be directed to the said city to preach the gospel of Christ."

Today the church dedicated to St. Trophime remains one of the masterpieces of Romanesque architecture in France. It replaced a smaller Carolingian church of the ninth century, and immediately became an important shrine for pilgrims, especially those gathering in Arles before setting off for Spain. Here was where those who had chosen this route to Santiago received their final blessing before departure. The sheer size of the new church reflects the numbers expected. The cult of Trophime as a saint with supposed links to Christ's apostles would undoubtedly have cemented the association of the new cathedral with the Santiago pilgrimage, which was likewise linked to one of the apostles. So, in Arles the two cults would have merged. The immense carved portal on the west front was completed later in the twelfth century, and here the figure of St. James is prominent on the right of the main door, while the portal itself is decorated with images of scallop-shells.[30] Then, behind the cathedral, hidden away behind high walls in the heart of the city, are the loveliest cloisters in Provence; and here on a panel depicting the Supper at Emmaus St. James is depicted wearing a scallop-shell on his conical hat, as though he was a pilgrim to his own shrine.

The Saint-Trophime portal also bears a message to pilgrims which would have been all too familiar from the carvings and paintings on countless newly built churches throughout the land. Right over the entrance to the cathedral, in the center of the tympanum, sits Christ as the Supreme Judge, while around him are arranged the symbols of the four evangelists whose gospels in the New Testament have spread his teaching far and wide. For the benefit of any remaining doubters the theme of Divine Judgment is

30. See Plate 21.

pressed home by threatening scenes of the Last Judgment, and these are complemented by further scenes illustrating the suffering of early Christian martyrs, all described with customary ghoulish relish.

Pilgrims would by now have grown accustomed to the punitive vision of human fate as preached by the medieval Church. To what extent that dark vision colored their experience of pilgrimage is hard to gauge, except that illuminated manuscripts of the period generally depict pilgrims enjoying the pleasures of travel and the conviviality of each other's company rather than laboring under the burden of sin and dread as promulgated by official Church teaching and in the sculpture which illustrated it. Today, too, how often do we find ourselves admiring the joyous architecture of Raoul Glaber's "white mantle of churches," yet recoiling from the relentlessly dark sermons in stone which decorate those same churches?

*

Dominating both the church and the cloisters of Saint-Trophime, and rising above the ancient city like a finger pointing triumphantly to heaven, is the majestic twelfth-century bell tower, the finest in the whole of Provence, and a landmark visible for miles across the flat countryside of the Rhône valley and delta. For nearly nine hundred years travelers have mounted the stairway to the highest gallery of the tower and gazed southwards over that intricate expanse of marshland and waterways which was once the granary of Roman Gaul, the Camargue. Just visible in the distance, by the shores of the Mediterranean, rises another bell tower, one that is quite different in shape, like a thin slab of golden stone set against the horizon. Beneath the tower is the hulk of a church jagged with battlements along the entire roofline, and clearly built for defense as much as for prayer.[31]

This is the medieval parish church of Les Saintes-Maries-de-la-Mer. Its very name—the Saint Maries of the Sea—identifies it as another special place of pilgrimage. It is not one that is listed in *The*

---

31. See Plate 22.

Pilgrim's Guide because it was not strictly speaking on the road to Santiago, and Aimery Picaud was not one to encourage diversions from the main track. Nonetheless a great many pilgrims would have known of it, and the story attached to it; and (like the detour to Rocamadour on the previous road from Le Puy) there would have been many keen to stretch their journey in order to visit the site of one of the most touching of all Christian legends of the Middle Ages.

The story of the Three Marys is as follows. After the Crucifixion the enemies of Christ continued their vendetta by placing members of his family and followers in an open boat, or raft, without sails or oars, and setting it adrift off the coast of Palestine. They consisted of Mary Salome, who was the mother of the apostles James and John, Mary Jacob who was the sister of the Virgin Mary, and Mary Magdalene. In addition to the three Marys the boat also contained the Magdalene's sister Martha and her brother Lazarus, Christ's friend whom he had brought back from the dead. Also on board the drifting vessel were two further disciples of Christ and saints, Maximin and Sidonius, along with a mysterious woman called Sara, who is believed to have been the Egyptian servant of Mary Jacob and Mary Salome. In all there was a complement of eight.

According to the legend, the only sign of habitation they found on this lonely stretch of coast was an abandoned Greek fortress, and here Mary Jacob and Mary Salome, together with their servant Sara, took refuge and established a small oratory, so creating the first Christian community in the western Mediterranean. Meanwhile the younger, more vigorous members of the party spread out to preach the gospel to the inhabitants of the nearby region. Lazarus headed eastwards for the Greek port of Massalia (Marseille), where he became its first bishop. Maximin and Sidonius traveled inland to what is now Aix-en-Provence, while Martha followed the River Rhône northwards to Tarascon, where she famously tamed a predatory dragon (whose sculpted effigy still decorates the town). As for Mary Magdalene, she was guided by an angel to a grotto where she remained in a state of continual prayer until her death thirty-five years later.

The appeal of the legend to Santiago pilgrims is clear enough. All the occupants of the magical craft, with the exception of the enigmatic Sara, were close associates of Jesus—which alone would have been a powerful magnet in a society yearning for close contact with events close to the life of Christ. Furthermore one of the Marys who stayed behind here to found the first Christian community was the mother of St. James, whose shrine was the ultimate goal of their pilgrimage. Equally important, according to the legend, this was where Mary Magdalene first arrived on these shores. She had been witness to both the Crucifixion and the Resurrection, and it was this closeness to the two central events in the Christian story which accounts for her supreme importance in the eyes of the early Church. She was his messenger, that "glorious Mary" whose shrine at the "large and most beautiful basilica" at Vézelay was richly praised in *The Pilgrim's Guide.*

Mention of the great Burgundian abbey of Vézelay is particularly appropriate in the context of Les Saintes-Maries. The story of the three Marys cast adrift by their persecutors in Palestine before being miraculously carried here is a natural coda to the parallel legend described in Chapter Seven in which the body of Mary Magdalene was later stolen from this region and reburied at Vézelay, so enabling that abbey to become one of the leading shrines in Christian Europe as well as the starting point of one of the four principal roads to Santiago.

The legend of Les Saintes-Maries is an invaluable preface to that fanciful and often disreputable saga which brought fame and wealth to one of the greatest of medieval abbeys, and led ultimately to its downfall.

Just how much of the story of the Three Marys would have been known to pilgrims in the twelfth century is open to dispute. The generally acknowledged source of the legend is that colorful account of the lives of the saints called *Legenda Aurea* (the Golden Legend), compiled in Italy by an eminent Dominican cleric and chronicler by the name of Jacobus (or Jacques) de Voragine in the thirteenth century. The work enjoyed immense popularity judging by the numerous copies made of it.

Jacobus de Voragine did not invent these "lives": his achievement was to gather together material from a variety of sources, much of it consisting of folktales passed down orally over the centuries or gleaned from accounts by earlier churchmen. And here we begin to see that the Three Marys' legend may have its roots far deeper than is generally believed. We know that as early as the fourth century a church was founded on the coast of the Camargue on the site of a Roman temple, and that in the sixth century the Bishop of Arles described a church there as bearing a dedication to "Sainte-Marie-de-Ratis"—St. Mary of the Raft. Which particular Mary is unclear, but the dedication certainly suggests that the story of the arrival of a saint by that name by sea was already current, and perhaps had been so ever since Roman times during the reign of Constantine when the emperor was living in Arles only a short distance away. In other words, long before the *Legenda Aurea* was compiled pilgrims would have been making their way to Les Saintes-Maries-de-la-Mer fully aware of a special shrine related to the magical arrival of a vessel carrying biblical figures from the Holy Land.

By the twelfth century pilgrims bound for Santiago who were making a detour here would have been greeted by a gaunt fortress of a church that had recently been rebuilt after a long history of plunder and destruction by Saracen invaders from Spain. It was this edifice that survives today as the parish church of Les Saintes-Maries—a large and impressive hulk of a building that clearly would have served a more important function than that of a parish church serving a small fishing village. It had been built of sandstone quarried far inland, then shipped down the Rhône at no doubt considerable cost. What is more, the elegant band of blind arches running around the entire building is evidence of those highly skilled itinerant stonemasons from Lombardy whose work we have already seen gracing many of the early churches in Burgundy. In short, no expense seems to have been spared. It was a place of prestige, a church that must already have been well established as a popular center of pilgrimage.

The story of the Three Marys did not end there. Three centuries later local curiosity led to another chapter being written. Hearsay and blind faith were no longer enough to satisfy a more skeptical society. Physical proof was required. In 1448 Count René of Provence (popularly known as Good King René since he was also the titular King of Naples) ordered an excavation beneath the church in the area of the high altar, its purpose being to put to the test the long-held belief that here was where Mary Salome and Mary Jacob would have founded their original oratory after their voyage from Palestine—and where they would ultimately have been buried. King René's conviction proved to be well-founded: bones were indeed discovered buried on either side of the present altar. Furthermore they were said to have given off a sweet odor as they were unearthed, considered to be conclusive proof that here were indeed the relics of the two Marys who had stayed behind here—one the sister of the Virgin Mary, the other the mother of the apostle St. James.

Today the celebrations to mark these discoveries of five and a half centuries ago have shifted from pilgrimage to pageant. In a touching ceremony held every May effigies of the two Marys in a model boat, along with a statue of their servant Sara, are carried on a symbolic journey back to the sea, traditionally led by the Archbishop of Aix-en-Provence. As the procession nears the water riders mounted on white Camargue horses form a protective semi-circle facing the shore as the effigies are borne slowly into the shallows. Then, in an informal ceremony of baptism, sea water is gently sprinkled over them. They have received the benediction of the sea which had brought them here.

The theme of the sea, and magical sea voyages, lies as close to the heart of the St. James legend as it does the legend of the Marys. Every pilgrim setting out for Santiago de Compostela in the twelfth century would have been familiar with the story of how the body of their apostle was carried by sea from the eastern Mediterranean out into the Atlantic before finally reaching the shores of northwest Spain. A miracle of the oceans feeds both legends almost in parallel.

Pilgrims who had made a detour to Les Saintes-Maries in medieval times may have returned to Arles before heading west on the next leg of their journey. Alternatively they may have taken the old Roman road northwest across the wilderness of the Camargue, crossing one of the branches of the Rhône by ferry before making their way to the next stopping place on the pilgrim road. Here was a place which carried a legend just as rich as those embracing Arles and Les Saintes-Maries. This was Saint-Gilles, with its historic abbey, and a wealth of miraculous events, about which *The Pilgrim's Guide* has a great deal to say.

*

The wild swamplands of the Camargue have always possessed a romantic mystique, quite apart from the touching legend of the Three Marys. There is a haunting magic about the place. Today, as yesterday, white horses roam freely among the reed-beds, the lagoons are speckled with scarlet flamingos in their thousands, and black bulls raise their heads imperiously as you pass, their lyre-shaped horns marking them out as a breed of great antiquity, once far more widespread. They date back at least as early as the third millennium BC, featuring in the Stone Age wall-paintings of the famous Hall of the Bulls in the Lascaux caves further north in central France.

Pilgrims setting out from Arles westwards by the direct route would have enjoyed at least a glimpse of the Camargue marshes before arriving at their next stopping place, Saint-Gilles. Here was a town awarded a very special accolade by the *Guide* on account of the fame of the saint who gave his name to the town. He was "the most Blessed Gilles" who was "extraordinarily famous in all parts of the world." The hagiography continues in full flow: "After the prophets and the apostles nobody among the other saints is more worthy, nobody more holy, nobody more glorious, nobody more prompt to lend help ... Oh, what a beautiful and valuable labor it is to visit his tomb."

The author, presumably Aimery Picaud, is writing about Saint-Gilles as if this was a travel diary. He is actually there on the spot, which may account for the breathless tone of his account. Whereas accounts of other saints and their shrines are respectfully impersonal, here is an impassioned eye-witness description. "I myself have verified what I am saying ... I regret indeed having to die before being able to report all the saint's feats worthy of veneration, these being so many and so great." Picaud's eulogy is accompanied by an account of some of the miracles attributed to Gilles, including one which apparently cured the Emperor Charlemagne of his (undisclosed) sins.

Picaud has introduced us to a shadowy figure who is nonetheless presented as special and remarkable; and it is intriguing to consider why. Gilles was by no means a "super-saint": he had not been a disciple of Christ or part of his family circle. He had not even been martyred. A mixture of legend and flimsy historical evidence establishes St. Gilles as having been born in Greece early in the seventh century. He received a divine calling to spread the gospel, leading him to set out on a raft which in due course brought him to the shores of Provence. (Similarities to the Three Marys legend are obvious, even with a gap of seven centuries, perhaps suggesting a long-held popular belief in this region that early Christianity was brought here miraculously by sea.) Why Gilles felt compelled to spread the gospel in southern France is puzzling since the area had already been converted to Christianity for three centuries. Gilles is then said to have lived the life of a Christian hermit befriended, so the story goes, by a deer he had saved by intercepting a huntsman's spear, and who proceeded to repay him by providing regular supplies of milk. The huntsman in question turned out to be the Visigoth (or Frankish) King Wamba. Amazed by what he had witnessed he undertook to found an abbey on the site, of which Gilles became the first abbot.

The cult of St. Gilles spread steadily, stimulated by the usual accounts of miracles attributed to him. It reached its zenith in the late eleventh and twelfth centuries, attracting followers far beyond the boundaries of Provence. There was, for instance, a nobleman in Poland towards the end of the eleventh century who narrowly survived a hunting accident. As a result of his escape from death he offered a prayer promising to visit the shrine of St. Gilles in thanksgiving—a promise he fulfilled. Clearly no saint closer to home would have been powerful enough to save him.

Such adulation from afar was no accident: rather, it was the natural outcome of an event which had taken place only a few years before the pilgrimage made by the grateful Polish nobleman. It was an event which in all probability also accounts for the effusiveness of *The Pilgrim's Guide*'s description of the place and its saint. What took place was this. In the year 1090 the abbey, which had been founded on modest lines four hundred years earlier, became one of a number of monasteries in this region to be placed under the reforming authority of the great Burgundian abbey of Cluny. This was a procedure we have witnessed many times before in earlier chapters as Cluny extended its empire of dependent abbeys and priories; except that the circumstances surrounding the take-over of Saint-Gilles were unusually propitious. The small town of Saint-Gilles was the principal seat of one of the most powerful and ambitious feudal lords in France, Count Raymond IV of Toulouse. Raymond would have recognized the huge benefit to his own position of becoming the patron of an abbey of international fame and prestige; and what more rewarding a step towards achieving such an ambition than to open negotiations with the most influential monastic house in the Christian world, Cluny? In addition, the Abbot of Cluny was a man who had held that position for forty years (and was to hold it for a further twenty years), and who already presided over many hundreds of monasteries right across Europe. Abbot Hugh, Hugh the Great, or St. Hugh as he became, besides being a pious churchman was a shrewd political animal, and would have seen the rewards of being supported by so powerful a figure as Raymond, particularly since Saint-Gilles was strategically placed on one of the main pilgrim routes to Spain. An important abbey here would greatly strengthen the pilgrimage movement leading to Spain, and in the process enhance the cause of Reconquest which lay close to the abbot's heart.

Altogether, an enduring agreement between the Counts of Toulouse and the Abbots of Cluny could become a formidable and lucrative partnership on both sides. And so it proved to be.

One further binding thread in the contract between Abbot Hugh and Count Raymond greatly boosted the prestige of their shared enterprise. Six years later, in 1096, as the enlarged Abbey of

Saint-Gilles was beginning to take shape on an ambitious scale, the new high altar was consecrated by no less a figure than the pope. Again it was Hugh who played the key role in bringing the ceremony about. Pope Urban II was a Frenchman who had spent his early years as a monk at Cluny, eventually appointed by Hugh to be the abbey's Grand Prior before his election to the papacy. It was another political triumph for Hugh. Cluny now had their own man as head of the Church.

Pope Urban's arrival at Saint-Gilles took place in the wake of an extensive tour, largely organized by Cluny, which the pope had made in order to promote an undertaking that was to become one of the seminal events of the Middle Ages, the First Crusade, aimed at liberating Jerusalem and the Holy Land from the hands of the Muslims. The crusade was launched within months of Urban consecrating the high altar here. It could scarcely have been a more auspicious moment for the fortunes of Saint-Gilles, both the abbey and the town. To complete the picture Count Raymond himself became one of the leaders of the crusade, launching the triumphant assault on Jerusalem in 1098.

Furthermore the success of the First Crusade opened up a rich traffic in goods from Palestine and the Middle East which no longer needed to reach France by way of Venice. They could arrive directly here in Languedoc in Raymond's own seat of power, and this lucrative business was further aided by trading privileges in Palestine which the count had obtained. Until the construction of Aigues-Mortes closer to the sea more than a century later Saint-Gilles became the principal port in this region for receiving luxury items of all kinds—silks, ivory, metalwork—brought from the Holy Land and from further east. The annual trade fairs held here as a result brought together merchants from north and south, east and west, to the handsome benefit of all, including of course the new abbey beneath whose expansive walls these colorful events were regularly held.

With its fame now spread across Europe, rich trade from the East, gifts from returning crusaders and ever more lavish donations from wealthy pilgrims heading for Spain, Saint-Gilles became in a very short time one of the brightest jewels in Cluny's crown. And the abbey church itself soon took on a splendor to match. By the second half of the twelfth century the ghosts of Abbot Hugh and Count Raymond would have been wearing satisfied smiles.

The prestige of the Abbey of Saint-Gilles abbey at that time can be measured by the scale and grandeur of the new church, or at least those parts of it which have survived, above all the west front. Here is one of the most glorious church facades in France, and a masterpiece of Romanesque architectural sculpture. Aimery Picaud was here too early to be able to break into his customary eulogies. And, unlike the shrine of St. Gilles in the crypt, the great west portal of the abbey church has mostly survived, even if somewhat battered by wars and religious conflict.

Set high above a flight of steps the facade consists of three elaborately carved portals which stretch majestically left and right across the entire width of the church. Within ranks of Corinthian columns each portal is decorated with wonderfully vigorous carvings on either side, accompanied by a frieze of further figures illustrating the life of Christ and his disciples. A tympanum characteristically crowns each portal, the central one devoted to the traditional theme of Christ in Glory. The whole facade bears strong similarities to the west front of Saint-Trophime in Arles, which pilgrims would have set eyes on only a short while before, and may well have been carved in the same stonemason's workshop. The entire mass of carvings is known to have been completed by the middle of the twelfth century, just a few decades after Picaud's visit. We even know the name of the principal sculptor involved, because inscribed on a statue of St. Jude are the Latin words BRUNUS ME FECIT—"Brunus made me." This is all we know about him. We are reminded of that celebrated Latin "signature," GISLEBERTUS HOC FECIT, which provides a shadowy identity to the authorship of an even finer masterpiece of medieval church carving, the great tympanum on the cathedral of Autun which was the subject of Chapter 10. In the overall anonymity of medieval artistry such glimpses of a living person are precious and rare. What kind of man, we would love to know, was Brunus, or Gislebertus—men who spent their lives creating some of the most noble works of Christian art to have come down to us?

The later history of the abbey would have saddened Picaud. The Hundred Years' War between France and England did Saint-Gilles no favors. Pilgrims dwindled in number. By the early fifteenth century donations to the abbey were so few that the monks were reduced to pleading to the Holy Roman Emperor that "the devotion of Christians to St. Gilles has altogether ceased and the faithful no longer come and visit his tomb. In former times the great affluence of pilgrims was a wonderful boon to the abbey and town of St. Gilles, but now the place is deserted and impoverished." Income was now so low that the monks claimed that they could no longer afford food and clothing.

The French religious wars in the following century led to a Protestant army occupying the town, slaughtering the local priests and choirboys. They used the abbey church itself as a fortress, and in their reforming zeal destroyed the celebrated tomb and reliquary of St. Gilles which Picaud had described so lovingly. At least they left the great west portal of the church more or less intact, satisfying themselves with hacking off the stone heads of a number of Old Testament prophets to whom they took offence.

As a result, by sheer good luck, we are still able to admire some of the surviving masterpieces of medieval church carving. At the same time, those aware of the dark history of this now rundown town might feel relieved to take the next step along the pilgrim road towards the relative peace and tranquillity of "the desert" which lay ahead.

16. Desert Songs

A FTER SAINT-GILLES THE PILGRIM ROAD CONTINUES to skirt the Camargue marshlands before reaching what later became the important Protestant city of Montpellier. The medieval road then headed briefly northwards into the rugged hill country known as the *garrigue*. This bleak and rocky landscape exerted an irresistible appeal to early Christian leaders in search of an ascetic life far from the madding crowd where they could contemplate their god in peace without the distraction of earthly pleasures. Two places close to one another came to prominence as early as the eighth century and exemplify this Christian yearning for solitude and self-deprivation. And as their fame became widespread and the pilgrimage movement gained momentum over the following centuries, both places became key shrines along the most southerly of the French pilgrim roads to Spain. Reaching them required a short meander into the wilds of the *garrigue*.

Both shrines have direct connections to the Emperor Charlemagne. The first of them, Aniane, only a short distance from Montpellier, subsequently fell victim to the fury of the French wars of religion between Protestants and Catholics in the late sixteenth century; and today there is nothing in Aniane to remind the traveler of how important a historical landmark this place once was. For medieval pilgrims Aniane may have been their first resting place after Saint-Gilles; and if there were any among them with some knowledge of monastic history, then here was a place demanding the deepest respect.

The story of Aniane can be read as a parable of the early days of Christianity. The man who became known as St. Benedict of Aniane (or "the second St. Benedict," the first being the founder of the order) spent his early years as a member of the court of Charlemagne, and subsequently rose to be one of the emperor's most successful field commanders during his campaigns against barbarian insurgents in northern Italy. Benedict then abandoned the military life and became a monk, and in about the year 780 he founded a small monastic community on his own estate here in the foothills, a location suitably remote from any town or village. The community was modelled on the isolated cells of Christian monks established during the third and fourth centuries in the Egyptian desert during the final decades of the Roman Empire.

Within a few years the Eastern model was replaced by a new, highly disciplined monastery at Aniane whose monks strictly followed the Rule set down by Benedict's illustrious namesake St. Benedict of Nursia at Monte Cassino in southern Italy two hundred years earlier. The new monastery flourished, attracting donations from wealthy benefactors and ever-increasing numbers of pilgrims.

The success of Benedict's venture at Aniane rested on the widespread reforms in monastic life which he instigated, at first here in Languedoc and later in abbeys and priories throughout the Christian world. The significance of his work can only be fully grasped by understanding just how crucial monasteries were to the fragile structure of early Christian society. Monasteries were not only houses of God and guardians of the true faith, they were virtually the only seats of learning and literacy: they represented moral order, social stability and the rule of law and justice. Outside their secure walls lay disorder and disruption. Anarchy was forever threatening. Accordingly, when those monastic pillars of society were seen to be crumbling, social order itself was at risk. Benedict saw around him a monastic world in which rules of worship and moral codes were being widely flouted. Abbots, as well as the monks below them, were often living openly with their wives and mistresses, enjoying a life of comfort and privilege given them by their status as supposed men of God and by the lands and worldly goods the monasteries had acquired from bequests and donations. Many monks had become parasites rather than servants of God.

This was the common state of affairs which Benedict did much to rectify. Having been Charlemagne's ideal soldier Benedict became his ideal reformer. He helped create the orderly Christian institutions which the emperor believed to be vital in an unstable world. When Charlemagne was crowned in Rome on Christmas Day in 800 it was an event which perfectly symbolized the emperor's ambition to recreate a coherent Christian empire as it had been under the first Roman Christian emperor Constantine five hundred years earlier. Charlemagne's vision was to re-establish a Christian world in which the former Roman Empire was reborn within the embrace of the Frankish dynasty he had founded (a vision realized to this day in the name of the principal territory he ruled, France).

After Charlemagne's death fourteen years later it was left to his son and heir, Louis the Pious, to put into practice many of the monastic reforms he had urged. And it was under Emperor Louis that Benedict of Aniane was placed in charge of all the monasteries throughout the empire. Benedict proved to be a zealous and uncompromising reformer; yet his successes turned out to be largely short-lived. With the break-up of the Carolingian empire shortly afterwards and the resulting anarchy of the ninth century many of the religious houses whose practices and rituals Benedict had reformed slipped back into their old ways, while many others suffered irreparably from the Saracen, Magyar and Viking invasions which wrecked so much of France throughout that darkest of centuries.

There is no mention of St. Benedict of Aniane in *The Pilgrim's Guide* even though the pioneer monastery he founded lay directly on the pilgrim road between Saint-Gilles and Toulouse, both of which are accorded much praise by the author, as is the celebrated shrine which pilgrims would soon be visiting barely a few miles beyond Aniane, Saint-Guilhem-le-Désert. We can only speculate on what would seem a surprising omission. Perhaps Aimery Picaud, whom we assume to have been the author, was unaware of the place, and indeed of its founder who also receives no mention in the text's catalog of saints.

Or maybe the omission was quite deliberate, on the grounds that a fulsome tribute to St. Benedict of Aniane and his monastic reforms would have detracted from the glory of Cluny, the abbey which had sponsored *The Pilgrim's Guide* and where, according to its final words, it was mainly written. As we have seen on so many occasions, the Santiago pilgrimage was Cluny's show, and not one to be easily shared.

Whatever the case, what remains true is that medieval travelers who benefitted from the charity of monks as they made their way from abbey to abbey, hospice to hospice, were welcomed into religious houses run on disciplined lines that owed much to the rigorous reforms first imposed by Benedict of Aniane. Weary after a long day on the open road they would have found themselves drawn into in atmosphere of calm and serenity, perhaps invited to participate in an elaborate church service that was rich in music and chanting: and this too was a debt to Benedict, who had laid such emphasis on liturgy and prayer in the daily life of a monk.

In many respects Benedict's reforms in monastic life foreshadowed those instituted by those indefatigable Abbots of Cluny several centuries later when the mood and resolve of Christian Europe was altogether more buoyant; at a time when, in the celebrated words of the Cluniac monk Raoul Glaber, "it was as if the world itself was casting aside its old age and clothing itself anew in a white mantle of churches."

＊

From Aniane the pilgrim road heads for the "desert," as the region is still described. The rough hills ahead are split by a gorge down which the River Hérault plunges towards the plains of Languedoc. The first intimation that here is a place of history is a bridge known as the Pont du Diable, the Devil's Bridge. It dates from the eleventh century, and was built by monks from the nearby Abbey of Saint-Guilhem, appropriately known as Saint-Guilhem-le-Désert.

We are again following in the footsteps of Charlemagne. Like St. Benedict of Aniane, St. Guilhem was one of the emperor's most successful generals, as *The Pilgrim's Guide* makes clear. "The most saintly Guilhem was an eminent standard-bearer of King Charles the

Great, a most valiant soldier and a great expert in war. It was he who by his valor, as it is told, brought under Christian rule the cities of Nîmes, Orange and many others." Guilhem was an aristocrat, a cousin and close friend of the emperor, and was brought up in the imperial court. As a soldier the victories enumerated in the *Guide* were against Saracen insurgents from Spain, whom he seems to have been responsible for expelling from much of Languedoc, temporarily at least.

On one of these campaigns his beloved wife died, and her death prompted him to abandon the life of a soldier in favor of the solitude of the barren hills to the north of Montpellier. Here, in 804 Guilhem built a monastery a short distance from where Benedict had established his monastic community at Aniane almost a quarter of a century earlier. And in recognition of Guilhem's services, and of his commitment to the monastic life, Charlemagne presented him with the most precious Christian relic in his possession, a fragment of the Holy Cross.

Unlike Benedict's community at Aniane, Guilhem's monastery survives today—at least part of it, namely the abbey church and fragments of the cloister. An enchanting village clusters around it, taking its name from the abbey's founder. Saint-Guilhem-le-Désert is a jewel in the desert.

The saint himself died in 812, just two years before the death of the emperor he had served. Processions and special services in the village still honor his saint's day (May 3) when the reliquary containing the fragment of the Holy Cross presented to him by Charlemagne is solemnly carried through the village to the church, those taking part in the procession traditionally carrying snails' shells as oil lamps through the narrow streets. Folklore has preserved touches of the Middle Ages in Saint-Guilhem, and everything about the immediate setting and the embracing hills around contribute to making a visit here a journey into a past era re-enacted before our eyes.

The church itself dates from the eleventh century, two and a half centuries after Guilhem's death. As so often with early Church leaders their fame, and the cults that grew up around them, blossomed long after their lifetime in response to a radically different social and political climate. St. Guilhem's fame became widespread throughout

Europe largely due to a new spirit of religious fervor in the eleventh and twelfth centuries which expressed itself as a cult of the saints and their shrines, and in the resulting popular urge to go on pilgrimages to visit those shrines.

A further accompaniment and stimulus to the pilgrimage movement onwards was the emergence of stories and songs—the *chansons de geste*—which dramatized the feats of Christian heroes of the past, and filled the ears and imaginations of pilgrims as they made their way to some chosen shrine. Again we have the example of Chaucer's band of pilgrims setting out for Canterbury and entertaining each other with tales and music to while away the time on the lonely road. They were making their way through the fertile garden of England rather than the "desert" of Languedoc; yet there is no reason to suppose they would have entertained each other in a different way.

The song recounting the exploits of Guilhem was among the most popular of these *chansons de geste*, largely on account of its bloodcurdling account of the hero's slaughter of Saracens invaders. Its bellicose theme would have found particular favor with the sponsors of the Santiago pilgrimage (notably Cluny) for whom the cult of St. James and the Christian Reconquest of Spain were inextricably linked. Notoriety brought wealth to Saint-Guilhem chiefly through donations by well-to-do pilgrims. As a result a new and much enlarged abbey church was built in the late eleventh and twelfth centuries, and which today stands handsomely in the heart of the village. A finely carved west door opens out on to a gracious square where a tangle of narrow alleys converge.

The abbey church survived the desecrations of the religious wars and the French Revolution by becoming the local parish church, which it still is. The abbey itself was less fortunate, and nothing of it remains. The cloister suffered only marginally less. In the twelfth century it was rebuilt as a two-storey cloister, which only the wealthiest religious houses tended to permit themselves, with Corinthian columns bearing traditional acanthus-leaf capitals combined with medieval imagery of flowers and vines. Here at Saint-Guilhem sadly little remains of what would have been one of the glories of Romanesque church architecture, though much of it has recently been recreated. After the French Revolution the cloister was transformed into a veritable shopping precinct, including a tannery. The entire abbey complex was then sold off as a stonemason's yard (a fate shared by the abbey church of Cluny). Much of the present village of Saint-Guilhem is built of it, and there is scarcely a farmhouse or inn within an hour's journey of the place that does not owe its limestone walls to the abbey founded by Charlemagne's legendary standard-bearer.

Late in the nineteenth century many of the carved columns from the cloister were discovered supporting a vine arbor belonging to a lawyer in nearby Aniane. A far-sighted American benefactor acquired the columns, and today the finest record of Saint-Guilham's majestic cloister graces the Cloisters Museum in New York.

*

It may seem a curious coincidence that two monasteries within a very short distance of one another, at Aniane and Saint-Guilhem, should each have been founded by a leading soldier serving under the Emperor Charlemagne, both of whom then abandoned a military career in favor of the solitary life of a hermit in the wilderness. Yet a closer look reveals a certain shared significance—in when the two monasteries were founded, and in the locations chosen. Wars and revolution have subsequently taken a heavy toll on both places, and today a great deal of reconstruction in the mind's eye is required to bring them to life; yet pilgrims arriving here in the twelfth century on this the most southerly road to Spain would have found themselves in a setting rich with echoes of the very beginnings of Christian monastic life. And the strongest of those echoes was the call of the wild—a belief in the healing power of the wilderness.

The appeal of the desert originated in the example of Christ himself, who had spent forty days and forty nights in the wilderness battling with Satan and overcoming him. For fervent believers here lay the only route to spiritual maturity, removing themselves from the marketplace of the world to live as hermits. Where Christ had led, others must follow. Hence, when persecution and social alienation contributed to driving early Christianity underground, it surfaced in places as far removed as possible from that busy world, in remote place reminiscent of Christ's wilderness, the chief of them being the Egyptian desert. It was here that the most uncompromising believers chose to remove themselves, convinced that only a life of solitude and the harshest asceticism could permit a profound understanding of God. The Ancient Greek philosophers had taken quite the opposite view, that civilization could only be found within the city walls: life outside the city gates was a Bacchanalian chaos, and if you lived there you were an outcast. But for the desert fathers towns and cities were "Babylon," the natural consequence of mankind after the Fall.

One of the most eminent of modern medievalists, Christopher Brooke, has aptly put the question: "How can one begin to characterize a movement of the human spirit so distant and so strange?" Yet, however distant it may seem, the hermitage ethic, and the extreme asceticism that accompanied it, lie at the very roots of western monasticism, and therefore inevitably part of the bloodstream of the medieval pilgrimage movement which sprang from it.

Solitary though these "desert fathers" initially were, many of them began to attract followers, so that soon small communities were established in the desert. Hence Christian monasticism was born. The man generally acknowledged to have been the founding father of these desert cells was St. Anthony, who was born about 251 AD, and who spent some twenty years in the Egyptian desert until his death in 356 at the age of over a hundred (which, if true, says a great deal for the rewards of an ascetic life). His was a lifespan which took in several of the most historic events in the early history of Christianity, two in particular: the final and most severe persecutions of the Christians under the Roman Emperor Diocletion at the very beginning of the fourth century, and—secondly—the sudden Roman conversion to Christianity under his successor, the Emperor Constantine, barely ten years later.

To have experienced two of the most cataclysmic events in Christian history clearly had a volcanic effect of St. Anthony's life and outlook: he became a passionate believer in the holy nature of martyrdom, allied with a corresponding horror of all emotions of joy and celebration, especially pleasures of the flesh. In doing so, he and those who followed him (such as St. Jerome) supplied the early Church with an ideal of extreme asceticism as the perfect way of serving and understanding God.

This combination of a cult of suffering with a loathing of all physical pleasures, born of these remote communities in the Egyptian desert, set the tone for much of the religious thought and teaching of the early Church, as well as providing a field-day for painters and sculptors required to illustrate the most lurid scene of carnal temptation. And though the extreme asceticism preached by St. Anthony became softened in the writings of St. Augustine and in particular St. Benedict, founder of the Benedictine order, it was nonetheless never far away from the ambitions of those zealots who gave up their worldly careers in order to found the earliest monasteries in Europe—men such as St. Guilhem who chose as his ideal setting a "desert" as close in spirit to the Egyptian desert as France could offer, as well as being safely removed from all distractions and temptations of human society. The spirit of St. Anthony was never far away from the Languedoc hills.

The dark legacy of the desert fathers led to a host of contradictions for the monasteries of Western Europe which grew out of that Eastern tradition, principally because medieval monasteries flourished in a very different world. And inevitably these contradictions were shared by the pilgrimage movement which depended to a large extent on those monasteries, without which there would have been no pilgrimages. So, we may wonder, was the medieval pilgrim making his way to a shrine in "the desert" here in Languedoc imagining that he was shunning the corrupt world of "Babylon?" Probably not; yet the monastery he was visiting had been founded in precisely that spirit.

This conflict of purpose illuminates a key question which has been asked ever since the beginning of Christianity: is the good Christian life compatible with earthly joys and pleasures? Or does that life in its truest form demand asceticism and a renunciation of such earthly pleasures? St. Anthony would have had no problem answering the second question in the affirmative, even though Christ himself would probably not have agreed, having been happy to turn water into wine to aid festivities at the wedding in Canaan. It would seem to be in the nature of human zeal that those who most vigorously point the way to God tend to be more disapproving of earthly pleasures than God's own son would have been. If we read Chaucer's *Canterbury Tales,* or look at mediaeval manuscripts illustrating scenes of pilgrims singing and dancing on their way to Santiago, we see little of St. Anthony's bleak certainty. The spirit of pilgrimage was a rare fusion of a renunciation of worldly comforts with a joyful embrace of the physical world through which the pilgrim boldly made his way.

At the same time the world the pilgrim encountered on his long journey through Western Europe was never free of that long-standing Christian conflict between body and spirit. Both in France and Spain the pilgrim roads led to hermitage after hermitage, many of them established by men who had built roads and bridges to aid travelers, and who were bravely dedicated to safeguarding their region against bandits, thieves and of course Saracens. As hermits they may have shunned the busy world, yet by their tireless exertions they still contrived to make it a better place. The hermit's life exemplifies one of the major contradictions that colored medieval Christianity— between the ideal of renunciation of the material world, and the equally powerful determination to improve it.

In short, one of the most compelling aspects of the medieval pilgrimage movement is how the long and winding road that pilgrims were compelled to take manages to draw together so many of the diverse religious impulses that drove Christians of the age to lead the life they chose. The pilgrim roads, today as yesterday, lead the traveler through an illustrated history of Christianity, from its darkest moments to its triumphant hours, from martyrs to heroes, masterpieces of art and architecture to personal messages scratched on the door of an inn. All life is there—even in the "desert" of Languedoc from where the next stage of this road takes the traveler westwards to the great city which gave this road its name, the *Via Tolosana.* That city is Toulouse.

17. A Cargo of Relics

FROM ST. GUILHEM-LE-DÉSERT PILGRIMS CONTINUED to be led from one local saint to another. The road was a narrow thread linking shrine to shrine through the rough countryside of the *garrigue*. Today's traveler, following in their footsteps across this primordial landscape, wanders from village to stone village, each of them having grown up over the centuries like a dependent family around one or other of these local shrines.

A recurring feature of early Christian saints is that reliable information about them is often thin on the ground. There are usually no more than a few bare facts, and the rest is folklore, changing color and content over the succeeding centuries. Then, with the pilgrimage movement under way there grew up a cult of these early saints whose shrines could now be venerated. In response disingenuous churchmen began to create fanciful biographies of these shadow figures from the past, making imaginative use of the few fragments of available material and thereby greatly enhancing the appeal of the saint's shrine to passing pilgrims and potential benefactors. The biography industry within the medieval church became a flourishing affair. History was widely re-written, or more often simply invented. Hagiography became a fine art, contributing much to the aura surrounding these early Christian heroes and martyrs, upon which much of the popular appeal of pilgrimages depended.

An example of just such an embroidered image confronted the medieval pilgrim a short distance after leaving Saint-Guilhem-le-Désert, at Lodève. Travelers today arriving here in May can still find themselves participating in a week-long celebration which has been conducted ever since pilgrims bound for Santiago came here in the

Middle Ages. The occasion is the annual feast of the local saint, St. Fulcran, who was bishop here in the mid-tenth century, and was responsible for building Lodève's first cathedral, long since replaced by the present Gothic pile dedicated to him.

The only written source of information about Fulcran is a pious biography by a successor as Lodève's bishop four centuries later. From this account we learn that Fulcran led a pure and holy life dedicated to aiding the poor and sick; that he founded and endowed hospices, was canonized for his good works and had undertaken the pilgrimage to Santiago. In this instance history may well have been re-written in the interests of the pilgrimage movement. If Fulcran had indeed made the journey to Santiago he must have been among the very first leading churchmen to do so, rivaling the claim generally made for the Bishop of Le Puy who is known to have made the pilgrimage in 950 (see Chapter 12). Perhaps the more likely scenario is that the saint's biographer, writing soon after the high noon of the pilgrimage movement when Fulcran's tomb had by now been venerated by millions of Santiago pilgrims over the past three centuries, concluded that the legendary bishop must surely have undertaken the pilgrimage himself, so making the saint an honorary recruit to the Santiago cause.

In reporting the lives of early saints any possible connection with Spain was welcomed, particularly as the pilgrim roads were now drawing closer to the Pyrenees. The goal of St. James' own city combined with the mission of the Reconquest gave the Spanish peninsula an air of the Promised Land. With the Pyrenees now on the southern skyline, pilgrims continued deeper into Languedoc, reaching the town of Castres which had grown up around one of the early abbeys reformed at the beginning of the ninth century by St. Benedict of Aniane. Castres had been a prominent Roman town, as its name suggests (*castrum,* a fortified place). To the medieval pilgrim, however, the Roman association had a very different significance. Later in the ninth century the abbey had been presented with the relics of one of the most celebrated of Christian martyrs in Spain during the final savage persecutions under the Roman

Emperor Diocletian. The martyr was St. Vincent of Saragossa. And over the following centuries his relics brought fame and wealth to the town and abbey of Castres to an extent which might seem out of proportion to St. Vincent's importance historically. Yet in a religious climate which held relics of the saints to be a spiritual link between humans and God a shrine such as this possessed much that a worshipper would travel far to venerate. Here was an early martyr who had been horribly tortured, whose body was subsequently protected by ravens from vultures, and finally spirited away from Saracen lands to a Christian country where pilgrims could pause to offer up their prayers and hopes for a reconquered Spain now only a few days ahead. It was a powerful and appealing narrative.

Castres flourished on the fame of St. Vincent until the fates turned against the town. It became a stronghold of the heretical Cathars only to be crushed in the Albigensian Crusade of the thirteenth century. Worse followed. The Black Death devastated the town in the following century, and it was then sacked by the English army under Edward the Black Prince during the Hundred Years' War. The sixteenth-century religious wars were no kinder to Castres either: as a predominantly Protestant town it was brought to heel, and its fortifications formally demolished. Even the relics of St. Vincent, revered by many centuries of pilgrims, perished in the course of the town's turbulent history. In the absence of surviving monuments to its gilded past the modern traveler may feel that the most appropriate comment lies with yet another, more recent connection to Spain. Here in the local museum devoted to the eighteenth-century Spanish artist Francisco Goya, his incomparable series of etchings, *The Disasters of War*, are displayed across its walls as a most telling *memento mori*. Goya's horrific images of human cruelty feel like a match for those gruesome depictions of early Christian martyrs on the face of so many churches lining the pilgrim roads to Santiago.

Castres suffered humiliatingly from the ruthless suppression of the heretical Cathars in the thirteenth century. Yet one outcome of that heartless campaign was a policy of vigorous propaganda by the triumphant Church authorities to "sell" those aspects of the orthodox faith which had been rejected or ignored by the Cathars. Prominent among the beneficiaries of this propaganda was the pilgrimage movement itself. Suddenly funds were made available for sophisticated artists to decorate churches with appropriate narratives of the saints, in particular St. James.

To the west of Castres, close to Toulouse, lies the village of Rabastens; and here the local parish church, formerly attached to a Benedictine priory, is covered wall to wall in frescoes illustrating the life of St. James. These robust and colorful paintings were executed soon after the suppression of the Cathars, and were later concealed behind layers of whitewash (which at least preserved them) not removed until the late nineteenth century. Here is a new world of official imagery from the thirteenth and early fourteenth centuries. Time has moved on from the early days when St. James was depicted simply as one of the twelve apostles. Now he is his own champion, promoting his own pilgrimage. He is a pilgrim to his own shrine. And he wears the full pilgrim's gear—scallop-shell on the brim of his hat, staff grasped in his hand, gourd for carrying water at his waist. Present-day pilgrims could even recognize him as one of their own. The identity is complete. At Rebastens the tourist industry of the present century can feel not so far away.

*

From here it is only a short distance to Toulouse, the city that gave this pilgrim road its name. Toulouse had long been one of the most important cities in France, politically and spiritually. It was the capital established by the Visigoths after the collapse of the Roman Empire in the fifth century. Then, from the mid-ninth century successive Counts of Toulouse ruled a virtually independent Languedoc with buccaneering panache for the next 350 years until losing most of their powers to the French crown as a result of the Albigensian Crusade against the Cathars, whom the Counts of Toulouse stubbornly refused to denounce, so bringing about their own downfall. By the thirteenth century pilgrims setting out on the

Via Tolosana some days earlier would have prayed in the great abbey church of Saint-Gilles where Raymond VI, Count of Toulouse, had been stripped naked and publicly whipped in the presence of twenty bishops and a representative of the pope.

For pilgrims earlier in the Middle Ages these dramatic events were still to come. To them the pervasive Cathar cult would have been merely an unorthodox local variant of the Christian faith, one to which some Church authorities were apparently content to turn a blind eye. Pilgrims came to Toulouse for quite other reasons. Here was a shrine with a special appeal since it contained the remains of the first Bishop of Toulouse from the third century who was martyred at the hands of the Romans in a spectacularly gruesome fashion even by the standards of the day, as *The Pilgrim's Guide* recounts with some relish. "On this route," the author insists, "one should visit the most worthy remains of the Blessed Saturninus, bishop and martyr who, when arrested by the pagans on the capital of the city of Toulouse, was tied to some furious and wild bull, and then dragged from the heights of the citadel down a flight of stone steps for a mile. His head crushed, his brains knocked out, his whole body torn to pieces, he rendered his worthy soul to Christ. He is buried in an excellent location close to the city of Toulouse where a large basilica was erected by the faithful in his honor."

St. Saturninus became abbreviated to St. Sernin (perhaps because this sounded less pagan), and the "excellent" basilica mentioned in *The Pilgrim's Guide* would have been the predecessor of the abbey church of Saint-Sernin which the author would have known soon after it was constructed, and which still stands today. It is one of the five huge "pilgrim churches" built to a similar pattern during the eleventh and twelfth centuries (as described in Chapter 3), all of them designed specifically to accommodate large numbers of pilgrims. As Kenneth Conant has stated, "the pilgrimage formula was brought to a climax at Saint-Sernin,"[32] the only other surviving church of comparable size in the group being the cathedral of Santiago de Compostela itself. It

---

32. Kenneth J. Conant, *Carolingian and Romanesque Architecture, 800-1200* (London: 1959)

has even been claimed by scholars that the architect of Saint-Sernin, or at least a pupil, may have been responsible for the design of the great Galician cathedral, built at much the same time.

Surprisingly, up until the tenth and early eleventh century the tomb of St. Sernin had been a more respected place of pilgrimage for Frenchmen than Santiago. This may partly have been due to geographical convenience, as well as to the relative safety of traveling to Languedoc; but it was also because of a bewildering quantity of holy relics which the abbey had come to possess. These included—so the abbey proudly claimed—no fewer than twenty-seven entire bodies of Christian martyrs. The most important of the relics were the contents of a veritable cargo of holy artifacts donated by the Emperor Charlemagne. Among the most popular of these was the celebrated Horn of Roland, which the emperor's nephew had blown with his dying breath at Roncesvalles (see Chapter 15). Many of the abbey's treasures met the fate of so much church property, looted or destroyed in the fury of the French Revolution. Among the survivors is the magnificent thirteenth-century reliquary containing the remains of St. Sernin to whom the abbey was dedicated, and which was its most venerated treasure.

Inventories show that in addition to the customary fragment claimed to be of the Holy Cross Saint-Sernin possessed a vast store of gold and silver reliquaries, caskets and statues, as well as fragments of the bodies of six of Christ's apostles. All these assorted relics were publicly displayed in the various radial chapels which worshippers could approach by means of the broad ambulatory beyond the high altar characteristic of these huge pilgrim churches.

One richly decorated reliquary no longer in existence must have created some confusion among pilgrims about to head for Santiago. We know from the abbey's medieval inventory that it was a casket displaying an image of St. James portrayed as a pilgrim to his own shrine, wearing a pilgrim's brimmed hat complete with scallop-shell. Since this was a reliquary it must have contained human remains, presumably believed to be those of St. James himself. In all probability this was one of the fragments of the six apostles known to have been donated to the abbey by Charlemagne. This had taken place around the year 800, several decades before the discovery of the supposed tomb of St. James in northwest Spain. So, at this point there were suddenly two bodies of the saint to be venerated on the same pilgrim road. Santiago had a rival. We may be reminded of the pilgrim in France a few centuries later who was proudly shown the head of John the Baptist at two successive shrines, only to be reassured by the priest at the second shrine that the first head must have been that of St. John as a young man.

It is generally believed that the discovery of the apostle's tomb in Spain barely a quarter of a century later cast a serious doubt over the authenticity of the Toulouse relics. (After all, the abbey still had the remains of five apostles to display.) It would also have been in the interest of those promoting the pilgrimage to Santiago to ignore the Saint-Sernin relic. Not surprisingly, *The Pilgrim's Guide* makes no mention of it. The abbey, on the other hand, clearly decided to have it both ways by commissioning a handsome new reliquary for the now-discredited relics, proudly decorated with the image of the apostle dressed as a pilgrim making the journey to his own "real" tomb in far-distant Spain. Were pilgrims to Saint-Sernin perplexed when they set eyes on the reliquary? We have no record.

*

The Pilgrim's Guide makes it clear that the abbey and Church of Saint-Sernin lay outside the city walls. With its enormous size it must have been an astonishing sight for pilgrims approaching Toulouse across the Languedoc plain. And it would have become an even more dominant landmark from the fourteenth century when the handsome octagonal bell tower was raised higher still, then even higher a century later with the addition of a tall spire.[33]

Saint-Sernin has hardly changed, and the magnificent bell tower still dominates the surroundings, except that the great pilgrim church no longer stands commandingly in the open countryside. Toulouse has grown around and beyond it. The church was begun in about

---

33. See Plate 23.

1070. While it conforms to the "formula" for pilgrim churches, one unusual feature is that a large part of it is constructed of brick, this being the most readily available building material in the alluvial plain in which the city is set. The proportions are huge, the nave being more than 350 feet in length, barrel-vaulted at a great height. From the beginning it was always an Augustinian house, as were so many abbeys and priories on the approaches to the Pyrenees, in many cases dependences, or "out-stations," of the great Augustinian monastery of Roncesvalles. Saint-Sernin, on the other hand, was always its own master, and being on the *Via Tolosana* it prepared pilgrims for a more demanding crossing of the High Pyrenees than the Pass of Roncesvalles, as *The Pilgrim's Guide* explains (see the following chapter).

As the Santiago pilgrimage grew in strength and numbers in the latter part of the eleventh century the role of Saint-Sernin as a key stopping place on one of the principal routes to Spain became increasingly important: hence the decision to build an enormous church. Accordingly it is no surprise that the Burgundian monastery of Cluny, sponsors of the pilgrimage in so many of its aspects, should have made a move to add it to the growing list of religious houses already under its authority. In the year 1082, at the height of Cluny's empire-building at the hands of the formidable Abbot Hugh, the Bishop of Toulouse, himself a Cluniac, took the bold step of dismissing the Augustinian canons of Saint-Sernin and replacing them with monks from Cluny. For a brief period the abbey, with its famous treasury of relics and its vast new pilgrim church in the process of construction, became the most valuable of all Cluny's possessions along the pilgrim roads. But for once Abbot Hugh's ambitions were thwarted. At first the papacy turned a blind eye—not surprisingly since Pope Urban II had himself been a monk at Cluny and been appointed by Hugh as the abbey's Grand Prior. But subsequently the papacy felt compelled to rule that the behavior of the Bishop of Toulouse was unacceptable, and Augustinian canons were restored.

Even so, relationships with Rome seem to have remained cordial since in 1096 Pope Urban made the journey to Toulouse in order to consecrate the new high altar; while a quarter of a century later his successor Calixtus II (the very pope whose name is quoted as part-author of *The Pilgrim's Guide*) presided over a ceremony dedicating an area of the new abbey church.

The main part of the high altar consecrated by Pope Urban survives. It is a magnificent slab of carved marble, and we even know who carved it since his name is engraved discreetly on the side, which is an indication of the esteem in which he was held. The craftsman was Bernard Gelduin, about whom we know very little except that he was a specialist in marble sculpture and a master of his art. Saint-Sernin contains some of the most refined and sophisticated carvings of this period anywhere in Europe, and since Gelduin was permitted to inscribe his own name on the most important object in the church it seems more than likely that he and his workshop were responsible for the other carvings, made at the same period, for the eastern end and crypt of the church. The finest of these is a series of seven marble plaques in low relief: they are of a feathery delicacy which makes us wish that a larger body of Gelduin's work had survived for us to admire. The theme of the seven plaques is the familiar one of Christ in Majesty accompanied by the customary symbols of the six evangelists—perhaps in this case a tacit reminder to pilgrims that the abbey claimed to possess remains of all six!

The plaques are among the glories of this abbey church. But there are other outstanding works of sculpture at Saint-Sernin, notably the carved figures round the south door of the church, known as the Porte Miègeville. And there on one side of the door is the figure of St. James, as if to remind us that this is one of the principal churches built specifically for pilgrims, many of whom would be heading off to the apostle's shrine in Spain. The Porte Miègeville carvings are a little later than the Gelduin plaques and high altar: we have now moved into the early years of the twelfth century. Together these works point to the existence of a major school of sculpture in southwest France, as we have also come across in Moissac, Conques and Saint-Gilles.

Significantly this was the very period when many of the new abbeys and priories were being established in northern Spain safeguarding the main pilgrim route to Santiago, the *camino*. It was also when Santiago's cathedral—a match for Saint-Sernin—was also being built. As we shall see in later chapters, it was these talented and successful workshops in France which came to play a key role in building and embellishing the new religious houses in northern Spain, both humble and grand. The skills of these architects, stonemasons and sculptors were taken across the Pyrenees to help create a new Christian Spain, often combining with the very different skills of Moorish craftsmen now working for their new Christian masters; the result was a compelling hybrid of creative achievements in northern Spain as if the Holy War had never existed.

Before long there were darker days for Toulouse and Saint-Sernin. When the Albigensian Crusade crushed the Cathars early in the thirteenth century Toulouse was besieged by the French armies of the north acting in the interest of religious orthodoxy combined with the acquisitive ambitions of the French crown. One incident which took place here at the pilgrim church during that siege may be seen as an appropriate act of retribution. The leader of the crusade, Simon de Montfort, had turned the campaign into a personal vendetta against the ruler of Languedoc, Count Raymond VI of Toulouse. It was Raymond who had been held responsible for the assassination of the papal legate near Saint-Gilles in 1208, an event which had sparked off the brutal campaign against the Cathars. Now, ten years after the murder, and with the Cathars all but destroyed, Raymond fled to Spain to avoid capture. Then, as de Montfort and his army were besieging Raymond's capital from outside the walls of the city Cathar supporters had taken refuge on the high roof of the new pilgrim church. One of them hurled a stone at the soldiers gathered below. The stone struck Simon de Montfort, killing him instantly.

A decade separated the two violent deaths. It was a decade which also saw the death of Count Raymond's political ambitions for the independence of Languedoc. As so often medieval pilgrims passed through territories fraught with dangers and disruption, protected only by their staff and recognizable clothing, and by the blessing (they hoped) of God.

Today's traveler walking the same road may be spared those hazards. Instead what the pilgrim roads offer is a journey through history—political history, religious, artistic, architectural, as well as the history of peoples and of social change. This road, the *Via Tolosana,* presents the traveler with an especially rich variety of these historical landmarks. It begins in Arles with the first flowering of Christianity in Roman Gaul, and the touching legend of the Three Marys arriving on the nearby shores of the Camargue. Then Saint-Gilles and one of the finest abbeys of Cluny's vast empire. Next, Saint-Guilhem-le-Désert and the legacy of Charlemagne's generals with their nostalgia for the wilderness and the life of the early desert fathers. Finally Toulouse and the death-knell of the Cathar heresy and all hopes of an independent Languedoc.

And tomorrow—the last trek southwards, towards the High Pyrenees, and into Spain at last.

18. Into Spain

THE FOUR PILGRIM ROADS ARE DRAWING CLOSER TOGETHER NOW as we approach the Pyrenees. The fourth road, the *Via Tolosana*, keeps a certain distance from the other three as it heads south, and before long will make its own crossing of the mountains before joining up with the others on the farther side, in Spain, or—more precisely—the region of Spain which was then the kingdoms of Aragon and Navarre.

At first the road from Toulouse continues westwards, keeping away from the mountains. There are frequent reminders that this has long been a route for Santiago pilgrims. The small town of L'Isle-Jourdain was the first halt after leaving Toulouse. Here the former hospice of Saint-Jacques is plentifully engraved with scallop-shells, while it also boasts a polychrome wooden statue of the saint in full pilgrim's outfit including multiple scallop-shells. The figure dates from many centuries later than the foundation of the hospice, after the French wars of religion at a time when the Catholic Church was reasserting its authority, keen to promote its long links with history including the heroic years of the Santiago pilgrimage and its contribution to the Reconquest of Spain. Soon the road leads to the city of Auch, the capital of Gascony, and here too there is a late tribute to the pilgrimage in the form of a window in the sixteenth-century cathedral with a colorful depiction of St. James dressed as a pilgrim *en route* to his own shrine. Again we are reminded that as the era of pilgrimage fervor passed, and lost its sense of mission, so the cult of St. James came to be celebrated almost sentimentally, with images that stressed the glamor of the pilgrimage movement rather than its spirit.

Only one major staging-post remained before the Pyrenees. And here at Oloron-Sainte-Marie the medieval church survives where pilgrims would have attended their last mass before heading for the mountains; and it possesses one of the finest carved portals in the whole of southwest France. Formerly the city's cathedral, the Church of Saint-Marie possesses a main doorway comparable to that of Moissac. It has classic Romanesque proportions, with its central column supporting a magnificent tympanum in marble framed by bands of figures above and on either side. The carvings include the customary twenty-four old men of the Apocalypse with their musical instruments and jars of perfume; yet unlike the usual stereotypes these are a humorous and joyful bunch, suggesting that the sculptor took pleasure in presenting them as local people of the town.[34] And this impression is reinforced by an accompanying frieze of carvings illustrating lively scenes of local peasant life—boar hunting, salmon fishing and filleting, cheese making, barrel making, preparing hams, plucking geese, and so on. Such glimpses of simple daily life are a refreshing rarity after so many didactic sermons in stone.

From Oloron the climb begins. The Somport Pass is at an altitude of almost six thousand feet. The Romans had used it to move their legions to and from northern Spain. The Carthaginian general Hannibal had crossed the Pyrenees here with his elephants on his way to attacking Italy. It was a popular trade route for merchants bringing goods northwards from Spain; and now it had become one of the two main crossing-points for pilgrims making for Santiago. They would normally be taking the pass in late spring, returning in late summer, so snow would not pose a problem. On the other hand, there were always winter travelers, and it was apparently customary for local mountain people to keep the pass clear of snow, and in return for such service they would be excused the customary border controls and taxes.

In the tenth and early eleventh centuries the Somport Pass seems to have been the favored crossing for pilgrims on this southern

34 See Plate 24.

route from Languedoc, Provence and northern Italy. But with the glamor of the Charlemagne legend cultivated by the popularity of the *Chanson de Roland*, by the late eleventh century the Pass of Roncesvalles was becoming more widely used, and many pilgrims reaching Oloron would then head westwards a mere day's journey to Saint-Michel-le-Vieux from where they would take the Col de Cize over the mountains to Roncesvalles, their imaginations fired by the sound of Roland's celebrated horn drifting on the wind.

The fame of the Somport Pass rested on the reputation of its hospice which welcomed travelers who had made the long climb from the plains of Gascony, or from the valleys of Aragon and Navarre to the south. This was the hospice of Santa Cristina de Somport; and the fourth chapter of *The Pilgrim's Guide* proudly nominates it as one of the "Three Hospices of the World," the other two being in Rome and Jerusalem. All three had been established "in places where they were very much needed ... for the restoring of saintly pilgrims, the resting of the weary, the consolation of the sick, the salvation of the dead, and assistance to the living." The exact circumstances of its foundation are obscure, but Santa Cristina seems to have been founded early in the eleventh century, and to have expanded greatly over the next hundred years as the numbers of pilgrims heading for Santiago hugely increased. The hospice was attached to a priory, and like Roncesvalles it was administered by Augustinian canons. Vulnerably isolated as it was, it enjoyed invaluable protection from the local Viscounts of Béarn on the French side of the Pyrenees and from the rulers of Aragon and Navarre on the Spanish side.

By the twelfth century Santa Cristina had become a major link in the long chain of hospices which were established across France and Spain for the benefit of pilgrims—in itself a measure of their numbers. Each hospice tended to be within a day's journey on foot from the next. The quality of the services on offer obviously varied considerable, but in those attached to richly endowed priories the spirit of Christian charity sometimes led to a fierce pride in being able to provide food and accommodation superior to those of rival establishments. Santa Cristina was certainly one of these, and developed a competitive rivalry with Roncesvalles even though they were both Augustinian houses. Lacking the glamorous appeal of the Charlemagne legend, Santa Cristina seems to have been keen to compensate in other ways. Endowments made it possible to provide hospitality on a relatively lavish scale: instead of the customary single crowded dormitory there were at least eight separate rooms for smaller numbers where pilgrims might actually enjoy a night's sleep. Not surprisingly, Santa Cristina earned a reputation for being among the most welcoming of hospices along the road to Santiago. Under the patronage of the Kings of Aragon and Navarre it provided a royal introduction to Spain.

The more normal experience of pilgrims seeking food and a bed for the night was a great deal less salubrious, unless they were able to pay for an inn, which few were in a position to afford. Many hospices offered no more than the basics, which might consist of a large straw mattress on the floor of a single dormitory; and this mattress would be shared by any number of travelers, often as many as twenty, together with an even larger host of fleas, lice, cockroaches, mosquitoes and other predatory creatures of the dark. There would have been plenty of pilgrims in the depths of the night nursing severe doubts about the sacredness of their mission.

More fortunate were travelers who found themselves welcomed by monasteries, or by hospices that were attached to monasteries and administered by monks, nuns or lay brothers, Here, as at Santa Cristina, the principle of free hospitality for pilgrims would generally prevail. This was a fundamental tenet of Christian charity in the medieval Church, and monasteries were even required to set aside a proportion of their revenues to provide it. Sometimes the abbot himself would undertake to wash the feet of pilgrims, a humble echo of the Last Supper and Christ washing the feet of his disciples. As *The Pilgrim's Guide* urged its readers to understand. "It should be known that the pilgrims of St. James, whether poor or rich, have the right to hospitality and to diligent respect." This is the final sentence of the *Guide* before the concluding words explaining that the book was written "mainly in Cluny."

Santa Cristina was not a Cluniac outpost unlike so many of the religious houses along the pilgrim roads. Nonetheless, in the context of the vast enterprise of the pilgrimage movement to which successive Abbots of Cluny were dedicated, Cluniacs and Augustinians were a single brotherhood, and according to the *Guide*, Santa Cristina's founders, whoever they were, "will partake without any doubt of the kingdom of God."

Today Santa Cristina, unlike Roncesvalles, is nothing but an archaeological site. Amid the turmoil of the French religious wars in the sixteenth century the Augustinian canons abandoned the place and sought refuge south of the Pyrenees safely away from France, in Aragon. They settled at the foot of the valley in the city of Jaca which—ironically—had in any case been the next stopping-place for Santiago pilgrims leaving Santa Cristina and the Somport Pass.

The track they would have taken survives, as old roads often do, being too narrow and winding to be transformed into modern highways. Instead, they become the favored paths for the modern traveler and pilgrim following in the footsteps of history. Here the old pilgrim road southwards from the ruins of Santa Cristina hugs the valley of the fast-flowing Rio Aragón as it flows towards Jaca. Pilgrims were now deep into the kingdom of Aragon, the neighboring kingdom to that of Navarre, where the road from Roncesvalles a short distance to the west was the combined route bringing pilgrims from Paris, Vézelay and Le Puy. Soon the fourth road would join them and become a single pilgrim road across northern Spain all the way to Santiago—the *camino*.

Meanwhile pilgrims who had crossed the Pyrenees by the Somport Pass made for Jaca, then the capital of the kingdom of Aragon. Strangely *The Pilgrim's Guide* has nothing to say about Jaca, or about the people of Aragon. This silence is in striking contrast to the author's salacious comments of the domestic and sexual habits of the neighboring Navarrese (alluded to in Chapter 15). All he writes, dispassionately, is that "having cleared the Pass of Somport, one finds the country of Aragon."

Gateway to Spain at the foot of the Pyrenees, Jaca was the first major stopping-place for pilgrims after crossing the mountains.

Today it remains a gem of a city, small and compact, and possessing one of the earliest cathedrals in Spain. In the late eleventh and twelfth centuries Jaca was not only the capital of the newly established kingdom of Aragon, but was also a bright symbol of a new Christian Spain that was at last rebuilding itself, materially and spiritually, after the long centuries of Saracen incursions from the south. The banner of the Reconquest was now flying proudly. With the aid of Burgundian knights Toledo had fallen to the Christians in 1085, soon to become the seat of the Spanish Church with a Cluniac monk from Languedoc as its first archbishop.

Jaca's Romanesque cathedral was begun at about this time in the new spirit of resurgent Christianity; and pilgrims arriving at the city from the Somport Pass would soon have found themselves in the company of those who were building it. By the entrance to Jaca, outside the city walls, a rapidly expanding community of foreigners was establishing itself. Some would have been former pilgrims who decided to stay: but mostly they were builders and craftsmen of all kinds, together with merchants and tradesmen, administrators, clerics and churchmen, the majority immigrants from southern France, Poitou and Burgundy, all of them welcomed here with special privileges by the rulers of Aragon, and all of them putting their shoulders to building a new and flourishing city and region free at last from the Saracen threat. They were people with all manner of skills that were urgently needed in the newly liberated kingdom of northern Spain. The passes over the Pyrenees, which since Roman times had seen the passages of armies, were now pilgrim roads, trade routes and channels for artisans, architects and craftsmen, as well as monks and ecclesiastics fired with a missionary zeal. Pilgrims had become so many Pied Pipers leading the way south. And as the area of the new Christian kingdoms expanded dramatically with the Reconquest, so this diverse migration from across the mountains helped fill the vacuum left by the departing Saracens. In the wake of the conquering Christian armies huge opportunities were suddenly opening up for churchmen and craftsmen alike.

When it came to creating the new churches and priories of northern Spain the inspiration and creative ideas came principally from north of the Pyrenees—from Languedoc, Poitou and of course from Cluny and Burgundy in general—with the grateful support of the Spanish rulers. But there were also local traditions and local skills on hand, and these were readily incorporated in the new building program. There was a long tradition of church building in Spain pre-dating the Saracen invasions. In addition there was the influence of Islamic building, in particular the Great Mosque at Córdoba completed in the tenth century. In fact in some respects Islamic building skills exceeded those of the Christian Church, particularly in areas involving geometrical design, including the construction of domes. Muslim craftsmen were forbidden by the Koran to represent living forms: hence the carved figures around the portals of the new Spanish churches tended to be the work of sculptors from north of the Pyrenees, whereas much of the architectural detailing is more likely to have been the work of Muslim craftsmen.

Jaca's cathedral displays precisely this hybrid character—a harmonious blend of several cultures. The tunnel-vaulted porch and carved portal carry strong echoes of Burgundy; the intricate capitals in the interior feel closer to southwest France, while the nave itself suggests the early Romanesque churches of the Loire, Saint-Benoît in particular. In all, pilgrims who had traveled from Paris, Vézelay or Le Puy may have felt they had brought their own world with them—at least until the architectural detailing told them that here were touches of Muslim Spain.

Suddenly, when it came to building a new Christian world in northern Spain, military Reconquest had succeeded in opening the way to a creative partnership. There would be a great many churches and priories along the *camino* all the way from here to Santiago which would remind the medieval pilgrim, as they do today's traveler, of that partnership between the Christian vision of worship and that of Islam.

*

From Jaca the fourth pilgrim road strikes westwards across the hilly countryside into Navarre. Everywhere they stopped for a night's rest, or to pray, travelers were made aware of those same links with places they had come from north of the Pyrenees. At Sangüesa the new rulers of Navarre founded a royal chapel here, Santa Maria la Real, celebrating the departure of the Saracens by erecting the most majestic of towers rising triumphantly above the town. They also commissioned a massive carved south portal packed with figures on the theme of the Last Judgment strongly reminiscent of so many churches in Burgundy under the shadow of Cluny. Those traveling workshops of skilled stonemasons and sculptors we detected at Jaca had moved westwards along the pilgrim road.

In the context of the pilgrimage a more important link with Burgundy lies a short distance away in the foothills nearby. The abbey of San Salvador de Leyre was established early in the eleventh century only to be largely destroyed by the Saracens. Early in the following century the abbey was resettled by monks from Cluny who substantially rebuilt and enlarged it, making it one of the key monasteries and hospices on the pilgrim road to Santiago. A further rebuilding of the abbey church took place in the thirteenth century at the hands of the Cistercians; but parts of the Cluniac church survive, in particular a superb west portal with its central column and bands of richly carved figures and writhing beasts crowded into the arches above. Here the Cluniac monks seem likely to have employed another of those traveling workshops, this one from western France. Aimery Picaud, had he come here, would have been happily reminded of those numerous jewel-box churches of the Saintonge in his native Poitou.

Further west, and another striking church catches the eye. It is a curious octagonal chapel set within an arcade of columns in the middle of open fields. This is Santa Maria de Eunate. Its connection with the pilgrim road has long been a matter of disagreement. It has sometimes been claimed that the Knights Templar built it, copying the shape of the church of the Holy Sepulchre in Jerusalem. But why would the Templars have chosen to build it right on the pilgrim road? A more likely explanation is that whoever built it intended it to be a funerary chapel for pilgrims who died along the way, and whose bones have been found buried here.[35] The chapel becomes a salutary

---

35. See Plate 25.

reminder that pilgrimages were not always the jovial outings evoked by Chaucer.

The Eunate chapel is also a reminder of why one of the longer chapters of *The Pilgrim's Guide* is devoted to "The Bitter and Sweet Waters to be found along the Road." The chapter is Aimery Picaud at his most personal and graphic. In the same chapter we have already heard his account of murderous boatmen in Gascony: now he offers a salutary warning of the fate which can befall pilgrims attempting to cross rivers here in Navarre. "There runs a river called the Rio Salado. Beware of drinking its waters, or watering your horse, for this river is deadly. While we were proceeding towards Santiago we came across two Navarrese seated on its banks and sharpening their knives. They make a habit of skinning the mounts of the pilgrims that drink from that water and die. To our question they answered with lies, saying that the water was indeed healthy and drinkable. Accordingly we watered our horses in the stream, and had no sooner done so than two of them died: these men then skinned them on the spot."

True or not, this is one of the few passages in the *Guide* which offers a glimpse of the author's personal journey to Santiago—which we know from a surviving letter he is reputed to have made with two companions (see Chapter 3). Unlike the great majority of pilgrims he traveled on horseback, not on foot, which as a privileged member of the Cluniac operation he was clearly entitled to do, even if his horse did end up skinned by the hated Navarrese before his eyes.

After the rural chapel of Eunate there was one short stretch of the *Via Tolosana* remaining. "Finally," the *Guide* states, "you arrive at Puente la Reina." Today, at the eastern edge of the town a statue representing the eternal pilgrim stands by the roadside at the point where the fourth pilgrim road joins the one from Roncesvalles which had combined the other three on the far side of the Pyrenees.

Here was the meeting place for all those pilgrims from every corner of Europe. The giant spider's-web which had spread across an entire continent had become a single thread. From now onwards there would be one principal road, the Spanish *camino*, known generally—since the majority of pilgrims using it were from north of the Pyrenees—as the *camino francés,* the Way of the French.

At that meeting point stand the medieval Church of the Crucifix. And here those pilgrims from all over Europe would celebrate their first mass together, as suddenly every language known to the Christian was subsumed by the unifying sound of Latin. Between the church and the former hospice where they had rested rises an arch over the porch, and here pilgrims would pass as they made their way into the first town of the *camino,* Puente la Reina. They made their way, as we do today, down the long corridor of a central street, the Calle Mayor. The Church of Santiago, St. James, still stands, and as an enduring echo of when pilgrims on the open road needed to be warned that town gates would soon close the church bell still rings out forty times at nightfall.

At the far end of town the Calle Mayor opens up to a sweep of the River Arga, and rising high over it the most famous bridge on the entire Spanish stretch of the pilgrim road. The magnificent hump-backed bridge of six Romanesque arches which gives its name to the town was built early in the eleventh century on the orders of the consort of King Sancho III of Navarre, Doña Mayor (although Doña Estefanía, the wife of his successor Garcia el de Nájera, has also been credited). And it was constructed principally for the benefit of pilgrims.[36]

There is a resonance, an aura, about this great medieval bridge carrying the pilgrim road westwards which makes it a symbol of the pilgrimage itself. It seems to tell a story of how it all became possible. It was built at a time when the Saracens still controlled most of the Iberian Peninsula. Nowhere was safe from their marauding bands. Roads were hazardous, if they existed at all, and bridges were even fewer. The Christian rulers of northern Spain clung to their fragile kingdoms hoping that the tide might soon turn in their favor. Even Pamplona, the capital of Navarre, had been briefly overrun by the Saracens; and only a little earlier Santiago itself has been sacked.

---

36. See Plate 26.

This was King Sancho's inheritance.

One thread of hope lay in the growing appeal of the Santiago legend and the crescendo of fervor within the pilgrimage movement it inspired. Since the discovery of the tomb St. James had been seen as the champion of the Reconquest, and the pilgrims undertaking the long trek to reach his shrine were his resolute standard-bearers. Hence creating a safe path for pilgrims through this hostile terrain became a vital weapon in the fight against the Infidel; and the great bridge of Puente la Reina was a mark of the Christian will to make that journey possible.

At the heart of this resurgence lay the relationship between the beleaguered ruler of Navarre, King Sancho III, and Christian leaders across the Pyrenees in France, above all Cluny. It is an intriguing anomaly that in the century following the building of the great bridge at Puente la Reina the Santiago pilgrimage should have flourished on this unlikely partnership—between a Burgundian abbey and a dynasty of Spanish monarchs. The Abbots of Cluny were men of God dedicated to the role of moral leadership and unwavering monastic discipline. The Christian monarchs of northern Spain, on the other hand, were for the most part unscrupulous political operators bent on the acquisition of territory and power by any available means, and who thought nothing of killing or robbing each other if necessary. Yet over a period of more than a century the two institutions formed a liaison that proved to be of enormous material benefit to both.

What made such a partnership workable was a shared ambition to reclaim the territories in northern Spain that had long been in Saracen hands. For Cluny this meant re-establishing the power of the Church in Spain. For the Spanish rulers it meant re-establishing their own dynastic power in those same lands. Their twin ambitions matched perfectly.

Queen Doña Mayor's great bridge which opened the way to Santiago stands as a monument to that partnership. It had begun early in the eleventh century when the young Sancho sought to establish a cordial relationship with Cluny's Abbot Odilo. Cluny was then beginning an extraordinary expansion of its monastic empire and influence throughout Europe. Lavish gifts of property and land were also making the abbey exceedingly wealthy. It was known to King Sancho that the reclaiming of Spain for Christendom was Odilo's most burning ambition. Contact between Navarre and Cluny became established, maintained by a regular exchange of letters and by visiting ambassadors. As a result Cluny began to employ its prestige, wealth and powerful connections (especially with the papacy) to sponsor military campaigns against the Saracens in collaboration with the Spanish kings.

It is unclear what direct support Cluny may have given to these campaigns, though the vital role played in the Reconquest by Burgundian knights suggests that the abbey's powerful feudal connections in Burgundy played an important part. The immediate result was a large expansion of Sancho's territories. In gratitude the Spanish monarch showered the abbey with gifts, many of them plundered from former Saracen lands. He also sent monks to be trained at Cluny, and invited Cluniac monks to Spain where they were soon installed in priories providing facilities for the ever-increasing flow of pilgrims. Military success, plunder, servicing the pilgrim road, the growth of monastic power in Spain: all these disparate elements came together.

So the pilgrim bridge at Puente la Reina pointed not only to Santiago, but to the continued success of that collaboration between monarchy and monastery in the decades to come. King Sancho maintained his relationship with Cluny right up to his death in 1035. It was then continued with increasing benefit to both sides by his son Ferdinand I, and finally by his grandson Alfonso VI, a relationship which culminated in the ultimate triumph of the capture of Toledo from the Saracens in 1085. The back of Saracen power in Spain had now been broken.

Three generations of Spanish kings had been the instruments of that success; while several hundred miles to the north, at Cluny, two generations of abbots had made that success possible. Odilo died in 1048 having been abbot for almost fifty-five years. He was succeeded by Hugh the Great, who remained abbot for a further sixty years.

Their combined period of office overlapped three centuries, from the end of the tenth century to the beginning of the twelfth. In the story of the Santiago pilgrimage this extraordinary chapter is far and away the most important.

*

Travelers today who stand on that Bridge of the Queen, where the four roads have become one, can follow the line of the old pilgrim road westwards as it climbs the far hill. Soon they will be enjoying the rich legacy of that partnership between France and Spain which has made the *camino*—the Way of St. James—one of the historic journeys of the world.

There are just four hundred miles to go.

V. Camino Francés:
The Spanish Road

19. The Way of St. James...
and Crowning Glory

As its name suggests, the Way was created specifically for pilgrims, together with a host of stonemasons, craftsmen, administrators, churchmen, engineers, merchants, peddlers, hangers-on and functionaries of every kind. All these diverse bands of travelers on the road contributed in one way or another to building and servicing what was in effect a new city, Santiago de Compostela, St. James' city. It had a new population drawn from all over northern Spain and beyond, new housing, new urban facilities and a vast and ambitious new cathedral set at its very heart. Without the shrine of the apostle the city would never have existed.

In Spain the Way is known as the *camino*; and today's pilgrims trekking towards Santiago may justifiably feel it to be their Way of St. James. By contrast, the pilgrim roads north of the Pyrenees were mostly former Roman roads long used for transporting armies to and from Spain, and later as busy trade routes. Building the new road, then servicing it for the benefit of pilgrims in their thousands, was always going to be a Herculean task for those on whose shoulders it fell. Until the late tenth century, much of this 400-mile stretch of northern Spain was intermittently under Saracen rule; and even when the beleaguered Christian rulers of the region managed to regain a degree of control of their lands it still remained a lawless area. Besides, there were numerous rivers to cross, and few bridges over them—which is why the Queen's Bridge at Puente la Reina was such a heroic achievement in the early days of the *camino*.

In the late eleventh century a great transformation took place. A distinctive road was established, hospices and inns set up, churches and priories founded. Cart tracks became roads. Bridges replaced fords, as at Hospital de Órbigo, where the thirteenth-century bridge is the longest on the *camino*.[37] Villages grew into towns. By the end of that era there would be a place for pilgrims to rest and be fed within a day's walk along the entire length of the *camino*, with lodgings of some description no more than twenty miles apart, most provided by the new monasteries.

In such an unsafe and unstable world, the monasteries were crucial for all travelers. They were havens of safety, and of physical and spiritual comfort. And at this time when a nation was being rebuilt after centuries of Saracen dominance, it was the monastic orders north of the Pyrenees that made the largest contribution to that rebuilding. Here lay the building skills, the craftsmanship, the organization and the wealth needed to undertake so massive a task.

Successive rulers in northern Spain gave every encouragement to these foreigners. French masons and stone carvers brought invaluable skills south of the Pyrenees. To a lesser extent this traffic in skills also worked the other way around, with Saracen craftsmen invited to employ their special carving skills on abbey portals and cloisters in southern France, such as Moissac. But in Spain the influx of skilled artisans from the north amounted to immigration on a widespread scale, encouraged by tempting offers of tax exemptions and other privileges if foreign craftsmen agreed to settle there. In the Navarre capital of Pamplona in the eleventh century it was reckoned that there were more French settlers than natives.

It was a similar story at Estella (Lizarra in Basque), the first town of any size along the *camino* after Puente la Reina. The Saracens had wrecked much of it in the tenth century, but now its reconstruction was to a large extent the work of French settlers— builders, craftsmen, merchants and traders—who enjoyed the same privileges as those given to foreigners in Pamplona, in addition to the unlimited opportunity for practicing their skills (which the

---

37. See Plate 27.

burgeoning city welcomed). The area around what is today the Plaza San Martín was virtually a French colony, and it was here that pilgrims naturally gravitated. The local church was even dedicated to a French saint, St. Martin of Tours.

Estella retains the look of a pilgrimage town. It bristles with Romanesque churches, and these bear evidence of immigrant craftsmen from north of the Pyrenees as well as from the Muslim south. From the Plaza San Martín a long flight of steps leads to San Pedro de la Rúa, one of the finest pilgrim churches on the *camino*. The magnificent portal is clearly to a large extent the work of Saracen stone carvers, while the most elegant of cloisters possesses carvings whose spiritual home is western France, Poitou and the land of those jewel-box churches of the Saintonge.

Echoes of pilgrimage are everywhere in Estella. In the Church of the Holy Sepulchre there is an exceptionally fine carving of St. James dressed as a pilgrim to his own shrine. The Church of San Miguel Archangel possesses a carved portal whose sculpted figures would have reminded any traveler from Burgundy of those powerful, elongated figures on the abbey churches of Vézelay, Cluny and Autun: evidence of yet another traveling workshop. And on the side of the twelfth-century palace of the Kings of Navarre, in the heart of the pilgrims' quarter, stands a carved stone capital depicting Roland killing a giant Saracen with his lance—one more instance of the Charlemagne legend being hitched to the Santiago cult.

Towns like Estella were key places for pilgrims to assemble after days on the road, and to feel at home in what was in effect a French colony. Yet on the bare stretches of road between such towns it was very different; and here it was the monasteries that came into their own. For the traveler they were oases in the desert—a desert, what is more, that all too recently had been patrolled by Saracen armies, and were still a hunting ground for bandits taking advantage of the vacuum left by the departing Muslim soldiers. A short distance to the southeast stands the monastery of San Juan de la Peña, built in the ninth century at the base of a protective cliff—which proved to be no protection at all, as the monastery was plundered by the

Saracens soon afterwards. The monastery limped on for a further century until, early in the eleventh century, it was colonized and rebuilt by monks from Cluny at the invitation of King Sancho, ruler of Navarre. This was about the time when the king's consort, Queen Doña Mayor, was apparently sponsoring the construction of the great bridge at Puente la Reina. These twin ventures—the pilgrim's bridge and the rebuilt monastery a short distance away—are among the first tangible achievements of that remarkable partnership between successive Abbots of Cluny and the rulers of northern Spain. Bridge and abbey: they were the foundation stones upon which the great enterprise of the Santiago pilgrim road came to be built.

It was the most powerful of the Christian rulers of northern Spain who made the most valuable contribution to the *camino*. He was King Alfonso VI, ruler of Aragon, Navarre, Castile, León and Galicia. His relationship with Abbot Hugh of Cluny led to lavish gifts to the abbey, including a chain of monasteries and priories in his own kingdom, most of them on the Spanish pilgrim road. These religious houses acted as service stations for pilgrims heading for Santiago as well as greatly strengthening the Church's grip on territories only recently in Saracen hands.

By far the most important of these monasteries was the abbey of Sahagún, midway between two of the major cities along the pilgrim road, Burgos and León. Today Sahagún is a hollow shell: little remains of the former abbey complex except a semi-ruined chapel that was once attached to the monastery. Yet in the eleventh and twelfth centuries at least fifty dependent priories were under the authority of its abbot. To have acquired Sahagún represented a huge expansion of the Cluniac monastic empire.

Today the impact of Cluny on the region is greatly reduced, since so many of its most important monasteries have long been destroyed: Sahagún, Nájera and Carrión de los Condes among them. The network of smaller religious houses along the pilgrim road remains the brightest legacy of Cluny. The modern traveler frequently comes face to face with small, elegant Romanesque churches that look as if they had been transplanted from north of the Pyrenees.

These echoes of Burgundy, and of Poitou and Languedoc, are among the small gems of the *camino*. They feel particularly precious because they date from the earliest days of the Santiago pilgrimage, a time when traveling the pilgrim road was still a hazardous undertaking. The Church of San Martín at Frómista, to the west of León, was one of the very first to be built along the *camino*, later drawn into the Cluniac empire. Its carvings tell of those traveling workshops from western France—Aimery Picaud's beloved Poitou. Even more French, with another portal indebted to Poitou, is the little pilgrim's chapel dedicated to St. James at Villafranca on the edge of the El Bierzo mountains, which form the final land barrier before Galicia and Santiago itself.

In Spain as in most of Western Europe, by the thirteenth century the Gothic style of church building had superseded Romanesque; this included the regions spanned by the *camino*. Accordingly the story of the Santiago pilgrimage, and what was created around it, is multi-layered, its early chapters often concealed beneath the grandeur that followed. The two key cities of Burgos and León exemplify this transformation from humble beginnings to splendor. Both cities are dominated by magnificent Gothic cathedrals which are among the glories of the Spanish pilgrim road. The great Gothic cathedral of Burgos has a mixed inheritance. The basic design was modelled on that of the French cathedral at Bourges. Its present spectacular appearance, bristling with spires and pinnacles, is more Germanic, derived from the Gothic cathedrals of the Rhineland, especially that of Cologne, where its architect, known as Juan de Colonia, came from.

León, too, possesses one of the loveliest cathedrals in Spain, but also with some of the most terrifying iconography of damnation.[38] Purely Gallic in style, it has the look of having been transported from northern France where it would have kept company with the early Gothic cathedrals of Reims and Amiens. Once inside, however, the colors of Spain and the burnished sierras take over. León cathedral

---

38. See Plate 28.

possesses the finest stained glass south of the Pyrenees. Generations of pilgrims on their way to Santiago have stood in the gloom of this nave, gazing up to marvel at how the sunlight seems to strike these windows with a burst of fire, transforming the expanse of glass into a revelation of light.[39]

The symbolism is timeless. At heart all pilgrimages are about traveling from a place of darkness in search of light. At the end of the *camino*, as legend has it, that point of light had been a star—Compostela, the Place of the Star. And the road leading to it has been under a canopy of stars, the Milky Way, La Voie Lactée, as French-speaking pilgrims have known it. What was created there, in that Place of the Star, is the subject of the final section of this book.

*

"Finally Compostela, the most excellent city of the Apostle ... the happiest and most spiritual of all the cities of Spain." Thus ends Chapter 3 of *The Pilgrim's Guide*.

The first Santiago cathedral had been a modest affair, built in the ninth century only decades after the apostle's tomb was claimed to have been found. This early building was severely damaged in 997 by the notorious Saracen warlord Al-Mansur. A century later the experience of seeing that church, still bearing the scars of Saracen assault, would have been deeply etched in the mind of a young man who was soon to become responsible for much of the cathedral that replaced it. He was Diego Gelmírez (c. 1069-1149), a key figure in the story of Santiago and the pilgrimage, and later the city's first archbishop. It was during his lengthy period in charge of operations—almost forty years—that a glorious phoenix rose from the ashes.

Work on the new cathedral began sometime between 1075 and 1078. According to *The Pilgrim's Guide*, "From the year the first stone was laid until the final one was in place 44 years passed." The claim that the entire cathedral was constructed in under half a century is only

---

39. See Plate 29.

partly true: much work remained to be done, particularly at the west end, which was not completed until late in the twelfth century. And the ultimate bravura touch of the twin Baroque towers overlooking the Praza do Obradoiro[40] was not added until the eighteenth. Nonetheless the main body of the basilica—the nave, the north and south portals, the choir and apsidal chapels were indeed finished, or nearly finished, in those 44 years—a truly remarkable achievement.

While Diego Gelmírez deservedly receives the credit for overseeing the bulk of the work, it was one of his predecessors who was responsible for commissioning the original design of the cathedral in the 1070s and supervising the earliest building work. He was Diego Peláez, Bishop of nearby Iria Flavia (today Padrón), Santiago not yet being a bishopric. Bishop Paláez recognized the need for a cathedral able to accommodate the ever-growing numbers of pilgrims who were arriving here now that the *camino* was relatively safe, with new bridges to facilitate travel and new monasteries and hospices to provide food and accommodation along the way.

By the year 1088 a great deal of the basic work on the east end of Santiago cathedral was completed. It would have been an extraordinary sight—this huge stone edifice towering high above the clusters of timber-built houses and shacks that are all that the city would have been at this time.

Then, in that year, a power struggle led to the deposition of Bishop Peláez. A long period of political unrest and insurrection overtook Santiago and Galicia as a whole, bringing building work on the cathedral to a halt for a number of years. The partly built cathedral remained an empty shell.

The deadlock was eventually broken in 1094 by the appointment of a monk from Cluny by the name of Dalmatius as the first Bishop of Santiago, now replacing Iria Flavia as the seat of the local bishop. The authoritative hand of Cluny was once again controlling events at the very heart of the Santiago pilgrimage.

At much the same time, and probably with the connivance of

40. Workshop Square, an allusion to the stonemasons' workshops located here during construction.

Cluny, Count Raymond of Burgundy came to Spain to marry the daughter and heir of King Alfonso VI, the eight-year-old Urraca, whose mother, Queen Constance, was Abbot Hugh's niece. On marrying Urraca Raymond received the title of Count of Galicia from King Alfonso, so becoming the administrator of the entire region, which included Santiago. Altogether it had been a multiple Burgundian triumph.

It was also a turning point in the fortunes of Santiago and its empty shell of a cathedral. In about the year 1092 the new man in charge of Galicia, Raymond, appointed Diego Gelmírez to be his secretary responsible for all building operations relating to the cathedral. An ambitious young man, able and industrious, Gelmírez immediately saw to it that construction work was resumed, now under the supervision of a new master mason known to us only as Stephen.

Before long Gelmírez was doubly in charge since Bishop Dalmatius had died only a year after taking office. There followed a hiatus of several years until 1100, when Gelmírez was elected Santiago's second bishop. Over the next quarter of a century, under his sharp eye, the bulk of the new cathedral was completed. And it was during this period, probably during the 1120s, that the description of the cathedral incorporated into *The Pilgrim's Guide* would have been recorded. "It is true to say," the author claims, that "you cannot find one single crack or defect in it. The basilica is wonderfully built, spacious, bathed in light, of excellent dimensions and proportions in width, length and height, and altogether of the most marvelous workmanship." The account, which goes on at great length, was evidently designed to be a tribute to Bishop Gelmírez, who in all probability commissioned it.

From the time of his election as bishop, Gelmírez' greatest ambition was to expand both the importance of his own office and the prestige of his city as the principal focus of pilgrimage in Christian Europe. Essential to these twin ambitions was the need to cement relationships with both the papacy and with Cluny.

Accordingly, early in his time as bishop he visited Rome twice (in 1100 and 1106), on both occasions traveling with his retinue along the pilgrim roads in France, visiting the monasteries of Toulouse, Moissac and possibly Conques, as well as spending time at the abbey of Cluny where he was able to meet the now-aged Abbot Hugh, his spiritual mentor and for so long a primary sponsor of the pilgrimage movement.

Not long after sealing his relationship with Cluny, Gelmírez lost the man who had first promoted him by making him his secretary and church administrator. Raymond of Burgundy, Count of Galicia, died in 1107. Two years later King Alfonso and Abbot Hugh also died. Within a short time three men who had done more than any others to further the cause of the Santiago pilgrimage were dead. Furthermore Gelmírez now found himself dealing with a new ruler in the form of Alfonso's tempestuous daughter Queen Urraca, who was also Raymond's widow.

Despite civil unrest and the insurrections that broke out in Santiago during Urraca's troubled reign, soon the stars began to shine brightly on Gelmírez and the fortunes of his city. In 1119 the late Count Raymond's brother, Guy of Burgundy, was elected pope as Calixtus II, the ceremony taking place at Cluny where the previous pope had died. Calixtus, a Burgundian aristocrat with strong links to Cluny, was the very pope who is credited with the authorship (or co-authorship) of several chapters of *The Pilgrim's Guide*, as well as being "one of the names of those who restored the pilgrim road."

Gelmírez himself became a leading beneficiary of these recent events. A year after the new pope was elected at Cluny, Santiago was elevated to the status of an archbishopric. And in the following year Pope Calixtus appointed Gelmírez as the city's first archbishop. In the same year, 1120, as if in celebration of this great moment, the nave of Santiago's cathedral was completed. Meanwhile, in Rome Pope Calixtus chose this time to canonize the late Abbot Hugh, the progenitor—together with his predecessor Odilo—of the Santiago pilgrimage movement.

There followed golden days. Elevation to the status of an archbishopric led to unheard-of prosperity for the city. Added to the flood of humble pilgrims arriving on foot Santiago was now attracting ever-growing numbers of wealthy benefactors keen to be associated with what Archbishop Gelmírez was happy to promote as "the new Rome." To facilitate matters further the network of Cluniac abbeys and priories along the *camino* made it possible for quantities of bullion to be transported to the city to pay for lavish new building works around the cathedral, monasteries being "safe houses" in more senses than one.

These years of triumph arrived too late to be recorded in *The Pilgrim's Guide*. Gelmírez lived on until the middle of the century, by which time the Romanesque cathedral he had supervised for forty years was complete in every detail except for the huge west portal. This remained in its partly restored state for at least twenty years after the archbishop's death, finally being replaced during the second half of the twelfth century by the present Pórtico da Gloria, set within the narthex of this most noble of cathedral entrances. The massive carved Pórtico is the most elaborate and probably the most celebrated of all the church portals along the pilgrim roads in France and Spain, matched only by that of Vézelay, to which in purely formal terms it bears some similarity. It is a stupendous achievement by a sculptor and master mason known simply as Maestro Mateo, or in his native France as Maître Mathieu, and about whom we know tantalizingly little beyond the fact that he and his team worked on this great portal for at least two decades.

In this late and magnificent flowering of Romanesque sculpture St. James is appropriately center-stage, high up on his commanding central pier, a figure at once modest and majestic. In his hand a scroll bears the Latin text *Misit me Dominus*—the Lord sent me. The apostle wears an expression of inscrutable serenity as he seems to be surveying the vast cathedral square below him where pilgrims from all over the world have been gathering in his honor for almost a thousand years.[41]

In 1974 I described the Pórticio da Gloria in *The Pilgrimage to Santiago*:

41. See Plate 30.

It is composed of three giant arches, each of them carved above, between and to the side. Dividing the central arch, which ids the main entrance to the cathedral itself, is the traditional central pillar, except that in the position normally occupied by the Madonna sits St. James himself perched above an elaborately carved Tree of Jesse. By tradition, the pilgrim to Santiago completed his journey by placing his fingers between the twisted stems of the tree, and this is a custom still observed: I have watched many a toddler lifted by his father after Sunday Mass so that he might squeeze rubbery fingers into the folds of Mateo's stone, now worn smooth by generations of hands.

On the inside of this central column, and partly obscured by the gloom of the nave, squats another carved figure. He is the stone image of Maestro Mateo himself, humble enough in appearance, though one feels he can scarcely have been that in life, and this figure has been another object of respect and reverence among pilgrims. On leaving the cathedral they would lower their heads and touch the brow of the master-builder and sculptor, in order, so it is said, that they might acquire some of the wisdom of the great man. The statue accordingly received the nickname of *O Santo dos Croques*, meaning the Saint of Skull-rappings...

Then on either side of Christ are the figures of St. John on his eagle, St. Luke on his bull, St. Mark on a lion and St. Matthew on his knees. And on the right-hand arch is a final reminder of that theme which had accompanied pilgrims all the way from Paris and Vézelay, Arles and Le Puy: the Torments of the Damned—and never have they looked more damned than here—with, nearby, other figures, who are enjoying the bliss of finding themselves on the other side of that barbed-wire fence, being conducted as if in a dream towards the Almighty by a squad of beautiful angels.

For travelers of all persuasions and interests, coming to the end of this long journey in which four roads finally merged into one, Santiago's Pórtico da Gloria, with its vision of heaven, presents itself to us as the final jewel in the crown. Now, as then, it is journey's end. It has been a journey inspired by a legend with its roots in the circle of Christ's closest companions. This legend has been so powerful over the course of many centuries that it gave rise to a network of pilgrim roads right across the continent of Europe. Accompanying these roads some of the finest architecture and works of art of Christian civilization have been created for the welfare of pilgrims, which make any journey along those "roads to heaven" an unforgettable experience for these who travel in even greater numbers today than in the Middle Ages. There are those, like Robert Louis Stevenson, who enjoy "travel for travel's sake. The great plan is to move." Others, as they wearily complete their journey in Santiago, may feel more in the spirit of St. Augustine: "The world is a book, and those who do not travel read only one page of it." For them *The Pilgrim's Guide* may stand as the first book of the world.

Selected Further Reading

This is a personal selection of books that have aided me in writing *The Four Roads to Heaven*. There are a great many others relating to the great pilgrimage of the Middle Ages, including numerous guidebooks to both the French pilgrim roads and to the Spanish *camino*. There are also many DVDs available, and these too can be found at many of the principal sites along the pilgrim roads, or in local bookshops.

Astbury, A., *Pilgrimage* (London: 2010)

Barraclough, G., *The Crucible of Europe* (London: 1975)

Bentley, J., *Restless Bones: The Story of Relics* (London: 1985)

Brooke, C.N.L., *Europe in the Central Middle Ages*
(London: 1964)
The Structure of Mediaeval Society (London: 1974)
The Age of the Cloister: The Story of Monastic Life in the Middle Ages (New York: 2001)

Brooke, R. and C., *Popular Religion in the Middle Ages*
(London: 1984)

Brown, P., *The Cult of the Saints: Its Rise and Function in Latin Christianity* (London: 1981)

Chaucer, G., *The Canterbury Tales* (London: 1997)

Conant, K.J., *Carolingian and Romanesque Architecture, 800-1200*
(London: 1959)

Cowdrey, H.E.J., *The Cluniacs and Gregorian Reform*
(Oxford: 1970)

Dunn, M. and Davidson. L.K., (eds), *The Pilgrimage to Compostela in the Middle Ages* (New York and London: 2000)

Evans, J., *Life in Mediaeval France* (London: 1969)
   Art in Mediaeval France (Oxford, 1969)
   (ed.), *The Flowering of the Middle Ages* (London: 1966)
Fletcher, R.A., *Saint James's Catapult: The Life and Times of Diego Gelmirez of Santiago* (Oxford: 1984)
Geary, P., *Furta Sacra: Thefts of Relics in the Central Middle Ages* (New York: 1978)
   Living with the Dead in the Middle Ages (New York: 1994)
Gittlitz, D.M., *The Pilgrimage Road to Santiago: The Complete Cultural Handbook* (New York: 2000)
Haskins, S., *Mary Magdalene: Myth and Metaphor* (London: 1993)
Huizinga, J., *The Waning of the Middle Ages* (London: 1967)
Jacobs, M., *The Road to Santiago de Compostela* (London: 1991)
Kendrick, T., *Saint James in Spain* (London: 1960)
Mâle, E., *Religious Art from the 12th Century in France* (London: 1949)
Mullins, E., *The Pilgrimage to Santiago* (Northampton, MA: 2001)
   In Search of Cluny: God's Lost Empire (Oxford: 2006)
Pirenne, H., *Mohammed and Charlemagne* (London: 1939)
Porter, A.K., *The Romanesque Sculpture of the Pilgrimage Roads* (New York: 1966)
Southern, R.W., *The Making of the Middle Ages* (Oxford: 1953)
Sumption, J., *The Age of Pilgrimage* (New York: 2003)
Trevor-Roper, H., *The Rise of Christian Europe* (London: 1966)
Ure, J., *Pilgrimage: The Great Adventure of the Middle Ages* (London: 2006)
Vieillard, J., *Le Guide du Pélerin de Saint-Jacques en Compostelle* (translation from the twelfth-century Latin) (Mâcon: 1938)
Zarnecki, G., (with Grivot, D.), *Gislebertus, Sculptor of Autun* (London: 1961)
   The Monastic Achievement (London: 1972)

Recommended websites

THE CONFRATERNITY OF ST. JAMES
www.csj.org.uk
Everything you need to know about the pilgrimage: an invaluable
resource, with practical advice and detailed route guides.

CAMINO ADVENTURES
www.caminoadventures.com
Useful information on the pilgrim routes and how to plan a
walking expedition.

AMERICAN PILGRIMS ON THE CAMINO
www.americanpilgrims.org
A good mix of history and practicalities aimed at a membership
organization.

Index